# Differential Equations, Discrete Systems and Control

# MATHEMATICAL MODELLING:
## Theory and Applications

VOLUME 3

This series is aimed at publishing work dealing with the definition, development and application of fundamental theory and methodology, computational and algorithmic implementations and comprehensive empirical studies in mathematical modelling. Work on new mathematics inspired by the construction of mathematical models, combining theory and experiment and furthering the understanding of the systems being modelled are particularly welcomed.

Manuscripts to be considered for publication lie within the following, non-exhaustive list of areas: mathematical modelling in engineering, industrial mathematics, control theory, operations research, decision theory, economic modelling, mathematical programming, mathematical system theory, geophysical sciences, climate modelling, environmental processes, mathematical modelling in psychology, political science, sociology and behavioural sciences, mathematical biology, mathematical ecology, image processing, computer vision, artificial intelligence, fuzzy systems, and approximate reasoning, genetic algorithms, neural networks, expert systems, pattern recognition, clustering, chaos and fractals.

Original monographs, comprehensive surveys as well as edited collections will be considered for publication.



*The titles published in this series are listed at the end of this volume.*

# Differential Equations, Discrete Systems and Control

## Economic Models

*by*

Aristide Halanay
*Faculty of Mathematics,*
*University of Bucharest,*
*Bucharest, Romania*

and

Judita Samuel
*Centre of Mathematical Statistics,*
*Romanian Academy,*
*Bucharest, Romania*

KLUWER ACADEMIC PUBLISHERS
DORDRECHT / BOSTON / LONDON

A C.I.P. Catalogue record for this book is available from the Library of Congress.

ISBN 978-90-481-4888-2

Published by Kluwer Academic Publishers,
P.O. Box 17, 3300 AA Dordrecht, The Netherlands.

Sold and distributed in the U.S.A. and Canada
by Kluwer Academic Publishers,
101 Philip Drive, Norwell, MA 02061, U.S.A.

In all other countries, sold and distributed
by Kluwer Academic Publishers,
P.O. Box 322, 3300 AH Dordrecht, The Netherlands.

*Printed on acid-free paper*

# Contents

# *Preface*

In the fall of 1992, Professor Dr. M. Altăr, chairman of the newly established department of Management with Computer Science at the Romanian–American University in Bucharest (a private university), introduced in the curriculum a course on Differential Equations and Optimal Control, asking us to teach such course.

It was an interesting challenge, since for the first time we had to teach such mathematical course for students with economic background and interests. It was a natural idea to start by looking at economic models which were described by differential equations and for which problems in decision making did arise. Since many of such models were described in discrete time, we decided to develop in parallel the theory of differential equations and that of discrete–time systems and also control theory in continuous and discrete time.

The present book is the result of our teaching experience with this course. It is an enlarged version of the actual lectures where, depending on the background of the students, not all proofs could be given in detail.

We would like to express our gratitude to the Board of the Romanian – American University, personally to the Rector, Professor Dr. Ion Smedescu, for support, encouragement and readiness to accept advanced ideas in the curriculum.

The authors express their warmest thanks to Mrs. Monica Stan - Necula for the excellent processing of the manuscript.

*The authors*

# *About the Notations*

The notations used in this book are fairly standard in mathematical literature (as well as in Mathematical Economics).

Nevertheless, we would like to point out some aspects:

If $f$ is a function of a real variable $t$, then the derivative at point $t$ is denoted by $f'(t)$.

If $f$ is a real–valued function of several variables, as $t$, $x$, then the derivative with respect to the second argument at point $(t, x)$ is denoted by $\frac{\partial f}{\partial x}(t, x)$.

We denote the second order partial derivatives of a function $f(x, y)$ by $\frac{\partial^2 f}{\partial x \, \partial y}$, $\frac{\partial^2 f}{\partial x^2}$ and $\frac{\partial^2 f}{\partial y^2}$, respectively.

If $f$ is a vector–valued function of a vector argument $x$, then we denote $\frac{\partial f}{\partial x}(x)$ the Jacobian matrix with entries $\frac{\partial f_i}{\partial x_j}$.

If $x$ is a vector in $R^n$, then we represent $x$ as a column vector and we denote by $x^*$ the row vector obtained by transposition; if the vector $x$ has complex elements (co–ordinates), then the vector $x^*$ is obtained by transposition and taking complex conjugates of the co–ordinates.

If $x$ is a vector in $R^n$, then $|x|$ denotes any usual norm, such as Euclidean norm $(x^* x)^{\frac{1}{2}}$ ,or $l^1$ – norm $\sum_{k=1}^{n} |x_k|$, or the $l^\infty$ – norm $\max_k |x_k|$.

If $x, y \in R^n$, then we denote their scalar product by $(x, y)$.

If $A$ is a matrix with real entries, then we denote by $A^*$ the matrix obtained by transposition; if the entries are complex numbers, then $A^*$ is obtained by transposition and by taking the complex conjugates of the entries.

For a matrix $A$, we denote by $|A|$ any usual matrix norm; the main connection between vector and matrix norms is expressed by the inequality $|Ax| \leq |A||x|$.

If $A$ is a square matrix, then we denote by $det\, A$ the corresponding determinant; nevertheless, in specific situations, for instance if $A = \begin{pmatrix} a_{11} & a_{12} \\ a_{21} & a_{22} \end{pmatrix}$, we denote the determinant by $\begin{vmatrix} a_{11} & a_{12} \\ a_{21} & a_{22} \end{vmatrix}$.

If $A$ is invertible, then we denote the inverse by $A^{-1}$.

If $A$ is a symmetric matrix $(A^* = A)$ then we write $A \geq 0$ if the corresponding quadratic form is positive semi–definite: $(Ax, x) \geq 0$ for all $x$.

We write $A > 0$ if the corresponding quadratic form is positive definite: $(Ax, x) > 0$ if $x \neq 0$.

We use the usual Landau notation: $O(\varepsilon)$ denotes a function with the property that $\left|\frac{O(\varepsilon)}{\varepsilon}\right| < M$ as $\varepsilon \to 0$, while $o(\varepsilon)$ denotes a function with the property that $\lim_{\varepsilon \to 0} \frac{o(\varepsilon)}{\varepsilon} = 0$.

We write $t \mapsto x(t)$ to denote the mapping from $t$ to $x(t)$.

# *Introduction*

Differential equations and their discrete counterpart represent the most usual and the simplest mathematical models for evolution.

Differential equations developed from the beginning of Mathematical Analysis, in connection with Geometry an Mechanics; they also have been used at an early stage in Physics and chemical processes. Discrete–time models have been used more recently in biological models, economic models and also in applications to technology. This is explained essentially by the fact that most natural processes take place in continuous time, while our observations, measurements and actions are usually performed at discrete moments of time.

Differential equations and discrete–time systems are used in economic models, if we consider evolutions of prices, capital and other economic quantitative elements as being produced by laws relating the speed of evolution to the state of evolving elements. At present, the quantitative study and control in economic science and practice play an increasing role and, as a consequence, differential equations, discrete–time systems and the corresponding control problems became part of curriculum in economic education, as they are in general in applied sciences.

This book grew up from a teaching experience for students in Business Administration, Management and Computer Science. It is not a book on economic models, it is a book on Mathematics, the mathematical stuff is presented in connection with economic models. It is because of this connection that we decided to discuss together the continuous time and the discrete time situations.

Having in mind the purpose of the book, there is considerable attention to the study of linear theory. The nonlinear theory is reduced to some fundamental facts about existence and uniqueness, numerical methods, stability and some optimization problems.

The book has two distinct parts: the first one is devoted to the study of free evolutions, described by differential equations or discrete–time systems, while the second one considers some control problems. Whereas the first part is kept at a more elementary level and can be used for an advanced undergraduate course, the second one is more suitable for a graduate course.

We start with the simplest case of first order equations, linear and affine, which allows to describe simple price models, economic growth models and also the deduction of the form of some production functions. The logistic equation is the simplest nonlinear example. The study goes further with second order linear equations with constant coefficients and then with linear systems with constant coefficients. In the discussion of models, even in this linear setting, attention is paid to develop the idea of stability.

As already mentioned, there is some nonlinear theory. We prove the fundamental existence and uniqueness theorem, describe the main concepts of stability theory and prove the main theorem on stability by the linear approximation. The numerical methods are aimed at giving an idea concerning the general philosophy; it is why only a few schemes are described, but with full proofs of the error estimate.

The first part ends by showing how differential equations can be used to solve problems concerning nonlinear equations and optimization. While all material in this first part is rather conventional, we believe that the selection, the style, the strong interaction between differential equations, discrete–time systems and economic models make it suitable for the purpose of education in applied mathematics, especially for students in economic science and applications.

The second part discusses some topics in control. It starts with a stabilization problem of linear systems, both in continuous time and discrete time. We discuss here, as an illustration, a model of stabilizing the inflation rate. We go further to the problem of optimal stabilization with respect to a quadratic cost, again in continuous and in discrete time. The theory is illustrated by the discussion of some price models.

The next topics is devoted to the linear–quadratic optimization on finite horizon. From here we pass to the classical problem of calculus of variation (as a free optimization problem) and end with the general theory of optimal control, providing a proof for Pontryagin's Maximum Principle.

It is clear that this second part is more advanced and contains more recent results. Here also we tried to remain connected to economic models. We hope that this second part is accessible to people having an undergraduate background in Advanced Calculus and Linear Algebra. We also hope that this text can be used by teachers, students and workers in Applied Mathematics, Business Administration, Management and Computer Science, in order to get information and skill in using differential equations and their discrete counterpart to solve their problems.

CHAPTER 1

# Linear and Affine Differential Equations. Equations with Separated Variables

## 1.1 Differential Equations Modelling Growth Processes

In order to understand the natural way in which differential equations occur in the study of evolutionary processes, we begin with some simple population growth models, that have also possible economic interpretations.

We consider a population that has a magnitude of $x_t$ at time $t$; we accept a natural evolution law (Malthus's law) that asserts a proportionality between the velocity of population's growth and the population's magnitude. If $x_{t+1}$ is the population's magnitude at time $t+1$, then the velocity of growth is $x_{t+1} - x_t$ (the population's growth in a time unit) and the law can be written as $x_{t+1} - x_t = \alpha x_t$, hence $x_{t+1} = (1+\alpha) x_t$, wherefrom it follows $x_t = (1+\alpha)^t x_0$. In other words, the law accepted leads to the conclusion that the population grows in a *geometric progression* .

The above simple model was a *discrete time* model, the moments of time being denoted by *natural numbers*. To be able to use the ideas of great efficiency of mathematical analysis, we can consider models in *continuous time* (in fact, all evolutions take place in continuous time; the variants depend on the manner in which we describe this evolution *mathematically*) accepting also the assumption that the population's magnitude may be described by *any real number*; such an assumption seems natural if we deal with a population of micro-organisms, for which counting of individuals is meaningless.

In such situations, the velocity of evolution at a given time is modelled by means of the *derivative* of the function describing the evolution. Thus, if $x(t)$ expresses the population's magnitude at time $t$, then the velocity of evolution at time $t$ is modelled by the derivative $x'(t)$ and the evolution law adopted becomes $x'(t) = \alpha\, x(t)$.

Taking into account the differentiation rules for the exponential function, we can give an explicit form for the evolution. Indeed, multiplying the relation $x'(t) - \alpha\, x(t) = 0$ by $e^{-\alpha t}$, we obtain the relation $e^{-\alpha t} x'(t) - e^{-\alpha t} \alpha\, x(t) = 0$, which can be written as $\frac{d}{dt}\left[e^{-\alpha t} x(t)\right] = 0$, leading to the conclusion that

$t \mapsto e^{-\alpha t} x(t)$ is a *constant function*,

hence $e^{-\alpha t} x(t) = e^{-\alpha t_0} x(t_0)$, therefore $x(t) = e^{\alpha(t-t_0)} x(t_0)$;

this relation shows that the evolution is completely determined when the value of the function $x$ at point $t_0$ is given (meaning the magnitude of the population at time $t_0$).

If $\alpha > 0$, then we conclude, from the properties of the exponential function, that $\lim_{t\to\infty} x(t) = \infty$. This property points out the limits of the Malthusian model of growth, since it is not possible to conceive indefinitely increasing populations, because of the limited character of resources.

The above remark suggests the need to replace the Malthusian model by a more adequate one, describing *limited growth* (Verhulst). In its continuous time variant, the velocity of evolution is still proportional to the population's magnitude, but the proportionality coefficient is no longer constant; it depends on the magnitude of the population and vanishes when a given threshold is reached. This law of limited growth is expressed by $x'(t) = \alpha\,[\beta - x(t)]\,x(t)$.

Let us give an explicit form of the function $x$. We assume, starting from the interpretation of the model, that $0 < x(t) < \beta$ (the population is a positive quantity, which does not exceed the threshold $\beta$).

Then $[\beta - x(t)]\,x(t) \neq 0$ and we can write

$$\frac{x'(t)}{[\beta - x(t)]\,x(t)} = \alpha .$$

We integrate in both sides from $t_0$ to $t$

$$\int_{t_0}^{t} \frac{x'(s)}{(\beta - x(s))\,x(s)}\,ds = \alpha\,(t - t_0).$$

We transform the integral by substitution; it follows

$$\int_{x(t_0)}^{x(t)} \frac{dx}{[\beta - x]\,x} = \alpha\,(t - t_0).$$

We compute the integral using a simple fractions decomposition:

$$\frac{1}{x(\beta-x)}=\frac{1}{\beta}\left[\frac{1}{x}+\frac{1}{\beta-x}\right].$$

Thus we have

$$\int_{x(t_0)}^{x(t)}\frac{d\,x}{x(\beta-x)}=\frac{1}{\beta}\int_{x(t_0)}^{x(t)}\left(\frac{1}{x}+\frac{1}{\beta-x}\right)dx=$$

$$=\frac{1}{\beta}\left\{\ln x(t)-\ln x(t_0)-\ln\left[\beta-x(t)\right]+\ln\left[\beta-x(t_0)\right]\right\}=$$

$$=\frac{1}{\beta}\ln\left[\frac{x(t)}{\beta-x(t)}\cdot\frac{\beta-x(t_0)}{x(t_0)}\right]=\ln\left\{\left[\frac{x(t)}{\beta-x(t)}\cdot\frac{\beta-x(t_0)}{x(t_0)}\right]^{\frac{1}{\beta}}\right\}.$$

Therefore

$$\ln\left\{\left[\frac{x(t)}{\beta-x(t)}\cdot\frac{\beta-x(t_0)}{x(t_0)}\right]^{\frac{1}{\beta}}\right\}=\alpha\left(t-t_0\right),$$

hence

$$\left[\frac{x(t)}{\beta-x(t)}\cdot\frac{\beta-x(t_0)}{x(t_0)}\right]^{\frac{1}{\beta}}=e^{\alpha(t-t_0)},$$

$$\frac{x(t)}{\beta-x(t)}\cdot\frac{\beta-x(t_0)}{x(t_0)}=e^{\alpha\beta(t-t_0)},$$

$$\frac{x(t)}{\beta-x(t)}=\frac{x(t_0)}{\beta-x(t_0)}\cdot e^{\alpha\beta(t-t_0)},$$

$$x(t)=\left[\beta-x(t)\right]\frac{x(t_0)}{\beta-x(t_0)}\,e^{\alpha\beta(t-t_0)},$$

$$\left[1+\frac{x(t_0)\,e^{\alpha\beta(t-t_0)}}{\beta-x(t_0)}\right]x(t)=\frac{\beta\,x(t_0)}{\beta-x(t_0)}\,e^{\alpha\beta(t-t_0)}.$$

Herefrom, eventually

$$x(t)=\frac{\dfrac{\beta\,x(t_0)}{\beta-x(t_0)}\,e^{\alpha\beta(t-t_0)}}{1+\dfrac{x(t_0)}{\beta-x(t_0)}\,e^{\alpha\beta(t-t_0)}}=\frac{\beta}{1+\dfrac{\beta-x(t_0)}{x(t_0)}\,e^{-\alpha\beta(t-t_0)}}.$$

The formula obtained shows that the value of $x$ at any point $t$ is uniquely determined by its value at a point $t_0$; it can also be seen that the values of $x$ remain smaller than $\beta$ for any $t$ and that $\lim_{t \to \infty} x(t) = \beta$, what confirms the limited character of the growth process.

## 1.2 Linear Differential Equations

The example of population growth according to Malthus's law can be generalized in a natural way, leading to the very important class of first order linear differential equations.

Let $a : I \subset \mathrm{R} \to \mathrm{R}$ be a continuous function.

We say that $x$ is a *solution of the linear differential equation* defined by $a$ if

1) $x : I \to \mathrm{R}$ is differentiable function;

2) for any $t \in I$, $x'(t) = a(t)\, x(t)$,

here $I$ is an open interval.

**Theorem 1.1** *For any* $t_0 \in I$ *and any* $x_0 \in \mathrm{R}$ *there exists a solution of the linear differential equation defined by* $a$ *such that* $x(t_0) = x_0$ *and such a solution is unique.*

***Proof*** Since $a$ is a continuous function, it is integrable; let $x$ be the function defined by $x(t) = e^{\int_{t_0}^{t} a(s)\,ds} x_0$. This function is differentiable, since $t \mapsto \int_{t_0}^{t} a(s)\, ds$ is differentiable, the exponential function is differentiable and the composition of two differentiable functions is differentiable.

If we denote $A(t) = \int_{t_0}^{t} a(s)\,ds$, then we have $A'(t) = a(t)$ and, from $x(t) = e^{A(t)}\,x_0$ it follows $x'(t) = A'(t)\,e^{A(t)}\,x_0 = a(t)\,x(t)$, hence $x$ is a solution of the linear differential equation defined by $a$. Obviously, $A(t_0) = 0$, hence $x(t_0) = x_0$. We have still to prove the uniqueness. To do so, let us consider $x: I \to \mathrm{R}$ differentiable, $x'(t) = a(t)\,x(t)$, $x(t_0) = x_0$. We have $x'(t) - a(t)\,x(t) = 0$ and, multiplying by $e^{-A(t)}$, we obtain

$$e^{-A(t)}\,x'(t) - e^{-A(t)}\,a(t)\,x(t) = 0,$$

or

$$e^{-A(t)}\,x'(t) - e^{-A(t)}\,A'(t)\,x(t) = 0, \quad \frac{d}{dt}\left[e^{-A(t)}\,x(t)\right] = 0,$$

hence the function $t \mapsto e^{-A(t)}\,x(t)$ is constant.

Therefore $e^{-A(t)}\,x(t) = e^{-A(t_0)}\,x(t_0) = x_0$ and $x(t) = e^{A(t)}\,x_0$; this shows that any solution with $x(t_0) = x_0$ is represented by the formula given.

## 1.3 Linear Affine Differential Equations

Let $a: I \subset \mathrm{R} \to \mathrm{R}$, $b: I \subset \mathrm{R} \to \mathrm{R}$ be continuous functions. We call *solution of the linear affine differential equation* defined by $a$ and $b$ a function $x: I \to \mathrm{R}$ such that

1) $x$ is differentiable;

2) $x'(t) = a(t)\,x(t) + b(t)$, for any $t \in I$.

**Theorem 1.2** *For any $t_0 \in I$ and $x_0 \in \mathrm{R}$ there exists a solution $x$ of the linear affine differential equation defined by $a$ and $b$ that satisfies $x(t_0) = x_0$; this solution is unique and it is represented by the formula*

$$x(t) = e^{\int_{t_0}^{t} a(s)\,ds} x_0 + \int_{t_0}^{t} e^{\int_{\tau}^{t} a(s)\,ds} b(\tau)\,d\tau .$$

***Proof*** We prove first the uniqueness and the representation formula of the statement.

Let $x$ be a solution of the equation, satisfying $x(t_0) = x_0$.

We denote, as above, $A(t) = \int_{t_0}^{t} a(s)\,ds$ and we write $x'(t) - a(t)\,x(t) = b(t)$. Multiplying by $e^{-A(t)}$, we obtain

$$e^{-A(t)}\,x'(t) - e^{-A(t)}\,a(t)\,x(t) = e^{-A(t)}\,b(t)$$

and, taking into account that $a(t) = A'(t)$, we see that

$$\frac{d}{dt}[e^{-A(t)}\,x(t)] = e^{-A(t)}\,b(t).$$

Integrating from $t_0$ to $t$, we obtain

$$e^{-A(t)}\,x(t) - e^{-A(t_0)}\,x(t_0) = \int_{t_0}^{t} e^{-A(\tau)}\,b(\tau)\,d\tau ,$$

hence

$$e^{-A(t)}x(t) = e^{-A(t_0)}x(t_0) + \int_{t_0}^{t} e^{-A(\tau)}\,b(\tau)\,d\tau =$$

$$= x_0 + \int_{t_0}^{t} e^{-A(\tau)}\,b(\tau)\,d\tau ,$$

wherefrom it follows

$$x(t) = e^{A(t)}x_0 + e^{A(t)}\int_{t_0}^{t} e^{-A(\tau)}\,b(\tau)\,d\tau .$$

Further, we can write

$$x(t) = e^{A(t)}x_0 + \int_{t_0}^{t} e^{A(t)} \cdot e^{-A(\tau)}\, b(\tau)\, d\tau =$$

$$= e^{A(t)}x_0 + \int_{t_0}^{t} e^{A(t)-A(\tau)}\, b(\tau)\, d\tau .$$

But $A(t) - A(\tau) = \int_{t_0}^{t} a(s)\, ds - \int_{t_0}^{\tau} a(s)\, ds = \int_{\tau}^{t} a(s)\, ds$ and, eventually

$$x(t) = e^{\int_{t_0}^{t} a(s)\, ds} x_0 + \int_{t_0}^{t} e^{\int_{\tau}^{t} a(s)\, ds} b(\tau)\, d\tau ;$$

this proves the uniqueness and the validity of the representation formula in the statement.

We have still to prove that a function given by this formula is indeed a solution.

The function $t \mapsto \int_{t_0}^{t} e^{-A(\tau)}\, b(\tau)\, d\tau$ is differentiable and its derivative at point $t$ is $e^{-A(t)}\, b(t)$. Using general results from Calculus, it follows that $x$ is differentiable and

$$x'(t) = A'(t)\, e^{A(t)}x_0 + A'(t)\, e^{A(t)} \int_{t_0}^{t} e^{-A(\tau)} b(\tau)\, d\tau + e^{A(t)}\left[e^{-A(t)} b(t)\right] =$$

$$= a(t)\left[e^{A(t)}x_0 + e^{A(t)} \int_{t_0}^{t} e^{-A(\tau)}\, b(\tau)\, d\tau\right] + b(t) = a(t)\, x(t) + b(t),$$

hence the function defined by the relation given in the statement is a solution of the equation. The fact that $x(t_0) = x_0$ is immediately checked and the theorem is thus completely proved.

This result is sometimes expressed in the form “the general solution of the linear affine equation is obtained as the sum of a particular solution and the general solution of the linear equation”.

The formula that provides the particular solution of the affine equation, which takes the value zero at $t_0$, shows that this solution is a linear function of $b$, in the sense that, if we have

$$x_1(t) = \int_{t_0}^{t} e^{\int_{\tau}^{t} a(s)\,ds} b_1(\tau)\,d\tau,$$

$$x_2(t) = \int_{t_0}^{t} e^{\int_{\tau}^{t} a(s)\,ds} b_2(\tau)\,d\tau,$$

then it follows

$$\alpha_1\, x_1(t) + \alpha_2\, x_2(t) = \int_{t_0}^{t} e^{\int_{\tau}^{t} a(s)\,ds} \left[\alpha_1\, b_1(\tau) + \alpha_2\, b_2(\tau)\right] d\tau,$$

hence the solution corresponding to $\alpha_1\, b_1 + \alpha_2\, b_2$ is $\alpha_1\, x_1 + \alpha_2\, x_2$. This property is called the *superposition principle.*

## 1.4 Simplest Models of Price Evolution in a Market Economy

We consider a commodity having the price $P\,(t)$ at time $t$. We admit that the evolution of the price is governed by the law of supply and demand, expressed in the form: the velocity of price evolution is proportional to the difference between demand and supply. Denoting by $D\,(t)$ demand at time $t$ and by $S\,(t)$ supply at time $t$, the law can be written as $P'(t) = \lambda\left[D(t) - S(t)\right]$.

Since the price increases when demand exceeds supply, it is natural to assume $\lambda > 0$. At the same time, demand, as well as supply depend on the price level and the simplest assumption is that of affine dependence: $D(t) = \alpha + a\,P(t)$, $S(t) = \beta + b\,P(t)$.

If we admit these assumptions, the following mathematical model of price evolution is obtained:

$$P'(t) = \lambda\left[\alpha + a\,P(t) - \beta - b\,P(t)\right] =$$

$$= \lambda\,(\alpha - \beta) + \lambda\,(a - b)\,P(t)\,.$$

The price evolution is described by a linear affine differential equation.

To obtain explicit expressions, let us start from the last remark concerning the representation of the solutions of affine equations and let us look for a simple particular solution.

The simplest such solution is a *constant solution*, expressing the *equilibrium* situation, when demand and supply are equal $(P' \equiv 0)$. Such a solution is given by $\hat{P} = \dfrac{\alpha - \beta}{b - a}$. It has an economic significance only if $\hat{P} > 0$ (prices have to be positive).

We remark that $\alpha$ is close to the level of demand when the price is small and $\beta$ to the level of supply when the price is small, so that the assumption $\alpha > \beta$ seems to be natural (we could even accept that $\beta = 0$). The coefficients $a$ and $b$ show the way in which the price level influences demand and supply; when the price increases, the demand decreases, what means that it is natural to assume $a < 0$; when the price increases, supply increases as well, so that we admit $b > 0$. It follows that $b - a > 0$, hence $\hat{P}$ has an economic meaning.

Taking into account the general remark, we can write $P(t) = \hat{P} + p(t)$, where $p$ is the general solution of the linear equation $p'(t) = \lambda(a - b)\,p(t)$. We have $p(t) = e^{\lambda(a-b)t}c$, hence

$$P(t) = \hat{P} + e^{\lambda(a-b)t}c, \quad c = e^{\lambda(a-b)t_0}\left[P(t) - \hat{P}\right].$$

Finally, $P(t) = \hat{P} + e^{\lambda(a-b)(t-t_0)}\left[P(t_0) - \hat{P}\right]$.

Since $a - b < 0$, it follows $\lim_{t\to\infty} e^{\lambda(a-b)(t-t_0)} = 0$, hence $\lim_{t\to\infty} P(t) = \hat{P}$, what expresses the tendency of prices to move towards the equilibrium value.

Thus, the model adopted corresponds to a "*stable*" market.

The very simplified character of the model is manifest, among other facts, in that we considered an isolated commodity, without taking into account the influence of all prices in a market on the price of a given commodity.

The models of this type can be presented in several variants. Thus, we can accept the idea that demand and supply depend not only on the instantaneous level of prices, but also on their evolution tendency, expressed by means of the derivative $P'$. In this way, we obtain a model of the form

$$P'(t) = \lambda\left[\alpha + a_0 P(t) + a_1 P'(t) - \beta - b_0 P(t) - b_1 P'(t)\right],$$

which also leads to an affine equation:

$$\left[1 + \lambda(b_1 - a_1)\right]P'(t) = \lambda(\alpha - \beta) - \lambda(b_0 - a_0)P(t)$$

or

$$P'(t) = \frac{\lambda(\alpha - \beta)}{1 + \lambda(b_1 - a_1)} - \frac{\lambda(b_0 - a_0)}{1 + \lambda(b_1 - a_1)}P(t).$$

We remark that the equilibrium value in this model is the same as in the previous one.

## 1.5 Discrete – Time Models for Price Evolution

In this case, the general law of price evolution, on the basis of demand and supply, is written as $P_{t+1} - P_t = \lambda(D_t - S_t)$.

If $D_t = \alpha + a\,P_t$, $S_t = \beta + b\,P_t$, it follows

$$P_{t+1} = P_t + \lambda(\alpha + a\,P_t - \beta - b\,P_t) =$$

$$= \lambda(\alpha - \beta) + [1 - \lambda(b - a)]\,P_t.$$

The equilibrium price corresponds, in this case as well, to the situation of equality between demand and supply, hence

$$\alpha + a\,\hat{P} = \beta + b\,\hat{P}, \text{ or } \hat{P} = \frac{\alpha - \beta}{a - b},$$

the same as in the continuous time case.

Denoting $p_t = P_t - \hat{P}$, we obtain

$$p_{t+1} = P_{t+1} - \hat{P} =$$

$$= p_t - \hat{P} + \lambda\left[\alpha + a\left(p_t + \hat{P}\right) - \beta - b\left(p_t + \hat{P}\right)\right] =$$

$$= p_t + \lambda(a - b)\,p_t = [1 - \lambda(b - a)]\,p_t;$$

we deduce

$$p_t = [1 - \lambda(b - a)]^{t - t_0}\,p_{t_0},$$

hence

$$P_t = \hat{P} + [1 - \lambda(b - a)]^{t - t_0}\left(P_{t_0} - \hat{P}\right).$$

We have $\lim\limits_{t\to\infty} P_t = \hat{P}$ if $-1 < 1 - \lambda(b - a) < 1$.

Accepting, as above, that $b > 0$, $a < 0$ and $\lambda > 0$, the second inequality is automatically verified. The first inequality reduces to $\lambda(b - a) < 2$, or $\lambda < \dfrac{2}{b - a}$.

We remark that a too strong influence of the difference between demand and supply on price increase, as well as a too strong dependence of demand and of supply on prices can have destabilizing effects on the price evolution.

The discrete-time variant allows for the description of several types of dependence of demand and supply on the level of prices. Thus, we can admit that supply depends not on the instantaneous level of prices, but on the previous one, while demand depends only on the present level. Under such assumptions, even the *equality* condition of demand and supply will determine a dynamics of prices.

Indeed, if $D_t = \alpha + a P_t$ and $S_t = \beta + b P_{t-1}$, then the equality $D_t = S_t$ will lead to $\alpha + a P_t = \beta + b P_{t-1}$, hence

$$P_t = \frac{1}{a}(\beta - \alpha + b P_{t-1}).$$

Essentially, we have an affine dependence of the same type as before. There exists a constant price satisfying the above relation, given by $\hat{P} = \frac{1}{a}\left(\beta - \alpha + b\hat{P}\right)$, hence $\hat{P} = \frac{\alpha - \beta}{b - a}$, the same as for the previous models.

Denoting again $p_t = P_t - \hat{P}$, we have now

$$p_t = \frac{1}{a}\left[\beta - \alpha + b\left(p_{t-1} + \hat{P}\right)\right] - \hat{P} = \frac{b}{a} p_{t-1},$$

hence $p_t = \left(\frac{b}{a}\right)^{t-t_0} p_{t_0}$ and $P_t = \hat{P} + \left(\frac{b}{a}\right)^{t-t_0}\left(P_{t_0} - \hat{P}\right)$.

In this case, stability of equilibrium, expressed by the condition $\lim_{t\to\infty} P_t = \hat{P}$, will be obtained if $\left|\frac{b}{a}\right| < 1$, hence if $b < |a|$.

We stress again that the above discussion corresponds to the equality of supply and demand. If not, we would obtain a connection between $P_{t+1}$, $P_t$ and $P_{t-1}$. Such equations will be discussed later on.

## 1.6 Simplest Models for Economic Growth

Let $Y(t)$ be the national product at time $t$ allotted for consumption and investment; investment, in its turn, has an *autonomous* component, $A(t)$, independent of the income (or national product), and another component, $I(t)$, which is related to income by the relation $I(t) = vY'(t)$.

Thus, we have the fundamental relationship

$$I(t) = C(t) + A(t) + I(t).$$

Accepting the simple relation $C(t) = cY(t)$ (meaning that a fixed ratio of income is allotted for consumption), we obtain the affine differential equation:

$$Y(t) = c\,Y(t) + v\,Y'(t) + A(t), \text{ i.e.}$$

$$Y'(t) = \frac{1}{v}(1-c)\,Y(t) - \frac{1}{v}A(t).$$

We denote $1-c$ by $s$, where $s$ has the meaning of "propensity to save"; the model takes the form $Y'(t) = \frac{s}{v}Y(t) - \frac{1}{v}A(t)$.

This model is known as the *Harrod-Domar model.*

If $A(t)$ is equal to a constant $A$, then we have an equilibrium level of the income $\hat{Y} = \frac{A}{s}$, and the evolution is given by

$$Y(t) = \frac{A}{s} + e^{\frac{s}{v}(t-t_0)}\left[Y(t_0) - \frac{A}{s}\right].$$

There exist several other models for the evolution of income (national product), leading to affine differential equations. Thus, we can define the "demand"

$$Z(t) = C(t) + A(t) = cY(t) + A(t)$$

and adopt the assumption that the velocity of evolution for the product $Y(t)$ is proportional to the deviation between its actual value and the "demand" (production is stimulated by demand):

$$Y'(t) = \lambda\left[Z(t) - Y(t)\right],\ \lambda > 0.$$

If we adopt again the assumption that $C(t) = cY(t)$, then it follows

$$Y'(t) = \lambda\left[cY(t) + A(t) - Y(t)\right] = -\lambda s Y(t) + \lambda A(t).$$

This model is known as the *Phillips model with multiplier*. If $A(t) = A$ is constant, we have again the equilibrium $\hat{Y} = \dfrac{A}{s}$ and a stable evolution towards the equilibrium, since $\lambda s > 0$.

## 1.7 Discrete – Time Models for Economic Growth

We write again the fundamental identity

$$Y_t = C_t + I_t + A_t.$$

We admit now that the level of consumption represents a given ratio of the previous level of income, hence $C_t = c\,Y_{t-1}$, and that investments at time $t-1$ induce the increase in income $Y_t - Y_{t-1} = \dfrac{1}{v} I_{t-1}$. Taking into account the relationship

$$Y_{t-1} = c\,Y_{t-1} + I_{t-1} + A_{t-1},$$

it follows

$$I_{t-1} = s\,Y_{t-1} - A_{t-1},$$

hence

$$Y_t = Y_{t-1} + \frac{1}{v}[s\,Y_{t-1} - A_{t-1}]$$

or

$$Y_t = \left[1 + \frac{s}{v}\right] Y_{t-1} - \frac{1}{v} A_{t-1}.$$

If we assume that $A_t = A$ is constant, we have the equilibrium given by $\hat{Y} = \dfrac{A}{s}$, as in the previous case, and the evolution is given by $Y_t = \hat{Y} + \left[1 + \dfrac{s}{v}\right]^{t-t_0} (Y_{t_0} - \hat{Y})$.

## 1.8 Production Functions

Production functions describe the connection between production factors (capital and labour) and the level of production. If we denote capital by $K$ and labour by $L$, and we consider a function $F : \mathrm{R}_+ \times \mathrm{R}_+ \to \mathrm{R}_+$, $F(K,L)$ is interpreted as the corresponding level of production. It is natural to admit the following properties:

$$F(0,0) = 0, \quad F(\lambda K, \lambda L) = \lambda F(K,L).$$

The last property, representing homogeneity of the first degree, expresses the fact that, if we multiply capital and labour by the same factor, then the result is multiplied by the same factor.

For a given production function, we associate in a natural way the following average indicators:

$y = \dfrac{Y}{L} = \dfrac{F(K,L)}{L}$, the average productivity of labour;

$\frac{Y}{K} = \frac{F(K,L)}{K}$, the average efficiency of capital;

$k = \frac{K}{L}$, the average capital–labour ratio.

We also associate the differential (marginal) indicators:

$$\eta_L(K,L) = \frac{\partial F(K,L)}{\partial L}; \quad \eta_K(K,L) = \frac{\partial F(K,L)}{\partial K}.$$

We have

$$y = \frac{1}{L} F(K,L) = F\left(\frac{K}{L},1\right) = F(k,1) = f(k);$$

$$\eta_L(K,L) = \frac{\partial}{\partial L}\left[L\, f\left(\frac{K}{L}\right)\right] = f\left(\frac{K}{L}\right) - L\, f'\left(\frac{K}{L}\right)\left(\frac{K}{L^2}\right) =$$

$$= f\left(\frac{K}{L}\right) - \left(\frac{K}{L}\right) f'\left(\frac{K}{L}\right) = f(k) - k\, f'(k);$$

$$\eta_K(K,L) = \frac{\partial}{\partial K}\left[L\, f\left(\frac{K}{L}\right)\right] = f'\left(\frac{K}{L}\right) = f'(k).$$

We also consider the elasticities:

$$E_L(K,L) = \frac{\frac{\partial F}{\partial L}(K,L)}{\frac{F(K,L)}{L}} = \frac{f\left(\frac{K}{L}\right) - \frac{K}{L} f'\left(\frac{K}{L}\right)}{f\left(\frac{K}{L}\right)} = 1 - \frac{k\, f'(k)}{f(k)};$$

$$E_K(K,L) = \frac{\frac{\partial F}{\partial K}(K,L)}{\frac{F(K,L)}{K}} = \frac{f'\left(\frac{K}{L}\right)}{f\left(\frac{K}{L}\right)\Big/\left(\frac{K}{L}\right)} = \frac{k\, f'(k)}{f(k)}.$$

We define the *rate of factor substitution* as

$$r(K,L) = \frac{\frac{\partial F}{\partial L}(K,L)}{\frac{\partial F}{\partial K}(K,L)} = \frac{f\left(\frac{K}{L}\right) - \frac{K}{L} f'\left(\frac{K}{L}\right)}{f'\left(\frac{K}{L}\right)} = \rho\left(\frac{K}{L}\right) = \rho(k),$$

and the *elasticity coefficient of substitution* as

$$\sigma\left(\frac{K}{L}\right) = \frac{\rho'\left(\frac{K}{L}\right)}{\rho\left(\frac{K}{L}\right)\Big/\left(\frac{K}{L}\right)} = k\,\frac{\rho'(k)}{\rho(k)}.$$

We can identify several types of production functions, starting from simple assumptions concerning the indicators defined above.

Thus, we can assume that the elasticities are constant. It can be seen that for such functions we have

$$k\frac{f'(k)}{f(k)} = 1-\alpha\text{, where } \alpha \text{ is a constant,}$$

hence $f$ is a solution of a linear differential equation

$$f'(k) = \frac{1-\alpha}{k} f(k).$$

The representation formula of the solution, given in **1.2**, leads to

$$f(k) = e^{\int_1^k \frac{1-\alpha}{t}dt} f(1).$$

We have $\displaystyle\int_1^k \frac{1-\alpha}{t}dt = (1-\alpha)\ln k = \ln k^{1-\alpha}$, hence

$$f(k) = e^{\ln k^{1-\alpha}} f(1) = k^{1-\alpha} f(1)$$

and therefore

$$F(K,L) = L f\left(\frac{K}{L}\right) = L\left(\frac{K}{L}\right)^{1-\alpha} F(1,1) = F(1,1)K^{1-\alpha}L^{\alpha}.$$

We found, in this way, the *Cobb-Douglas type production functions*, if $0 < \alpha < 1$. If $\alpha = 0$, then the function is proportional to capital and independent of labor, and if $\alpha = 1$, then $F(K,L)$ does not depend upon $K$ and is proportional to $L$.

Let us look now for the production functions having constant elasticity of factor substitution.

If $\sigma\left(\frac{K}{L}\right) = \alpha$, then we deduce $k\rho'(k) = \alpha\rho(k)$, hence $\rho$ is a solution of a linear differential equation: $\rho(k) = \rho(1)k^{\alpha}$. Further,

$$\rho(k) = -k + \frac{f(k)}{f'(k)},\quad \frac{f(k)}{f'(k)} = k + \rho(k)$$

and $f$ is a solution of the linear differential equation

$$f'(k) = \frac{1}{k + \rho(1)\,k^{\alpha}}\, f(k), \text{ hence } f(k) = f(1)\, e^{\int_1^k \frac{dt}{t+\rho(1)\,t^{\alpha}}}.$$

We can compute $\displaystyle\int \frac{dt}{t + \beta\, t^{\alpha}} = \int \frac{dt}{t\left(1 + \beta\, t^{\alpha-1}\right)}$.

We denote $t^{\alpha-1} = \tau$; $(\alpha - 1)\, t^{\alpha-2}\, dt = d\tau$.

$$\frac{dt}{t\left(1 + \beta\, t^{\alpha-1}\right)} = \frac{d\tau}{(\alpha - 1)\, t^{\alpha-2}\, t\left(1 + \beta\, t^{\alpha-1}\right)} =$$

$$= \frac{d\tau}{(\alpha - 1)\, \tau\, (1 + \beta\, \tau)} = \frac{1}{(\alpha - 1)}\left[\frac{d\tau}{\tau} - \frac{\beta\, d\tau}{1 + \beta\, \tau}\right],$$

hence

$$\int \frac{dt}{t + \beta t^{\alpha}} = \ln\left(\frac{t^{\alpha-1}}{1 + \beta\, t^{\alpha-1}}\right)^{\frac{1}{\alpha-1}};$$

$$\int_1^k \frac{dt}{t + \beta\, t^{\alpha}} = \ln\left(\frac{k^{\alpha-1}\,(1 + \beta)}{1 + \beta\, k^{\alpha-1}}\right)^{\frac{1}{\alpha-1}},$$

what leads to

$$f(k) = f(1)\left(\frac{k^{\alpha-1}\,(1 + \beta)}{1 + \beta\, k^{\alpha-1}}\right)^{\frac{1}{\alpha-1}} = f(1)\left(\frac{1 + \beta}{\beta + k^{1-\alpha}}\right)^{\frac{1}{\alpha-1}};$$

$$F(K,L) = L\, f\!\left(\frac{K}{L}\right) = f(1)\, L\left(\frac{1 + \beta}{\left(\frac{K}{L}\right)^{1-\alpha} + \beta}\right)^{\frac{1}{\alpha-1}} = A\left[K^{1-\alpha} + \beta\, L^{1-\alpha}\right]^{\frac{1}{1-\alpha}},$$

where $A$ is a constant equal to $f(1)\,(1 + \beta)^{\frac{1}{\alpha-1}}$ and $\beta = \rho(1)$.

The production function obtained in this way is a *CES type production function* (Constant Elasticity of Substitution).

## 1.9 Equations with Separated Variables

The differential equation modelling limited growth is a special example of a differential equation with separated variables.

Let $a: I \subset \mathbb{R} \to \mathbb{R}$ and $b: J \subset \mathbb{R} \to \mathbb{R}$ be continuous functions, $I$ and $J$ open intervals. A function $x: I_0 \subset I \to J$ is called a *solution of the differential equation with separated variables* defined by $a$ and $b$ if

1) $x$ is differentiable;
2) $x'(t) = a(t)\, b(x(t))$, for any $t \in I_0$.

**Theorem 1.3** *If $b(x) \neq 0$ for any $x \in J$ then for any $t_0 \in I$, $x_0 \in J$ there exist an interval $I_{(t_0, x_0)} \subset I$ and a function $x: I_{(t_0, x_0)} \to J$ which is a solution of the differential equation with separated variables defined by $a$ and $b$ and satisfies $x(t_0) = x_0$.*

***Such a solution is unique, in the sense that two solutions that coincide at $t_0$ coincide on the intersection of their domains.***

***Proof*** **We prove first uniqueness, by giving a representation formula for the solution, starting from the functions $a$ and $b$.**

**Since $a$ is continuous, it is integrable; we denote**

$$A(t) = \int_{t_0}^{t} a(s)\, ds.$$

Since $b$ is continuous and $b(x) \neq 0$, for any $x \in J$, the function $x \mapsto \dfrac{1}{b(x)}$ is continuous, hence integrable.

We denote $B(x) = \int_{x_0}^{x} \frac{d\xi}{b(\xi)}$ and notice that $B'(x) = \frac{1}{b(x)}$.

Since $b(x) \neq 0$ for any $x \in J$, it follows that $b$ does not change sign in $J$ (the Darboux property of a continuous function shows that if $b$ changed its sign then it would have to vanish at a point in $J$). We conclude that $x \mapsto \frac{1}{b(x)}$ preserves sign, hence $B$ is a strictly monotonous function. It follows that $B$ is invertible; we denote by $B^{-1}$ the inverse (satisfying $B^{-1}(B(x)) = x$, whatever $x \in J$).

Let us now consider a solution $x$ of the differential equation, satisfying $x(t_0) = x_0$.

We have $x'(t) = a(t)\,b(x(t))$ and, since $b(x(t)) \neq 0$, it follows $\frac{x'(t)}{b(x(t))} = a(t)$. We integrate on the interval $(t_0, t)$ and we obtain

$$\int_{t_0}^{t} \frac{x'(s)\,ds}{b(x(s))} = \int_{t_0}^{t} a(s)\,ds\text{, hence } \int_{x(t_0)}^{x(t)} \frac{d\xi}{b(\xi)} = \int_{t_0}^{t} a(s)\,ds.$$

With the above notations, the relation obtained can be written as $B(x(t)) = A(t)$, $x(t) = B^{-1}(A(t))$.

We have to find out for what values of $t$ is this formula valid. The function $B$ is strictly monotonous and continuous, hence it maps the interval $J$ bijectively on $B(J)$. The inverse function $B^{-1}$ is defined on $B(J)$ and its image is $J$. The function $B^{-1}(A(t))$ will be defined for those $t$ for which $A(t)$ is in $B(J)$; since $B(x_0) = 0$, it follows that $0 \in B(J)$ and since $B(J)$ is an open interval, we conclude that $B(J)$ will contain a neighbourhood of 0.

On the other hand, $A(t_0) = 0$ and $A$ is continuous, hence there exists a neighbourhood of $t_0$ such that, for $t$ in this neighbourhood, $A(t)$ is in $B(J)$ and hence, for $t$ in this neighbourhood, we can define $B^{-1}(A(t))$.

The representation formula obtained shows that the solution is <u>*unique*</u>, any solution satisfying $x(t_0) = x_0$ being equal, at any point of its domain, to $B^{-1}(A(t))$.

Let us show now that the function defined by $x(t) = B^{-1}(A(t))$ is a solution of the problem. We notice first that

$$x(t_0) = B^{-1}(A(t_0)) = B^{-1}(0) = x_0 .$$

Using the theorem on the differentiability of the inverse function and the chain rule, it follows that $x$ is differentiable.

We write $B(x(t)) = A(t)$ and apply the chain rule.

We infer $B'(x(t))x'(t) = A'(t)$, hence

$$\frac{x'(t)}{b(x(t))} = a(t), \quad x'(t) = a(t)\, b(x(t))$$

and this ends the proof.

## 1.10 Notes and References

This chapter is introductory and quite elementary. The matter may be found in standard books on differential or difference equations. We mention as closest references Allen [1] and Brand [3]. For production functions, we may refer to [2].

# CHAPTER 2

# Linear Differential Equations with Constant Coefficients

## 2.1 Second Order Differential Equations with Constant Coefficients

We start with the simple remark that, if $x(t) = e^{\alpha t}$, then $x'(t) = \alpha e^{\alpha t}$, $x''(t) = \alpha^2 e^{\alpha t}$, hence $x''(t) = \alpha^2 x(t)$; in the same way, if $x(t) = e^{-\alpha t}$, then $x'(t) = -\alpha e^{-\alpha t}$, $x''(t) = \alpha^2 e^{-\alpha t}$ and again $x''(t) = \alpha^2 x(t)$. We deduce that the functions $t \mapsto e^{\alpha t}$ and $t \mapsto e^{-\alpha t}$ satisfy the equation $x'' = \alpha^2 x$; it can be immediately seen that, for any real numbers $C_1$ and $C_2$, the function $x$ defined by $x(t) = C_1 e^{\alpha t} + C_2 e^{-\alpha t}$ satisfies the same differential equation.

We have $x'(t) = \alpha C_1 e^{\alpha t} - \alpha C_2 e^{-\alpha t}$ and, if we put $x(0) = x_0$, $x'(0) = v_0$, it follows $C_1 + C_2 = x_0$, $\alpha(C_1 - C_2) = v_0$,

$$C_1 = \frac{1}{2}\left(x_0 + \frac{v_0}{\alpha}\right),\ C_2 = \frac{1}{2}\left(x_0 - \frac{v_0}{\alpha}\right),$$

hence

$$x(t) = \frac{e^{\alpha t} + e^{-\alpha t}}{2} x_0 + \frac{e^{\alpha t} - e^{-\alpha t}}{2} \frac{v_0}{\alpha} = x_0 \operatorname{ch} \alpha t + \frac{v_0}{\alpha} \operatorname{sh} \alpha t;$$

$$\left(\operatorname{ch} \alpha t = \frac{e^{\alpha t} + e^{-\alpha t}}{2}; \ \operatorname{sh} \alpha t = \frac{e^{\alpha t} - e^{-\alpha t}}{2}\right).$$

The above developments led to the explicit form of the solution for the differential equation $x'' - \alpha^2 x = 0$, satisfying

$$x(0) = x_0, \ x'(0) = v_0.$$

Analogously, if $x(t) = \cos \alpha t$, then it follows $x'(t) = -\alpha \sin \alpha t$, $x''(t) = -\alpha^2 \cos \alpha t$, hence $x''(t) + \alpha^2 x(t) = 0$.

If $x(t) = \sin \alpha t$, then $x'(t) = \alpha \cos \alpha t$, $x''(t) = -\alpha^2 \sin \alpha t$, and again $x''(t) + \alpha^2 x(t) = 0$.

Since $t \mapsto \cos \alpha t$ and $t \mapsto \sin \alpha t$ satisfy the equation $x''(t) + \alpha^2 x(t) = 0$, it can be seen immediately that, for any real constants $C_1$ and $C_2$, the function defined by

$$x(t) = C_1 \cos \alpha t + C_2 \sin \alpha t$$

is a solution of the equation.

We have $x'(t) = -\alpha C_1 \sin \alpha t + \alpha C_2 \cos \alpha t$ and we determine the constants $C_1$ and $C_2$ from the conditions $x(0) = x_0$, $x'(0) = v_0$.

We have $x(0) = C_1 = x_0$ and $x'(0) = \alpha C_2 = v_0$, hence

$$x(t) = x_0 \cos \alpha t + \frac{v_0}{\alpha} \sin \alpha t.$$

In this way, we obtained the general form of the solution of equation $x'' + \alpha^2 x = 0$, satisfying $x(0) = x_0$, $x'(0) = v_0$.

It is useful to write this solution in another form, too. We denote

$$A = \sqrt{x_0^2 + \frac{v_0^2}{\alpha^2}}, \quad \frac{x_0}{\sqrt{x_0^2 + \frac{v_0^2}{\alpha^2}}} = \cos \varphi, \quad \frac{x_0}{\sqrt{x_0^2 + \frac{v_0^2}{\alpha^2}}} = -\sin \varphi.$$

With these notations, we have

$$x(t) = A(\cos\alpha t\cos\varphi - \sin\alpha t\sin\varphi) = A\cos(\alpha t + \varphi).$$

The function $x$ describes an oscillation with period $\dfrac{2\pi}{\alpha}$, amplitude $A$ and phase $\varphi$; it can be seen that the period of the oscillation does not depend on the initial conditions $(x_0, v_0)$, while the amplitude and the phase depend on the initial conditions.

We are now in the position of studying the general situation

$ax'' + bx' + cx = 0$ with $a \neq 0$,

which is equivalent to

$$x'' + \frac{b}{a}x' + \frac{c}{a}x = 0.$$

We shall try to reduce this general situation to the cases studied above, by performing a change of variables.

Let us consider a solution $x$ of the equation and define a new function $y$ by $y(t) = e^{\beta t}x(t)$. It follows $x(t) = e^{-\beta t}y(t)$,

$$x'(t) = e^{-\beta t}y'(t) - \beta e^{-\beta t}y(t),$$

$$x''(t) = e^{-\beta t}y''(t) - 2\beta e^{-\beta t}y'(t) + \beta^2 e^{-\beta t}y(t)$$

and, if

$$ax''(t) + bx'(t) + cx(t) = 0,$$

then we have

$$a\left[e^{-\beta t}y''(t) - 2\beta e^{-\beta t}y'(t) + \beta^2 e^{-\beta t}y(t)\right] +$$

$$b\left[e^{-\beta t}y'(t) - \beta e^{-\beta t}y(t)\right] + c\,e^{-\beta t}y(t) = 0.$$

Simplifying by $e^{-\beta t}$, we deduce

$$a\,y''(t) + (b - 2\beta a)\,y'(t) + (a\beta^2 - b\beta + c)\,y(t) = 0.$$

In order to find ourselves in one of the situations studied

above, we have to put $\beta = \frac{b}{2a}$, what leads to

$$y''(t) + \left(\beta^2 - \frac{b}{a}\beta + \frac{c}{a}\right)y(t) = 0,$$

hence to

$$y''(t) + \left(\frac{c}{a} - \frac{b^2}{4a^2}\right)y(t) = 0.$$

In what follows, we distinguish several cases.

The simplest one is that in which $\frac{c}{a} = \frac{b^2}{4a^2}$, hence $b^2 - 4ac = 0$. In this case, we have $y''(t) = 0$, hence $y(t)$ is a polynomial function of the first degree:

$$y(t) = C_1 + C_2 t, \text{ therefore } x(t) = e^{-\frac{b}{2a}t}(C_1 + C_2 t).$$

We can determine directly the constants $C_1$ and $C_2$:

$C_1 = y(0)$, $C_2 = y'(0)$; but $y(0) = x(0) = x_0$, $y'(0) = x'(0) + \beta y(0)$, hence $C_2 = v_0 + \frac{b}{2a}x_0$ and, finally

$$x(t) = e^{-\frac{b}{2a}t}\left[x_0 + \left(v_0 + \frac{b}{2a}\right)t\right].$$

Let us assume now that $b^2 - 4ac > 0$.

We denote $\alpha^2 = \frac{b^2 - 4ac}{4a^2}$ and obtain $y''(t) - \alpha^2 y(t) = 0$, hence $y(t) = C_1 e^{\alpha t} + C_2 e^{-\alpha t}$ is a solution; $x(t) = C_1 e^{(-\beta+\alpha)t} + C_2 e^{(-\beta-\alpha)t}$.

We have

$$-\beta + \alpha = -\frac{b}{2a} + \frac{\sqrt{b^2 - 4ac}}{2a},$$

$$-\beta - \alpha = -\frac{b}{2a} - \frac{\sqrt{b^2 - 4ac}}{2a},$$

and we notice that $-\beta+\alpha$ and $-\beta-\alpha$ are the roots of the second degree equation $ar^2+br+c=0$. We denote these roots by $r_1$ and $r_2$ and we have

$$x(t)=C_1e^{r_1t}+C_2e^{r_2t},\quad x'(t)=r_1C_1e^{r_1t}+r_2C_2e^{r_2t}.$$

We can determine the constants $C_1$ and $C_2$ by means of the initial conditions $x(0)=x_0$, $x'(0)=v_0$, from the relations

$$C_1+C_2=x_0,\quad r_1C_1+r_2C_2=v_0.$$

We consider now the last case, when $b^2-4ac<0$; in this case, we denote $\alpha^2=\dfrac{4ac-b^2}{4a^2}$ and we obtain $y''(t)+\alpha^2y(t)=0$, hence $y(t)=C_1\cos\alpha t+C_2\sin\alpha t$. Thus, we deduce

$$x(t)=e^{-\frac{b}{2a}t}\left[C_1\cos\frac{\sqrt{4ac-b^2}}{2a}t+C_2\sin\frac{\sqrt{4ac-b^2}}{2a}t\right].$$

We notice that in this case we can also have a connection with the roots of the second degree equation; if $b^2-4ac<0$, then the roots are of the form $r_1=\rho+i\alpha$, $r_2=\rho-i\alpha$, $\rho=-\dfrac{b}{2a}$, $\alpha=\dfrac{\sqrt{4ac-b^2}}{2a}$ and the solution can be written as

$$x(t)=e^{\rho t}\left(C_1\cos\alpha t+C_2\sin\alpha t\right).$$

We also notice that the first case we considered, when $b^2-4ac=0$, corresponds to the situation in which the roots of the equation $ar^2+br+c=0$ are real and equal: $r_1=r_2=r$.

The structure of the solution is $x(t)=e^{rt}\left(C_1+C_2t\right)$.

Thus, we find out that the structure of the solutions of a second order linear differential equation is obtained immediately by means of a second degree algebraic equation $ar^2 + br + c = 0$, called characteristic equation associated to the differential equation.

Let us make some important remarks concerning the behaviour of solutions for $t \to \infty$.

If the roots of the characteristic equation are real, then the behaviour is given by the exponentials $e^{r_1 t}$ and $e^{r_2 t}$. The solutions converge to zero for $t \to \infty$ if the roots $r_1$ and $r_2$ are negative. The same conclusion is valid if $r_1 = r_2 < 0$.

If the roots are not real, the behaviour is given by their real part $\rho$. If $\rho < 0$, then the solutions converge to zero for $t \to \infty$; if $\rho > 0$, then the solutions are oscillatory and unbounded, and if $\rho = 0$, then the solutions are oscillatory and bounded.

## 2.2 Discrete – Time Second Order Linear Equations

In what follows, we consider the recurrence equations of the form $ax_{t+2} + bx_{t+1} + cx_t = 0$, with $a \neq 0$.

In this case too, we denote the argument by $t$, although here $t$ is an integer number; we mark this fact by writing it as an index.

We look for solutions of the form $x_t = q^t$; substituting into the equation, we obtain $aq^{t+2} + bq^{t+1} + cq^t = 0$ and, after dividing by $q^t$, we have $aq^2 + bq + c = 0$.

If this equation has distinct real roots $q_1$, $q_2$, then we obtain for the recurrence equation the real solutions $q_1^t$, $q_2^t$ and the general solution will be of the form $x_t = C_1 q_1^t + C_2 q_2^t$.

To determine completely the solution, the values $x_0$ and $x_1$ should be given; the subsequent values $x_t$ , $t \geq 2$ follow then recurrently; if $c \neq 0$, then the sequence can be defined also for $t < 0$. From the conditions $C_1 + C_2 = x_0$, $C_1 q_1 + C_2 q_2 = x_1$ it follows that $C_1$ and $C_2$ are uniquely determined, since $q_1 \neq q_2$.

If the roots of the second degree equation are equal, then we perform a change of variable: $x_t = q^t u_t$, where $q$ is the multiple root of the second degree equation. It follows

$$a\, q^{t+2} u_{t+2} + b\, q^{t+1} u_{t+1} + c\, q^t u_t = 0\,,$$

$$a\, q^2 u_{t+2} + b\, q u_{t+1} + c\, u_t = 0\,.$$

Since the roots are equal, it follows $q = -\dfrac{b}{2a}$, hence

$$\frac{b^2}{4a} u_{t+2} - \frac{b^2}{2a} u_{t+1} + c\, u_t = 0\,,$$

$$b^2 u_{t+2} - 2b^2 u_{t+1} + 4ac\, u_t = 0\,.$$

But $b^2 - 4ac = 0$, hence

$$b^2 u_{t+2} - 2b^2 u_{t+1} + b^2 u_t = 0\,,\ \text{or}\ u_{t+2} - 2\, u_{t+1} + u_t = 0\,.$$

This can also be written as $u_{t+2} - u_{t+1} = u_{t+1} - u_t$.

Denoting $v_t = u_t - u_{t-1}$, we deduce $v_t = v_{t+1}$, hence the sequence $v_t$ is constant; writing $u_t - u_{t-1} = \hat{v}$, we deduce successively

$$u_1 = \hat{v} + u_0\,,\quad u_2 = \hat{v} + u_1\,, \ldots , u_t = \hat{v} + u_{t-1}$$

and, by adding, we obtain $u_t = t\hat{v} + u_0$, what leads to the solution

$$x_t = \left(-\frac{b}{2a}\right)^t (t\,\hat{v} + u_0)$$ of the recurrence equation.

If $b \neq 0$, then we can determine the constants $u_0$ and $\hat{v}$ from the conditions $x_0 = u_0$, $x_1 = \left(-\dfrac{b}{2a}\right)(\hat{v} + u_0)$.

If $b = 0$ then, taking into account that $b^2 - 4ac = 0$ and $a \neq 0$, we get $c = 0$ too, the equation reducing to $a\, x_{t+2} = 0$, which has the unique solution $x_t = 0$ for $t \geq 2$.

If the roots are not real, we take into account that for a complex number $z$, $z^t = |z|^t (\cos\theta t + i \sin\theta t)$ and that the roots of the second degree equation are complex conjugate. We obtain

$$x_t = C_1 |q|^t (\cos\theta t + i \sin\theta t) + C_2 |q|^t (\cos\theta t - i \sin\theta t) =$$

$$= |q|^t \left[(C_1 + C_2)\cos\theta t + i\,(C_1 - C_2)\sin\theta t\right].$$

The conditions for determining the constants $C_1$ and $C_2$ are now

$$C_1 + C_2 = x_0,\ |q|\left[x_0 \cos\theta + i\,(C_1 - C_2)\sin\theta\right] = x_1$$

and, since $\sin\theta \neq 0$, we can compute $C_1 - C_2$; knowing $C_1 + C_2$ and $C_1 - C_2$, we can determine $C_1$ and $C_2$. Since $x_0$ is real, it follows that $C_1$ and $C_2$ are complex conjugate. Denoting $C_1 = \alpha + i\beta$, $C_2 = \alpha - i\beta$, we have

$$C_1 + C_2 = 2a,\ C_1 - C_2 = 2i\,\beta;$$

hence

$$x_t = 2|q|^t (\alpha \cos\theta t - \beta \sin\theta t).$$

We see that $2\alpha = x_0$ and $2\beta |q| \sin\theta = |q| x_0 \cos\theta - x_1$.

We can analyze the behaviour of the sequence $x_t$ when $t \to \infty$.

In the last situation discussed, when the associated second degree equation has no real roots, it can be seen that the sequence converges to zero if $|q| < 1$, it is unbounded if $|q| > 1$ and bounded (oscillatory) if $|q| = 1$.

If the roots are real and equal, we also have convergence to zero for $|q|<1$. If $|q|=1$, we have in general an unbounded sequence, because of the factor $t\hat{v}+u_0$.

If the roots are real and distinct, then the solution $x_t$ converges to zero if both rots are strictly smaller than 1 in modulus. If the roots $q_1$ and $q_2$ satisfy $|q_1|\le 1$, $|q_2|\le 1$, then the sequence $x_t$ is bounded; if at least one of the roots is strictly greater than 1 in modulus, then there exist unbounded solutions.

## 2.3 Price Evolution in the Presence of Inventories

In the simple price evolution models discussed earlier, only two agents were present, one of them representing demand (consumption) and the other – supply (production). In the models with inventories a third agent occurs, which keeps the commodity in stores (trade) and which essentially influences price evolution.

If we denote by $Q_t$ the level of inventories at the end of time interval $t$, we have the natural relationship $Q_t - Q_{t-1} = S_t - D_t$, expressing the fact that the increase of inventories during the time interval $t$ is due to the difference between supply and demand.

**2.3.1** We adopt first as evolution law for prices the principle of proportionality between the variations of prices and of inventories:

$$P_t - P_{t-1} = -\lambda\left(Q_{t-1} - Q_{t-2}\right),\ \lambda > 0,$$

(when there is an excess of inventories, then prices tend to decrease).

If we accept again the relations

$$D_t = \alpha + aP_t, \; S_t = \beta + bP_t,$$

then we have

$$Q_{t-1} - Q_{t-2} = S_{t-1} - D_{t-1} = \beta + bP_{t-1} - \alpha - aP_{t-1} =$$

$$= \beta - \alpha + (b - a)P_{t-1}$$

and we obtain the model

$$P_t = [1 - \lambda(b - a)]\, P_{t-1} + \lambda(\alpha - \beta).$$

This is a first order affine recurrence.

The equilibrium price $\hat{P}$ corresponds to $\hat{P} = \dfrac{\alpha - \beta}{b - a}$, and the price evolution is given by $P_t = \hat{P} + [1 - \lambda(b - a)]^t \left(P_0 - \hat{P}\right)$.

We have $b - a > 0$ and $P_t$ will converge to the equilibrium price if $-1 < 1 - \lambda(b - a) < 1$, hence if $\lambda(b - a) < 2$.

If $\lambda(b - a) > 2$, then we have unbounded oscillations of prices. This situation would occur for great values of $\lambda$, expressing a strong and fast reaction of trade to inventory variation; such reaction generates instability.

**2.3.2** We consider now the case when there exists a "normal" level $\overline{Q}$ of inventories and we admit that price variation is proportional to the deviation of the inventories level from the normal one:

$$P_t - P_{t-1} = -\lambda\left(Q_{t-1} - \overline{Q}\right), \; \lambda > 0.$$

Writing the analogous relationship for the time interval $t - 1$:

$$P_{t-1} - P_{t-2} = -\lambda\left(Q_{t-2} - \overline{Q}\right)$$

and subtracting, we get

$$P_t - 2P_{t-1} + P_{t-2} = -\lambda\left(Q_{t-1} - Q_{t-2}\right) = -\lambda\left(S_{t-1} - D_{t-1}\right) =$$

$$= -\lambda(\beta - \alpha) - \lambda(b - a)P_{t-1}.$$

The model obtained in this way is described by a second order recurrence:

$$P_t - [2 - \lambda(b - a)]P_{t-1} + P_{t-2} = \lambda(\alpha - \beta).$$

The equilibrium level of prices is given by the same value $\hat{P}$ as above. If we denote $p_t = P_t - \hat{P}$, then we obtain

$$p_t - [2 - \lambda(b - a)]p_{t-1} + p_{t-2} = 0.$$

The associated second degree equation (characteristic equation) is $q^2 - [2 - \lambda(b - a)]q + 1 = 0$.

The discriminant of this equation is

$$[2 - \lambda(b - a)]^2 - 4 = \lambda^2(b - a)^2 - 4\lambda(b - a) =$$

$$= \lambda(b - a)[\lambda(b - a) - 4].$$

If $\lambda(b - a) > 4$, then the roots are real and distinct, their product being 1 and their sum negative, hence the roots are negative and one of them has modulus greater than 1, what corresponds to an unbounded oscillation of the price.

If $\lambda(b - a) < 4$, then the roots are not real, their product is equal to the square of the modulus, hence the modulus is 1 and a bounded, oscillatory evolution around the equilibrium price results.

We obtain second order recurrences for price evolution also in the case of the simple law of demand and supply, studied in

**1.5**, when the supply depends not on the current price, but on the previous one.

**2.3.3** Let us discuss now the continuos time analogues of the models with inventories.

We consider the velocity of inventory evolution to be proportional to excess demand: $Q'(t) = -\mu\left[D(t) - S(t)\right]$, $\mu > 0$.

In a first model, the velocity of price evolution will be assumed proportional to that of inventory evolution:

$$P'(t) = -\lambda Q'(t) = \lambda\mu\left[D(t) - S(t)\right],$$

what reduces to the simple model based on supply and demand only, which was analyzed in **1.4**.

For the model in which the "normal" level of inventories is present, we consider

$$P'(t) = -\lambda\left[Q(t) - \overline{Q}\right],$$

wherefrom it follows

$$P''(t) = -\lambda\, Q'(t) = \lambda\,\mu\left[D(t) - S(t)\right].$$

Taking into account the relations expressing the dependence of supply and of demand on prices, we obtain the second order affine differential equation

$$P''(t) + \lambda\,\mu\,(b - a)\,P(t) = \lambda\,\mu\,(\alpha - \beta).$$

The equilibrium price $\hat{P}$ is the same as above; denoting $p(t) = P(t) - \hat{P}$, we have $p''(t) + \lambda\mu(b - a)\,p(t) = 0$, hence

$$p(t) = C_1 \cos\sqrt{\lambda(b - a)}\,t + C_2 \sin\sqrt{\lambda(b - a)}\,t,$$

$$P(t) = \hat{P} + C_1 \cos\sqrt{\lambda(b - a)}\,t + C_2 \sin\sqrt{\lambda(b - a)}\,t.$$

Again, we have price oscillation around the equilibrium level.

## 2.4 Economic Growth Models

**2.4.1** We consider a continuous time model, in which the connection between the product $Y$ and investment $I$ is described by

$$I'(t) = -k\left[I(t) - vY'(t)\right],\ k > 0,$$

and the evolution velocity of the product is proportional to the deviation of the product from demand:

$$Y'(t) = -\lambda\left[Y(t) - Z(t)\right],\ \lambda > 0.$$

The demand $Z(t)$ is assumed to have three components: consumption $C(t)$, autonomous investment $A(t)$ and induced investment $I(t)$: $Z(t) = C(t) + A(t) + I(t)$.

Consumption, in its turn, is assumed to be a fixed ratio of the product (income): $C(t) = cY(t)$, $0 < c < 1$.

This model is known as *Phillips model with accelerator*.

Assuming that $A(t) = A$ is constant, we have

$$Y''(t) = -\lambda\left[Y'(t) - cY'(t) - I'(t)\right],$$

$$Y'(t) = -\lambda Y(t) + \lambda\left[cY(t) + I(t) + A\right],$$

$$I(t) = \frac{1}{\lambda}Y'(t) + sY(t) - A,$$

where $s = 1-c$ is the propensity to save.

$$I'(t) = -\frac{k}{\lambda}Y'(t) - ksY(t) + kA + kvY'(t),$$

$$Y''(t) = -\lambda sY'(t) + \lambda I'(t) =$$

$$= -\lambda sY'(t) - kY'(t) - \lambda ksY(t) + \lambda kA + \lambda kvY'(t).$$

Finally, we obtain the differential equation

$$Y''(t) + (\lambda s + k - \lambda k v) Y'(t) + \lambda k s Y(t) = \lambda k A .$$

The equilibrium is given by $\hat{Y} = \frac{A}{s}$; if we denote $y(t) = Y(t) - \hat{Y}$, then we have

$$y''(t) + [\lambda (s - k v) + k] y'(t) + \lambda k s y(t) = 0 .$$

The associated characteristic equation is

$$r^2 + [\lambda (s - k v) + k] r + \lambda k s = 0 .$$

The discriminant is

$$[\lambda (s - k v) + k]^2 - 4\lambda k s = (k + \lambda s - \lambda k v)^2 - 4\lambda k s =$$

$$= (k + \lambda s)^2 - 2\lambda k v (k + \lambda s) + \lambda^2 k^2 v^2 - 4\lambda k s =$$

$$= (k - \lambda s)^2 - 2\lambda k v (k + \lambda s) + \lambda^2 k^2 v^2 =$$

$$= (k - \lambda s)^2 - 2\lambda k v (k - \lambda s) + \lambda^2 k^2 v^2 - 4\lambda^2 k v s =$$

$$= (k - \lambda s - \lambda k v)^2 - 4\lambda^2 k v s .$$

The roots will be real if $2\lambda \sqrt{k v s} \le |k - \lambda s - \lambda k v|$.

The deviation $y$ from equilibrium will converge to zero if the real parts of both roots are strictly negative; this happens for

$$\lambda (s - k v) + k > 0 .$$

**2.4.2** We consider now some discrete – time models.

The assumptions of the model are $S_t = s Y_{t-1}$ and $C_t = Y_t - S_t$. Savings are allotted for autonomous investment $A_t$ and induced investment $I_t$ : $S_t = I_t + A_t$.

In a first variant, we put $I_t = v(Y_t - Y_{t-1})$. Thus, we have

$$v(Y_t - Y_{t-1}) = S_t - A_t = s Y_{t-1} - A_t ,$$

hence

$$vY_t = (v+s)Y_{t-1} - A_t,\ Y_t = \left(1+\frac{s}{v}\right)Y_{t-1} - \frac{1}{v}A_t.$$

This model is known as the discrete - time variant of the Harrod – Domar model.

If $A_t$ is constant, equal to $A$, then we obtain the equilibrium level $\hat{Y} = \dfrac{A}{s}$ and the deviation $y_t = Y_t - \hat{Y}$ satisfies $y_t = \left(1+\dfrac{s}{v}\right)^t y_0$, hence

$$Y_t = \frac{A}{s} + \left(1+\frac{s}{v}\right)^t \left(Y_0 - \frac{A}{s}\right).$$

The quotient $\dfrac{s}{v} = \rho$ is called warranted growth rate.

If $Y_0 > \dfrac{A}{s}$, then we have a monotonous growth in geometric progression, but if $Y_0 < \dfrac{A}{s}$, then $Y_t$ decreases; this last situation is unsatisfactory and shows that, if we have a small initial value of income, we can not accept a large level of autonomous investment and a small propensity to save.

**2.4.3** In a second variant, we assume that the variation of income equals the difference $U$ between planned investment $\tilde{I}$ and saving (actual investment); thus

$$Y_t - Y_{t-1} = \tilde{I}_{t-1} - S_{t-1} = v(Y_{t-1} - Y_{t-2}) - sY_{t-2}.$$

In this model, autonomous investment does not occur.

We obtain the recurrence: $Y_t - (1+v)Y_{t-1} + (s+v)Y_{t-2} = 0$.

The associated characteristic equation is

$$q^2 - (1+v)q + s + v = 0,$$

and we can write the general form of the solution by means of the roots of this equation.

The discriminant is $(1+v)^2 - 4(s+v) = (1-v)^2 - 4s$;

if $s < \left(\frac{1-v}{2}\right)^2$, then the roots are real while, for $s > \left(\frac{1-v}{2}\right)^2$, the roots are not real; in this last case, the modulus of the roots is greater than 1 (what indicates growth) if $s + v > 1$. Such a situation is obtained if $v = 1$.

Let us consider also the variant $Y_t = (1+\rho)\, Y_{t-1} + U_{t-1}$, $\rho = \frac{s}{v}$, which leads to the recurrence

$$Y_t - (1+\rho+v)\, Y_{t-1} + (s+v)\, Y_{t-2} = 0\,.$$

The characteristic equation is now $q^2 - (1+\rho+v)q + s + v = 0$, having the roots $q_1 = 1 + \frac{s}{v}$, $q_2 = v$. We obtain

$$Y_t = C_1 v^t + C_2 (1+\rho)^t,$$

which implies that the growth rate of $Y_t$ is given by the larger of the numbers $v$ and $1+\rho$.

**2.4.4** Let us return to the continuous – time model given by

$$Y' = \frac{s}{v} Y - \frac{1}{v} A(t) \text{ (Harrod – Domar model)}$$

and analyze the case when $A(t) = e^{rt} A_0$ (autonomous investment grows exponentially, with rate $r$).

We denote again $\rho = \frac{s}{v}$ and write the solution by means of the general representation formula

$$Y(t) = e^{\rho(t-t_0)}\, Y(t_0) - \frac{A_0}{v} \int_{t_0}^{t} e^{\rho(t-s)}\, e^{rs}\, ds =$$

$$= e^{\rho(t-t_0)}\,Y(t_0) - \frac{A_0}{v}\,e^{\rho t}\int_{t_0}^{t} e^{(r-\rho)s}\,ds =$$

$$= e^{\rho(t-t_0)}\,Y(t_0) - \frac{A_0}{(r-\rho)v}\left(e^{rt} - e^{rt_0}\,e^{\rho(t-t_0)}\right) =$$

$$= e^{\rho(t-t_0)}\left[Y(t_0) + \frac{A_0\,e^{rt_0}}{(r-\rho)v}\right] - \frac{A_0}{(r-\rho)v}\,e^{rt}.$$

If $r > \rho$, then the negative term is dominant and the evolution is unfavorable. If $r < \rho$, then we have growth, provided that $Y(t_0) > \dfrac{A_0}{(\rho - r)v}\,e^{rt_0}$.

**2.4.5** Let us discuss, finally, a discrete time growth model, known as the *Samuelson – Hicks model*. Its assumptions are:

$$Y_t = C_t + I_t + A_t,$$

$$I_t = v\left(Y_{t-1} - Y_{t-2}\right),$$

$$C_t = \gamma + c_1\,Y_{t-1} + c_2\,Y_{t-2},$$

with $0 < c_1 < 1$, $0 < c_2 < 1$, $c_1 + c_2 = c < 1$.

We deduce

$$Y_t = \gamma + c_1\,Y_{t-1} + c_2\,Y_{t-2} + v\left(Y_{t-1} - Y_{t-2}\right) + A_t,$$

$$Y_t - (v + c_1)\,Y_{t-1} + (v - c_2)\,Y_{t-2} = A_t + \gamma.$$

If $A_t$ is constant, then the equilibrium is given by

$$A + \gamma = (1 - c)\,\hat{Y};\quad \hat{Y} = \frac{A+\gamma}{1-c} = \frac{A+\gamma}{s}.$$

We denote $v - c_2 = w$. We have

$$v + c_1 = w + c_1 + c_2 = w + c = 1 + w - s.$$

The characteristic equation of the above model is

$$P(q) = q^2 - (1 + w - s)\,q + w = 0.$$

Its discriminant is

$$\Delta = (w + 1 - s)^2 - 4w = w^2 + [2(1 - s) - 4]w + (1 - s)^2 =$$

$$= w^2 - 2(s + 1)w + (1 - s)^2 .$$

The roots of the above equation (looked at as a function of $w$) are given by

$$w_{12} = 1 + s \pm \sqrt{(1+s)^2 - (1-s)^2} = 1 + s \pm 2\sqrt{s} = \left(1 \pm \sqrt{s}\right)^2 .$$

If $\left(1 - \sqrt{s}\right)^2 < w < \left(1 + \sqrt{s}\right)^2$, then we have $\Delta < 0$, the roots of the characteristic equation are not real, hence the evolution will be oscillating, damped for $w < 1$ and explosive for $w > 1$.

If $w < \left(1 - \sqrt{s}\right)^2$, then $\Delta > 0$, the roots of the characteristic equation are real. We have $P(1) = s > 0$ and

$$P(-1) = 2(1 + w) - s = c + 1 + 2v - 2c_2 =$$

$$= 2v + c_1 + c_2 + 1 - 2c_2 = 2v + c_1 + 1 - c_2 > 0 .$$

The product of the roots, $w$, satisfies the inequality

$$-1 < w < 1 \text{, since } -1 < -c_2 < w < \left(1 - \sqrt{s}\right)^2 < 1 .$$

The above conditions guarantee that the roots of the characteristic equation are located in the open interval $(-1, 1)$, hence the solution of the recurrence equation converges to the equilibrium.

If $w > \left(1 + \sqrt{s}\right)^2$, then the roots of the characteristic equation are real and positive, their product is greater than 1, hence at least one of the roots is greater than 1; it follows that there are solutions of the recurrence equation tending to infinity.

## 2.5 Second Order Linear Affine Equations

We consider first the simple situation $ax'' + bx' + cx = d$, where $a \neq 0$ and $d$ is constant. We assume $c \neq 0$.

We have the constant particular solution $\hat{x} = \frac{d}{c}$; denoting $\tilde{x} = x - \hat{x}$, we obtain $a\tilde{x}'' + b\tilde{x}' + c\tilde{x} = 0$, $x(t) = \hat{x} + \tilde{x}(t)$.

Knowing the structure of the solution $\tilde{x}$, we obtain the general form of the solution $x$.

In the case $c = 0$, the equation becomes $ax'' + bx' = d$; denoting $y = x'$, it follows $ay' + by = d$.

If $b \neq 0$, then we have the constant solution $\hat{y} = \frac{d}{b}$, hence $\hat{x}' = \hat{y} = \frac{d}{b}$ and $\hat{x}(t) = \frac{d}{b}t + k$ is a particular solution of the equation considered.

Since the characteristic equation of the linear differential equation $ax'' + bx' = 0$ is $r(ar + b) = 0$, we have $r_1 = 0$ and $r_2 = -\frac{b}{a}$ and the general solution of the affine equation will be

$$x(t) = C_1 + C_2 e^{-\frac{b}{a}t} + \frac{d}{b}t.$$

The constants $C_1$ and $C_2$ can be determined by means of the initial conditions $x(0) = x_0$, $x'(0) = v_0$.

Finally, if $b = 0$ too, then the equation reduces to $ax'' = d$, $x'' = \frac{d}{a}$, hence $x(t) = \frac{d}{2a}t^2 + C_2 t + C_1$, the constants $C_1$ and $C_2$ being determined by means of the initial conditions.

Thus, we have seen that, if the right - hand side is constant and if $r = 0$ is not a root of the characteristic equation, then the differential equation has a constant solution; if $r = 0$ is a simple root of the characteristic equation, then the differential equation has a polynomial solution of first degree and if $r = 0$ is a multiple root of the of the characteristic equation, then the differential equation has a polynomial solution of second degree.

Let as consider now the general case, $ax'' + bx' + cx = p(t)$, where $p$ is a polynomial function of degree $k$. It can be seen immediately that the superposition principle works in this case too, so that it is sufficient to analyze the case when $p(t) = t^k$.

We look first for a polynomial solution of degree $k$, hence $\hat{x}(t) = \sum_{j=o}^{k} \xi_j t^j$. We have

$$\hat{x}'(t) = \sum_{j=0}^{k} j\xi_j t^{j-1}, \quad \hat{x}''(t) = \sum_{j=0}^{k} j(j-1)\xi_j t^{j-2};$$

substituting into the equation, we get

$$\sum_{j=0}^{k} \left(aj(j-1)\xi_j t^{j-2} + bj\xi_j t^{j-1} + c\xi_j t^j\right) = t^k .$$

We get, first, the relation $\xi_k = \dfrac{1}{c}$, if $c \neq 0$, then

$$kb\xi_k + c\xi_{k-1} = 0, \ \xi_{k-1} = -\frac{kb\xi_k}{c};$$

$$ak(k-1)\xi_k + b(k-1)\xi_{k-1} + c\xi_{k-2} = 0,$$

wherefrom we get $\xi_{k-2}$; in general, for any $j$, we have

$$aj(j+1)\xi_{j+2} + b(j+1)\xi_{j+1} + c\xi_j = 0$$

and, knowing $\xi_{j+1}$ and $\xi_{j+2}$, we determine $\xi_j$.

In conclusion, if $c \neq 0$, then the system obtained admits a polynomial solution of the same degree as that of $t^k$, namely $k$.

If $c = 0$, then the equation reduces to $ax'' + bx' = t^k$; denoting $x' = y$, we obtain $ay' + by = t^k$. If $b \neq 0$, we conclude the existence of a polynomial solution $\hat{y}$ of degree $k$, hence the initial equation has a solution $\hat{x}$ such that $\hat{x}' = \hat{y}$ and this solution will be a polynomial of degree $k + 1$.

Finally, if $c = 0$, $b = 0$, $a \neq 0$, then the equation reduces to $ax'' = t^k$ and it admits a polynomial solution of degree $k + 2$.

We conclude the validity of the following assertion: the equation $ax'' + bx' + cx = p(t)$, where $p$ is a polynomial function of degree $k$, admits a polynomial solution of degree $k$, if $r = 0$ is not a root of the characteristic equation, of degree $k + 1$, if $r = 0$ is a simple root of the characteristic equation and of degree $k + 2$, if $r = 0$ is a multiple root of the characteristic equation.

We can consider now the more general situation

$$ax'' + bx' + cx = e^{\alpha t} p(t),$$

where $\alpha$ is real and $p$ is a polynomial function. We look for a particular solution of the form $\hat{x}(t) = e^{\alpha t} y(t)$. We have

$$\hat{x}'(t) = \alpha e^{\alpha t} y(t) + e^{\alpha t} y'(t),$$

$$\hat{x}''(t) = \alpha^2 e^{\alpha t} y(t) + 2\alpha e^{\alpha t} y'(t) + e^{\alpha t} y''(t).$$

Substituting into the equation, it follows

$$a\left[\alpha^2 e^{\alpha t} y(t) + 2\alpha e^{\alpha t} y'(t) + e^{\alpha t} y''(t)\right] +$$

$$+ b\left[\alpha e^{\alpha t} y(t) + e^{\alpha t} y'(t)\right] + c e^{\alpha t} y(t) = e^{\alpha t} p(t)$$

and, after simplification by $e^{\alpha t}$, we obtain for $y$ the equation:

$$a\, y''(t) + (2\alpha a + b)\, y'(t) + (a\alpha^2 + b\alpha + c)\, y(t) = p(t).$$

We are now in the case studied above.

If $a\alpha^2 + b\alpha + c \neq 0$, hence $\alpha$ is not a root of the characteristic equation, then we obtain a polynomial solution $y$ of the same degree as $p$, therefore the initial equation admits a solution $\hat{x}(t) = e^{\alpha t}\, y(t)$, where $y$ is a polynomial of the same degree as $p$.

If $a\alpha^2 + b\alpha + c = 0$, but $2a\alpha + b \neq 0$, hence if $\alpha$ is a simple root of the characteristic equation, then we obtain a polynomial solution $y$, whose degree exceeds by one unit the degree of $p$; if $a\alpha^2 + b\alpha + c = 0$ and $2a\alpha + b = 0$, hence if $\alpha$ is a multiple root of the characteristic equation, then we obtain a polynomial solution $y$ whose degree exceeds by two units the degree of $p$.

Thus, we discussed the case when $\alpha$ is real. In order to include also the case when $\alpha$ is a complex number, it is important to define the exponential function in the case of a complex argument.

We start from the remark that, for $x$ real, we have the expansion $e^x = 1 + \dfrac{x}{1!} + \dfrac{x^2}{2!} + \ldots + \dfrac{x^n}{n!} + \ldots.$

This expansion suggests the definition of $e^z$ as

$$e^z = 1 + \frac{z}{1!} + \frac{z^2}{2!} + \ldots + \frac{z^n}{n!} + \ldots.$$

The series of complex numbers, defined in this way, is *absolutely convergent* for any $z$, since the series

$$1 + \frac{|z|}{1!} + \frac{|z|^2}{2!} + \ldots + \frac{|z|^n}{n!} + \ldots$$

is convergent for any $z$.

Herefrom, using the properties of absolutely convergent series, we deduce the fundamental property of the exponential function: $e^{z_1+z_2} = e^{z_1} \cdot e^{z_2}$.

Herefrom, writing $z = \mu + i\nu$, we have $e^z = e^{\mu+i\nu} = e^{\mu} \cdot e^{i\nu}$ and

$$e^{i\nu} = 1 + \frac{i\nu}{1!} - \frac{\nu^2}{2!} - \frac{i\nu^3}{3!} + \frac{\nu^4}{4!} + \frac{i\nu^5}{5!} - \frac{\nu^6}{6!} - \frac{i\nu^7}{7!} + \frac{\nu^8}{8!} + \ldots.$$

Taking again into account the absolute convergence, we can perform a grouping of the terms, obtaining

$$e^{i\nu} = 1 - \frac{\nu^2}{2!} + \frac{\nu^4}{4!} - \frac{\nu^6}{6!} + \frac{\nu^8}{8!} + \ldots +$$

$$+ i\left(\frac{\nu}{1!} - \frac{\nu^3}{3!} + \frac{\nu^5}{5!} - \frac{\nu^7}{7!} + \ldots\right) =$$

$$= \cos\nu + i\sin\nu.$$

In this way, we obtained *Euler's* fundamental *formula*

$$e^{i\nu} = \cos\nu + i\sin\nu,$$

wherefrom it follows $e^{-i\nu} = \cos\nu - i\sin\nu$, hence

$$\cos\nu = \frac{e^{i\nu} + e^{-i\nu}}{2}, \ \sin\nu = \frac{e^{i\nu} - e^{-i\nu}}{2i}.$$

Let us remark now that, denoting

$$\varphi(t) = e^{(\mu+i\nu)t} = e^{\mu t}(\cos\nu t + i\sin\nu t),$$

we have

$$\varphi'(t) = \mu e^{\mu t}(\cos\nu t + i\sin\nu t) + e^{\mu t}(-\nu\sin\nu t + i\nu\cos\nu t) =$$

$$= \mu e^{\mu t}(\cos\nu t + i\sin\nu t) + i\nu e^{\mu t}(\cos\nu t + i\sin\nu t) =$$

$$= (\mu + i\nu)\, e^{(\mu+i\nu)t},$$

hence for $z = \mu + i\nu$, the function $t \mapsto e^{zt}$ is differentiated according to the same rule as in the case of a real $z$.

Starting from this remark, we can consider the general situation of the equation $ax'' + bx' + cx = e^{\alpha t}p(t)$, where $p$ is a polynomial function and $\alpha$ is a <u>*complex number*</u>.

In this case, the solution has to be a function defined on the real axis, but complex-valued; looking for a solution of the form $\hat{x}(t) = e^{\alpha t}y(t)$, we find out that things happen just as in the case of a real $\alpha$, hence the same conclusions are valid.

This case allows us to consider a right-hand side of the form $f(t) = e^{\alpha t}\cos\beta t$, or more generally, of the form

$$f(t) = p(t)\,e^{\alpha t}\,(A\cos\beta t + B\sin\beta t),$$

where $p$ is a polynomial function. Using the Euler formulae, we can write

$$f(t) = p(t)\,e^{\alpha t}\left[A\,\frac{e^{i\beta t} + e^{-i\beta t}}{2} + B\,\frac{e^{i\beta t} - e^{-i\beta t}}{2i}\right] =$$

$$= p(t)\left[\frac{1}{2}(A - iB)\,e^{(\alpha+i\beta)t} + \frac{1}{2}(A + iB)\,e^{(\alpha-i\beta)t}\right],$$

which leads us to the case discussed above when the right-hand side is of the form $f(t) = e^{(\alpha+i\beta)t}p(t)$, with $p$ being a polynomial function.

Although this procedure provides a complex-valued solution, when we determine a particular solution by means of real initial conditions, we get a real-valued solution.

Let us consider, in the end of this section, an important special case, namely $ax'' + bx' + cx = A\cos\alpha t$.

According to the general idea discussed above, we write

$\cos \alpha t = \frac{1}{2}\left(e^{i\alpha t} + e^{-i\alpha t}\right)$ and we obtain solutions of the form

$$\hat{x}(t) = q_1(t)\, e^{i\alpha t} + q_2(t)\, e^{-i\alpha t},$$

where $q_1$ and $q_2$ are constant if $\underline{i\,\alpha}$ is not a root of the characteristic equation, or first degree polynomials, if $\underline{i\,\alpha}$ is a root of the characteristic equation (since the characteristic equation admits the root $-i\,\alpha$ too, we deduce that such a root cannot be multiple, if $\alpha \neq 0$).

Therefore, if $i\,\alpha$ is not a root of the characteristic equation, then we have an oscillatory solution having the same period as the function $f$, while if $i\,\alpha$ is a root of the characteristic equation, then we have oscillations whose amplitude grows indefinitely when $t$ increases.

Let us analyze the circumstances in which this last situation occurs. The characteristic equation is $ar^2 + br + c = 0$.

If it admits the root $i\,\alpha$, then $-a\alpha^2 + ib\alpha + c = 0$, hence $b = 0$ (since $\alpha \neq 0$) and $c = a\,\alpha^2$.

Therefore, the differential equation has the form $a\,(x'' + \alpha^2 x) = A\cos \alpha t$ and it can be seen that the equation $x'' + \alpha^2 x = 0$ has solutions (free oscillations) of the form $C_1 \cos \alpha t + C_2 \sin \alpha t$, having the same period $\frac{2\pi}{\alpha}$ as the function in the right-hand side.

This situation is called <u>*resonance*</u>: the response of the system to oscillations having the same period as the free oscillations is an oscillation whose amplitude tends to infinity when $t \to \infty$.

If $a > 0$, $b > 0$, $c > 0$ (since $b > 0$, $i\,\alpha$ is not a root of the characteristic equation), then the real parts of the roots of the characteristic equation are strictly negative, the solutions of the associated linear equation converge to zero for $t \to \infty$.

It follows that any solution of the affine equation, which is a sum of $\hat{x}(t) = k_1 e^{i\alpha t} + k_2 e^{-i\alpha t}$ (the periodic solution) and a solution of the linear equation, will differ from $\hat{x}$ by some terms which vanish for $t \to \infty$.

This fact justifies the terms *permanent regime* for the solution $\hat{x}$ and *transients* for the solutions of the linear equation.

## 2.6 The Phillips Model with Several Types of Autonomous Investment

We discussed in **2.4** the Phillips model with accelerator assuming that autonomous investment is constant. In this section, we shall consider several other forms of the autonomous investment.

In order to deduce the differential equation for the cases when $A(t)$ is not constant, we recall the main assumptions of the model:

$$Z(t) = C(t) + A(t) + I(t); \tag{2.1}$$

$$I'(t) = -k\left[I(t) - vY'(t)\right],\ k > 0; \tag{2.2}$$

$$Y'(t) = -\lambda\left[Y(t) - Z(t)\right],\ \lambda > 0; \tag{2.3}$$

$$C(t) = cY(t),\ 0 < c < 1. \tag{2.4}$$

Substituting **(2.1)** and **(2.4)** into **(2.3)**, we have

$$Y'(t) = -\lambda\left[sY(t) - A(t) - I(t)\right], \text{ where } s = 1 - c. \tag{2.5}$$

By differentiating **(2.5)**, we get

$$Y''(t) = -\lambda\left[s\,Y'(t) - A'(t) - I'(t)\right]. \tag{2.6}$$

From **(2.5)**, we also have

$$I(t) = \frac{1}{\lambda}Y'(t) + s\,Y(t) - A(t). \tag{2.7}$$

By substituting **(2.7)** into **(2.2)**, we get

$$I'(t) = -k\left[I(t) - v\,Y'(t)\right] = \tag{2.8}$$
$$= k\left(v - \frac{1}{\lambda}\right)Y'(t) - ks\,Y(t) + k\,A(t).$$

Substituting **(2.8)** into **(2.6)**, we deduce

$$Y''(t) = (\lambda kv - \lambda s - k)\,Y'(t) - \lambda ks\,Y(t) + \lambda\left[k\,A(t) + A'(t)\right], \tag{2.9}$$

or

$$Y''(t) + (\lambda s + k - \lambda ks)\,Y'(t) + \lambda ks\,Y(t) = \lambda\left[k\,A(t) + A'(t)\right]. \tag{2.10}$$

**2.6.1** $A(t) = A\,e^{rt}$; the right-hand side of the above affine equation becomes

$$\lambda\left[k\,A(t) + A'(t)\right] = \lambda\,A\,e^{rt}\,(k + r) = \overline{A}\,e^{rt},$$

where we denoted $\overline{A} = \lambda\,A\,(k + r)$.

**a)** If the characteristic equation

$$\rho^2 + \left[\lambda\,(s - kv) + k\right]\rho + \lambda ks = 0$$

has no real roots, or if $r \neq \rho_1$, $r \neq \rho_2$ (in the case of real roots $\rho_1$, $\rho_2$), then we look for a particular solution of the form $\hat{x}(t) = M\,e^{rt}$. Substituting into equation **(2.10)**, we have

$$Y''(t) + (\lambda s + k - \lambda ks)\,Y'(t) + \lambda ks\,Y(t) = \overline{A}\,e^{rt}; \tag{2.11}$$

$$Y'(t) = rM\,e^{rt};\ Y''(t) = r^2M\,e^{rt};$$

$$M\left[r^2 + (\lambda s + k - \lambda kv)\,r + \lambda ks\right] = \overline{A}.$$

We remark that

$$r^2 + (\lambda s + k - \lambda k v) r + \lambda k s = P(r),$$

where $P(\cdot)$ is the characteristic polynomial of the equation. Since we assumed that $r$ is not a root of this polynomial, we have $P(r) \neq 0$ and $M = \dfrac{\bar{A}}{P(r)}$.

The general solution of the equation is

$$Y(t) = C_1 e^{\rho_1 t} + C_2 e^{\rho_2 t} + \frac{\bar{A}}{P(r)} e^{rt}.$$

The constants $C_1$ and $C_2$ can be determined from the initial conditions:

$$Y(0) = y_0 = C_1 + C_2 + \frac{\bar{A}}{P(r)}$$

$$Y'(0) = v_0 = \rho_1 C_1 + \rho_2 C_2 + r \frac{\bar{A}}{P(r)}.$$

**b)** We consider now that $r$ is a simple root of the characteristic equation, say $r = \rho_1 \neq \rho_2$.

Then we have $P(r) = 0$, $P'(r) = 2r + (\lambda s + k - \lambda k v) \neq 0$.

We look for a particular solution of the form $\hat{Y}(t) = (M t + N) e^{rt}$.

We have

$$\hat{Y}'(t) = M e^{rt} + r (M t + N) e^{rt};$$

$$\hat{Y}''(t) = 2r M e^{rt} + r^2 (M t + N) e^{rt}.$$

Substituting into equation **(2.11)**, we get

$$r^2 (M t + N) e^{rt} + 2r M e^{rt} +$$

$$+ (\lambda s + k - \lambda k v)\left[M e^{rt} + r (M t + N) e^{rt}\right] +$$

$$+ \lambda k s (M t + N) e^{rt} = \bar{A} e^{rt}$$

or, simplifying by $e^{rt}$,

$$(Mt + N)\left[r^2 + (\lambda s + k - \lambda kv)r + \lambda ks\right] +$$

$$+ M\left[2r + (\lambda s + k - \lambda kv)\right] = \overline{A}.$$

Taking into account that $P(r) = 0$, we obtain

$$M\left[2r + (\lambda s + k - \lambda kv)\right] = M\,P'(r) = \overline{A},$$

hence $M = \dfrac{\overline{A}}{P'(r)}$.

The general solution of the affine equation is

$$Y(t) = C_1 e^{rt} + C_2 e^{\rho_2 t} + \left(\frac{\overline{A}}{P'(r)}t + N\right)e^{rt} =$$

$$= e^{rt}\left[\frac{\overline{A}}{P'(r)}t + N + C_1\right] + C_2 e^{\rho_2 t}.$$

If we denote $C = N + C_1$, we can write

$$Y(t) = \left(\frac{\overline{A}}{P'(r)}t + C\right)e^{rt} + C_2 e^{\rho_2 t}.$$

$$Y'(t) = \frac{\overline{A}}{P'(r)}e^{rt} + r\,e^{rt}\left(\frac{\overline{A}}{P'(r)}t + C\right) + \rho_2 C_2 e^{\rho_2 t}$$

and we can determine $C$ and $C_2$ from the initial conditions:

$$Y(0) = y_0 = C + C_2$$

$$Y'(0) = v_0 = rC + \rho_2 C_2 + \frac{\overline{A}}{P'(r)}.$$

**c)** Let us assume now that $r$ is a multiple root of the characteristic equation: $r = \rho_1 = \rho_2$.

In this case, we have $P(r) = 0$, $P'(r) = 0$ and the general solution of the linear equation is $\widetilde{Y}(t) = (C_1 t + C_2)e^{rt}$.

We look for a particular solution of the affine equation,

assuming the form $\hat{Y}(t) = \left(M\,t^2 + N\,t + P\right)e^{rt}$. We have

$$\hat{Y}'(t) = (2t\,M + N)\,e^{rt} + r\left(M\,t^2 + N\,t + P\right)e^{rt};$$

$$\hat{Y}''(t) = 2\,r\,(2t\,M + N)\,e^{rt} + 2\,M\,e^{rt} + r^2\left(M\,t^2 + N\,t + P\right)e^{rt}.$$

Substituting into equation **(2.11)** and simplifying by $e^{rt}$, we get

$$\left(M\,t^2 + N\,t + P\right)\left[r^2 + r\,(\lambda s + k - \lambda kv) + \lambda ks\right] +$$

$$+(2\,t\,M + N)\left[2\,r + (\lambda s + k - \lambda kv)\right] + 2\,M = \overline{A}.$$

Taking into account that $P(r) = 0$, $P'(r) = 0$, we obtain $2\,M = \overline{A}$, hence $M = \dfrac{\overline{A}}{2}$.

The general solution of the affine equation will be

$$Y(t) = (C_1\,t + C_2)\,e^{rt} + \left(\frac{\overline{A}}{2}\,t^2 + N\,t + P\right)e^{rt} =$$

$$= \left(\frac{\overline{A}}{2}\,t^2 + \tilde{C}\,t + \overline{C}\right)e^{rt},$$

where we denoted $\tilde{C} = C_1 + N$, $\overline{C} = C_2 + P$. We have

$$Y'(t) = r\left(\frac{\overline{A}}{2}\,t^2 + \tilde{C}\,t + \overline{C}\right)e^{rt} + \left(\overline{A}\,t + \tilde{C}\right)e^{rt}$$

and we can determine the constants $\tilde{C}, \overline{C}$ from the initial conditions:

$$Y(0) = y_0 = \overline{C}$$

$$Y'(0) = v_0 = r\,\overline{C} + \tilde{C}.$$

**2.6.2** We assume $A(t) = A\cos\alpha t$, $\alpha \neq 0$.

The right-hand side of equation **(2.10)** becomes:

$$\lambda\left[k\,A(t) + A'(t)\right] = \lambda\left[k\,A\cos\alpha t - A\,\alpha\sin\alpha t\right] =$$

$$= \overline{A}\cos\alpha t + \overline{B}\sin\alpha t,$$

where we denoted $\overline{A} = \lambda A k$ and $\overline{B} = -\lambda A \alpha$.

We have to solve the differential equation

$$Y''(t) + (\lambda s + k - \lambda k v) Y'(t) + \lambda k s Y(t) = \tag{2.12}$$

$$= \overline{A} \cos \alpha t + \overline{B} \sin \alpha t .$$

We distinguish two cases, depending on whether $i\,\alpha$ is a root of the characteristic equation or not.

If $i\,\alpha$ is a root of the characteristic equation, then this equation must be of the form $r^2 + \alpha^2 = 0$, wherefrom we deduce the conditions

$$\lambda k s = \alpha^2 , \tag{2.13}$$

$$\lambda s + k - \lambda k v = 0 . \tag{2.14}$$

From **(2.13)** we have $\lambda s = \dfrac{\alpha^2}{k}$; substituting this relation into **(2.14)**, we get $\alpha^2 = k^2 (\lambda v - 1)$ or $\alpha = k\sqrt{\lambda v - 1}$.

Taking into account that $s = \dfrac{\alpha^2}{k\lambda} = \dfrac{k}{\lambda}(\lambda v - 1) = k\left(v - \dfrac{1}{\lambda}\right)$

and that $0 < s < 1$, we obtain the inequalities:

$$\frac{1}{\lambda} < v < \frac{1}{\lambda} + \frac{1}{k} . \tag{2.15}$$

Therefore, $i\alpha$ can be a root of the characteristic equation only if **(2.15)** is satisfied.

**a)** We assume that $i\alpha$ is not a root of the characteristic equation

Then we look for a particular solution of the form

$$\hat{Y}(t) = M \cos \alpha t + N \sin \alpha t .$$

We have

$$\hat{Y}'(t) = -\alpha M \sin \alpha t + \alpha N \cos \alpha t,$$

$$\hat{Y}''(t) = -\alpha^2 M \cos \alpha t - \alpha^2 N \sin \alpha t.$$

Substituting into equation **(2.12)**, we get

$$- \alpha^2 \left[M \cos \alpha t + N \sin \alpha t\right] +$$

$$+ \alpha (\lambda s + k - \lambda k v)\left[- M \sin \alpha t + N \cos \alpha t\right] +$$

$$+ \lambda k s\left[M \cos \alpha t + N \sin \alpha t\right] = \overline{A} \cos \alpha t + \overline{B} \sin \alpha t.$$

Identifying the coefficients of $\cos \alpha t$ and $\sin \alpha t$, respectively, we obtain the system:

$$\begin{cases} - \alpha^2 M + \alpha N (\lambda s + k - \lambda k v) + \lambda k s M = \overline{A} \\ - \alpha^2 N - \alpha M (\lambda s + k - \lambda k v) + \lambda k s N = \overline{B} \end{cases}$$

or

$$\begin{cases} \left(\lambda k s - \alpha^2\right) M + \alpha (\lambda s + k - \lambda k v) N = \overline{A} \\ - \alpha (\lambda s + k - \lambda k v) M + \left(\lambda k s - \alpha^2\right) N = \overline{B}. \end{cases}$$

The determinant of this system is

$$\Delta = \left(\lambda k s - \alpha^2\right)^2 + \alpha^2 (\lambda s + k - \lambda k v)^2.$$

We remark that this determinant would vanish only if conditions **(2.13)** and **(2.14)** were simultaneously satisfied, hence if $i\alpha$ were a solution of the characteristic equation. Since we assumed that this is not the case, we conclude that $\Delta \neq 0$ and we can determine $M$ and $N$:

$$M = \frac{\begin{vmatrix} \overline{A} & \alpha (\lambda s + k - \lambda k v) \\ \overline{B} & \lambda k s - \alpha^2 \end{vmatrix}}{\Delta} =$$

$$= \frac{\begin{vmatrix} \lambda k A & \alpha (\lambda s + k - \lambda k v) \\ - \lambda \alpha A & \lambda k s - \alpha^2 \end{vmatrix}}{\Delta} =$$

$$= \frac{\lambda^2 A\left(k^2 s + \alpha^2 s - kv\alpha^2\right)}{\left(\lambda ks - \alpha^2\right)^2 + \alpha^2 (\lambda s + k - \lambda kv)^2}$$

and

$$N = \frac{\begin{vmatrix} \lambda ks - \alpha^2 & \lambda k A \\ -\alpha(\lambda s + k - \lambda kv) & -\lambda\alpha A \end{vmatrix}}{\Delta} =$$

$$= \frac{\lambda\alpha A\left(\alpha^2 + k^2 - \lambda k^2 v\right)}{\left(\lambda ks - \alpha^2\right)^2 + \alpha^2 (\lambda s + k - \lambda kv)^2}.$$

The general solution of the equation is:

$$Y(t) = C_1 e^{\rho_1 t} + C_2 e^{\rho_2 t} + M \cos \alpha t + N \sin \alpha t.$$

We have

$$Y'(t) = \rho_1 C_1 e^{\rho_1 t} + \rho_2 C_2 e^{\rho_2 t} - \alpha M \sin \alpha t + \alpha N \cos \alpha t$$

and we can determine $C_1$, $C_2$ from the initial conditions:

$$Y(0) = y_0 = C_1 + C_2 + M$$

$$Y'(0) = v_0 = \rho_1 C_1 + \rho_2 C_2 + \alpha N,$$

or

$$C_1 + C_2 = y_0 - M$$

$$\rho_1 C_1 + \rho_2 C_2 = v_0 - \alpha N.$$

**b)** We assume that $i\alpha$ is a root of the characteristic equation.

Then the differential equation **(2.12)** becomes

$$Y'' + k^2 (\lambda v - 1) Y = \overline{A} \cos \alpha t + \overline{B} \sin \alpha t \tag{2.16}$$

or

$$Y'' + \alpha^2 Y = \overline{A} \cos \alpha t + \overline{B} \sin \alpha t \tag{2,16'}$$

and the characteristic equation is

$$r^2 + k^2 (\lambda v - 1) = 0 \text{ or } r^2 + \alpha^2 = 0.$$

The general solution of the linear equation is

$$\tilde{Y}(t) = C_1 \cos \alpha t + C_2 \sin \alpha t, \text{ where } \alpha = k\sqrt{\lambda v - 1}.$$

A particular solution of the affine equation will be of the form

$$\hat{Y}(t) = t\left(P\cos \alpha t + Q\sin \alpha t\right).$$

We have

$$\hat{Y}'(t) = P\cos\alpha t + Q\sin\alpha t + \alpha\, t\left(-P\sin\alpha t + Q\cos\alpha t\right),$$

$$\hat{Y}''(t) = 2\alpha\left(-P\sin\alpha t + Q\cos\alpha t\right) - \alpha^2\, t\left(P\cos\alpha t + Q\sin\alpha t\right).$$

Substituting into **(2,16')**, we obtain

$$2\alpha\left(-P\sin\alpha t + Q\cos\alpha t\right) - \alpha^2\, t\left(P\cos\alpha t + Q\sin\alpha t\right) +$$

$$+\,\alpha^2\, t\left(P\cos\alpha t + Q\sin\alpha t\right) = \overline{A}\cos\alpha t + \overline{B}\sin\alpha t,$$

or

$$-\,2\alpha P\sin\alpha t + 2\alpha Q\cos\alpha t = \overline{A}\cos\alpha t + \overline{B}\sin\alpha t.$$

Identifying the coefficients, we get

$$-\,2\alpha P = \overline{A} = \lambda k A\ ;\ P = -\frac{\lambda k A}{2\alpha}$$

$$2\alpha Q = \overline{B} = -\lambda\alpha A\ ;\ Q = -\frac{\lambda A}{2}.$$

We have

$$Y(t) = C_1\cos\alpha t + C_2\sin\alpha t - \frac{\lambda A}{2\alpha}t\left(k\cos\alpha t + \alpha\sin\alpha t\right) =$$

$$= \cos\alpha t\left(C_1 - \frac{\lambda A k}{2\alpha}t\right) + \sin\alpha t\left(C_2 - \frac{\lambda A}{2}t\right),$$

$$Y'(t) = -\alpha\sin\alpha t\left(C_1 - \frac{\lambda A k}{2\alpha}t\right) - \frac{\lambda A k}{2\alpha}\cos\alpha t +$$

$$+\,\alpha\cos\alpha t\left(C_2 - \frac{\lambda A}{2}t\right) - \frac{\lambda A}{2}\sin\alpha t.$$

We can determine $C_1$, $C_2$ from the initial conditions:

$$Y(0) = y_0 = C_1$$
$$Y'(0) = v_0 = -\frac{\lambda A k}{2\alpha} + \alpha C_2 . \tag{2.17}$$

We have, form **(2.17)**, $C_2 = \frac{1}{\alpha}\left(v_0 + \frac{\lambda A k}{2\alpha}\right)$.

We remark that case **b)** illustrates the resonance phenomenon described in **2.5**. It can be seen that the amplitude of oscillations corresponding to this case is

$$\sqrt{\left(y_0 - \frac{\lambda A k}{2\alpha} t\right)^2 + \left(\frac{v_0}{\alpha} + \frac{\lambda A k}{2\alpha^2} - \frac{\lambda A}{2} t\right)^2}$$

and that it tends to infinity for $t \to \infty$.

## 2.7 Higher Order Linear Differential Equations with Constant Coefficients

We shall illustrate the structure of solutions of higher order linear differential equations with constant coefficients by considering the case of a third order equation

$$ax''' + bx'' + cx' + dx = 0 .$$

We look for solutions of the form $x(t) = e^{rt}$; no distinction is made between the case when $r$ is real and the one in which $r$ is not real.

We have $x'(t) = re^{rt}$, $x''(t) = r^2e^{rt}$, $x'''(t) = r^3e^{rt}$, hence, if $x$ is a solution of the equation, we get

$$ar^3e^{rt} + br^2e^{rt} + cre^{rt} + de^{rt} = 0 ,$$

or

$$ar^3 + br^2 + cr + d = 0 .$$

The above equation is the *characteristic equation* associated to the given differential equation.

If the characteristic equation has distinct roots $r_1$, $r_2$, $r_3$, then we obtain, for the differential equation, the solutions

$$x_1(t) = e^{r_1 t},\ x_2(t) = e^{r_2 t},\ x_3(t) = e^{r_3 t}$$

and it can be seen immediately that, for any complex constants $C_1$, $C_2$, $C_3$, the function $x$ defined by

$$x(t) = C_1 e^{r_1 t} + C_2 e^{r_2 t} + C_3 e^{r_3 t}$$

is a solution of the equation.

We can determine the constants $C_1$, $C_2$, $C_3$, by means of the initial conditions $x(0) = x_0$, $x'(0) = v_0$, $x''(0) = w_0$. We have

$$x'(t) = r_1 C_1 e^{r_1 t} + r_2 C_2 e^{r_2 t} + r_3 C_3 e^{r_3 t}$$

$$x'(t) = r_1^2 C_1 e^{r_1 t} + r_2^2 C_2 e^{r_2 t} + r_3^2 C_3 e^{r_3 t}$$

and, for $t = 0$, we get, the system

$$\begin{cases} C_1 + C_2 + C_3 = x_0 \\ r_1 C_1 + r_2 C_2 + r_3 C_3 = v_o \\ r_1^2 C_1 + r_2^2 C_2 + r_3^2 C_3 = w_0 \end{cases}$$

The determinant of this system is the Vandermonde determinant

$$\begin{vmatrix} 1 & 1 & 1 \\ r_1 & r_2 & r_3 \\ r_1^2 & r_2^2 & r_3^2 \end{vmatrix},$$

which does not vanish if the numbers $r_1$, $r_2$, $r_3$ are distinct; this proves that, if the characteristic equation has distinct roots, then the constants $C_1$, $C_2$, $C_3$, are uniquely determined by requiring that the solution satisfy the given initial conditions.

Let us consider now the case when the characteristic

equation admits a multiple root $r_1 = r_2 = \rho$ and a simple root $r_3 = \sigma$. Since $\rho$ is a multiple root, we have

$$a\rho^3 + b\rho^2 + c\rho + d = 0,$$

$$3a\rho^2 + 2b\rho + c = 0.$$

We perform a change of variable, defined by $x(t) = e^{\rho t}y(t)$ and get:

$$x'(t) = \rho e^{\rho t}y(t) + e^{\rho t}y'(t),$$

$$x''(t) = \rho^2 e^{\rho t}y(t) + 2\rho e^{\rho t}y'(t) + e^{\rho t}y''(t),$$

$$x'''(t) = \rho^3 e^{\rho t}y(t) + 3\rho^2 e^{\rho t}y'(t) + 3\rho e^{\rho t}y''(t) + e^{\rho t}y'''(t).$$

Since $x$ has to be a solution of the differential equation, we have

$$a\left[\rho^3 e^{\rho t}y(t) + 3\rho^2 e^{\rho t}y'(t) + 3\rho e^{\rho t}y''(t) + e^{\rho t}y'''(t)\right] +$$

$$+ b\left[\rho^2 e^{\rho t}y(t) + 2\rho e^{\rho t}y'(t) + e^{\rho t}y''(t)\right] +$$

$$+ c\left[\rho e^{\rho t}y(t) + e^{\rho t}y'(t)\right] + d\,e^{\rho t}y(t) = 0,$$

hence

$$a\,y'''(t) + (3a\rho + b)\,y''(t) + \left(3a\rho^2 + 2b\rho + c\right)y'(t) +$$

$$+ \left(a\rho^3 + b\rho^2 + c\rho + d\right)y(t) = 0,$$

where the coefficients of $y'(t)$ and of $y(t)$ vanish, therefore

$$a\,y'''(t) + (3a\rho + b)\,y''(t) = 0.$$

This equation is satisfied if $y''(t) = 0$, hence $y$ is a polynomial function of first degree.

We deduce that $(C_1 + C_2 t)e^{\rho t}$, as well as $C_3 e^{\sigma t}$ are solutions, hence the function $x$, defined by $x(t) = (C_1 + C_2 t)e^{\rho t} + C_3 e^{\sigma t}$ is a solution of the differential equation.

Further, we have

$$x'(t) = \rho(C_1 + C_2 t)e^{\rho t} + C_2 e^{\rho t} + C_3 \sigma e^{\sigma t},$$

$$x''(t) = \rho^2(C_1 + C_2 t)e^{\rho t} + 2\rho C_2 e^{\rho t} + C_3 \sigma^2 e^{\sigma t}$$

and the conditions $x(0) = x_0$, $x'(0) = v_0$, $x''(0) = w_0$ lead to

$$C_1 + C_3 = x_0, \; \rho C_1 + C_2 + \sigma C_3 = v_0,$$

$$\rho^2 C_1 + 2\rho C_2 + \sigma^2 C_3 = w_0,$$

wherefrom

$$C_1 = x_0 - C_3, \; C_2 + \rho(x_0 - C_3) + \sigma C_3 = v_0,$$

$$C_2 = v_0 - \rho x_0 + (\rho - \sigma)C_3.$$

Substituting these values for $C_1$ and $C_2$ into the last equation of the system, we get

$$\rho^2(x_0 - C_3) + 2\rho(v_0 - \rho x_0) + 2\rho(\rho - \sigma)C_3 + \sigma^2 C_3 = w_0.$$

The coefficient of $C_3$ is

$$\sigma^2 - \rho^2 + 2\rho(\rho - \sigma) = \sigma^2 + \rho^2 - 2\rho\sigma = (\rho - \sigma)^2,$$

which does not vanish if $\sigma \neq \rho$, hence $C_3$ is uniquely determined from the above equation, and so will be $C_1$ and $C_2$. We would have reached the same conclusion by computing the determinant of the system:

$$\begin{vmatrix} 1 & 0 & 1 \\ \rho & 1 & \sigma \\ \rho^2 & 2\rho & \sigma^2 \end{vmatrix} = \begin{vmatrix} 1 & 0 & 0 \\ \rho & 1 & \sigma - \rho \\ \rho^2 & 2\rho & \sigma^2 - \rho^2 \end{vmatrix} =$$

$$= (\sigma - \rho)\begin{vmatrix} 1 & 1 \\ 2\rho & \sigma + \rho \end{vmatrix} = (\sigma - \rho)^2.$$

The case when $r_1 = r_2 = r_3 = r$ can be dealt with in the same way. The change of variable $x(t) = e^{rt}y(t)$ leads to $ay'''(t) = 0$, hence $y$ is a second degree polynomial function and the initial equation has a solution of the form

$$x(t) = e^{rt}\left(C_3 t^2 + C_2 t + C_1\right).$$

Since

$$x'(t) = re^{rt}\left(C_3 t^2 + C_2 t + C_1\right) + e^{rt}\left(2C_3 t + C_2\right),$$

$$x''(t) = r^2 e^{rt}\left(C_3 t^2 + C_2 t + C_1\right) + 2re^{rt}\left(2C_3 t + C_2\right) + 2e^{rt} C_3,$$

the initial conditions give:

$$C_1 = x_0,\ C_2 + rC_1 = v_0,\ 2C_3 + 2rC_2 + r^2 C_1 = 0,$$

and $C_1$, $C_2$, $C_3$ are uniquely determined.

We can describe now the corresponding results in the general case of an n-th order linear differential equation with constant coefficients.

Let us consider the *differential* equation

$$a_0 x^{(n)} + a_1 x^{(n-1)} + \ldots + a_{n-1} x' + a_n x = 0.$$

We associate the *algebraic* equation

$$a_0 r^n + a_1 r^{n-1} + \ldots + a_{n-1} r + a_n = 0.$$

Let $r_1, r_2, \ldots, r_s$ be the distinct roots of this equation, with multiplicities $n_1, n_2, \ldots, n_s$, $\sum_{j=1}^{s} n_j = n$.

Then a solution of the differential equation can be written as

$$x(t) = \sum_{j=1}^{s} e^{r_j t} p_j(t),$$

where $p_j(t)$ is a polynomial function of degree $n_j - 1$. The coefficients of these polynomials are determined from the initial conditions $x(0) = x_0^1$, $x'(0) = x_0^2, \ldots, x^{(n-1)}(0) = x_0^n$.

It is worth mentioning that, in this case too, it follows from the structure of solutions that, if the real parts of all roots $r_j$ are strictly negative, then all solutions converge to zero for $t \to \infty$.

Let us consider now the affine equation

$$a_0 x^{(n)} + a_1 x^{(n-1)} + \ldots + a_{n-1} x' + a_n x = f(t).$$

The <u>*superposition principle*</u> works: if $f(t) = \sum_{j=1}^{k} \alpha_j f_j(t)$ and if $x_j$ is a solution of the equation:

$$a_0 x^{(n)} + a_1 x^{(n-1)} + \ldots + a_{n-1} x' + a_n x = f_j(t),$$

then $x(t) = \sum_{j=1}^{k} \alpha_j x_j(t)$ is a solution of the initial equation.

We also remark that if $x_1$ and $x_2$ are solutions of the affine equation, then their difference $x_2 - x_1$ satisfies the associated linear equation, hence the general solution of the affine equation can be obtained from a particular solution, by adding the general solution of the linear (homogeneous) equation.

In the case when $f(t) = e^{\alpha t} p(t)$, where $p(t)$ is a polynomial function of degree $k$, we obtain a particular solution of the form $x(t) = e^{\alpha t} q(t)$, where $q$ is a polynomial of degree $m + k$, $m$ being the multiplicity of $\alpha$ as a root of the characteristic equation ($m = 0$ if $\alpha$ is not a root of the characteristic equation).

We shall not prove here all these assertions; they have been discussed in detail in the case $n = 2$ and will be treated again in the general case of systems of differential equations.

## 2.8 Discrete – Time Linear Affine Equations

We consider a discrete-time equation of the form

$$ax_{t+2} + bx_{t+1} + cx_t = f_t$$

and we look for particular solutions for several types of sequences $f_t$.

The case when $f_t$ is constant was already discussed, but we resume it in short. If $a + b + c \neq 0$, then there exists a constant solution $\hat{x} = \dfrac{f}{a+b+c}$; we remark that $a + b + c = 0$ if the characteristic equation $aq^2 + bq + c = 0$ admits the root $q = 1$.

We assume that the characteristic equation admits the root $q = 1$ $(a + b + c = 0)$. Then $c = -(a + b)$ and the recurrence equation has the form

$$ax_{t+2} - ax_t + bx_{t+1} - bx_t = f_t.$$

We denote $x_{t+1} - x_t = y_t$ and we have

$$x_{t+2} - x_t = x_{t+2} - x_{t+1} + x_{t+1} - x_t = y_{t+1} + y_t.$$

Thus, we obtain for $y_t$ the equation

$$a(y_{t+1} + y_t) + by_t = f \quad \text{or} \quad ay_{t+1} + (a + b)y_t = f.$$

We have a constant solution $\hat{y} = \dfrac{f}{2a+b}$ if $2a + b \neq 0$, hence if 1 is a simple root of the characteristic equation; in this case, $x_{t+1} - x_t = \dfrac{f}{2a+b}$, hence

$$x_1 - x_0 = \frac{f}{2a+b}, \; x_2 - x_1 = \frac{f}{2a+b} \text{ a.s.o.},$$

therefore $x_t = t\dfrac{f}{2a+b} + x_0$.

If $q = 1$ is a multiple root of the characteristic equation, then $a + b + c = 0$, $2a + b = 0$, hence $c = a$, $b = -2a$ and the recurrence equation takes the form

$$a\left(x_{t+2} - 2x_{t+1} + x_t\right) = f.$$

We denote again $x_{t+1} - x_t = y_t$ and deduce $y_{t+1} - y_t = \frac{f}{a}$, hence

$$y_t = t\frac{f}{a} + y_0 \quad \text{or} \quad x_{t+1} - x_t = t\frac{f}{a} + y_0;$$

herefrom

$$x_1 - x_0 = y_0,\ x_2 - x_1 = \frac{f}{a} + y_0,\ x_3 - x_2 = 2\frac{f}{a} + y_0,$$

$$x_t - x_{t-1} = (t-1)\frac{f}{a} + y_0.$$

By summation, we obtain

$$x_t - x_0 = \left[1 + 2 + \ldots + (t-1)\right]\frac{f}{a} + ty_0,$$

hence

$$x_t = x_0 + ty_0 + \frac{t(t-1)}{2}\frac{f}{a} = x_0 + t(x_1 - x_0) + \frac{t(t-1)}{2}\frac{f}{a},$$

which gives the general solution of the equation.

We consider now the case when $f_t$ is a polynomial function with respect to $t$; for the sake of simplicity, we assume that $f_t = At + B$. We look for a solution of the form $x_t = \alpha t + \beta$; substituting into the equation, we get

$$a(\alpha t + 2\alpha + \beta) + b(\alpha t + \alpha + \beta) + c(\alpha t + \beta) = At + B;$$

by identifying the coefficients, we obtain

$$\alpha(a + b + c) = A,$$

$$(2a + b)\alpha + (a + b + c)\beta = B.$$

If $a + b + c \neq 0$, then $\alpha = \dfrac{A}{a+b+c}$ and

$$\beta = \frac{B}{a+b+c} - \frac{(2a+b)A}{(a+b+c)^2},$$

therefore we can obtain a solution of the form required.

If $a+b+c = 0$ and $2a+b \neq 0$, then we look for a solution of the form $x_t = \alpha t^2 + \beta t$; substituting into the equation, we have

$$a\left[\alpha(t+2)^2 + \beta(t+2)\right] + b\left[\alpha(t+1)^2 + \beta(t+1)\right] +$$

$$+ c(\alpha t^2 + \beta t) = At + B.$$

The coefficient of $t^2$ is $\alpha(a+b+c) = 0$; the coefficient of $t$ is $2(2a+b)\alpha + (a+b+c)\beta$, hence $\alpha = \dfrac{A}{2(2a+b)}$;

identifying the terms of degree zero, we get

$$4a\alpha + 2a\beta + b\alpha + b\beta = B,$$

wherefrom

$$(2a+b)\beta = B - (4a+b)\alpha$$

and we can determine $\beta$, too.

Finally, if $a+b+c = 0$ and $2a+b = 0$, then we look for a solution of the form $x_t = \alpha t^3 + \beta t^2$.

Substituting into the equation, we get

$$a\left[\alpha(t+2)^3 + \beta(t+2)^2\right] + b\left[\alpha(t+1)^3 + \beta(t+1)^2\right] +$$

$$+ c(\alpha t^3 + \beta t^2) = At + B.$$

The coefficient of $t^3$ is $\alpha(a+b+c) = 0$ and that of $t^2$ is

$$a(6\alpha + \beta) + b(3\alpha + \beta) + c\beta = (a+b+c)\beta + 3\alpha(2a+b) = 0.$$

For the terms of first degree, we have

$$12a\alpha + 4a\beta + 3b\alpha + 2b\beta = A,$$

or

$$3\alpha(4a+b)+2\beta(2a+b)=A$$

and, since $2a+b=0$, we get $\alpha=\dfrac{A}{3(4a+b)}$.

Last, the terms of degree zero lead to $8a\alpha+4a\beta+b\alpha+b\beta=B$.

The coefficient of $\beta$ being $4a+b\neq 0$, $\beta$ can be determined in a unique way. We remark that $4a+b$ cannot vanish, since the relations $a+b+c=0$, $2a+b=0$, $4a+b=0$ imply

$$a=b=c=0.$$

We can treat in the same way the cases when the polynomial terms are of an arbitrary degree. The degree of the polynomial solution is the same as that of $f_t$ (the polynomial in the right-hand side) if $a+b+c\neq 0$; it raises with one unit if $a+b+c=0$, $2a+b\neq 0$ and raises with two units if

$$a+b+c=0 \text{ and } 2a+b=0.$$

Let us consider now the case when $f_t=\beta\alpha^t$, where $\beta$ is constant. We look for a solution having the same structure:

$$x_t=\gamma\alpha^t;$$

substituting into the equation, it follows

$$a\gamma\alpha^{t+2}+b\gamma\alpha^{t+1}+c\gamma\alpha^t=\beta\alpha^t.$$

Simplifying by $\alpha^t$, we get $\left(a\alpha^2+b\alpha+c\right)\gamma=\beta$, hence

$$\gamma=\frac{\beta}{a\alpha^2+b\alpha+c},$$

if $\alpha$ *is not a root of the characteristic equation*.

If $\alpha$ is a root of the characteristic equation, then we look for a solution of the form $x_t=\gamma t\alpha^t$; in this case, we have

$$a\gamma(t+2)\,\alpha^{t+2} + b\gamma(t+1)\,\alpha^{t+1} + c\gamma t\,\alpha^{t} = \beta\,\alpha^{t};$$

Simplifying by $\alpha^t$, we get

$$\gamma\left[a(t+2)\,\alpha^2 + b(t+1)\,\alpha + ct\right] = \beta\,.$$

The coefficient of $t$ is $a\alpha^2 + b\alpha + c$; it vanishes, since $\alpha$ is a root of the characteristic equation. The term of degree zero is

$$\gamma\,\alpha\,(2a\alpha + b) = \beta\,;$$

herefrom we get

$$\gamma = \frac{\beta}{\alpha\,(2a\alpha + b)}\,, \text{ if } 2a\alpha + b \neq 0,$$

hence if $\alpha$ is a simple root of the characteristic equation.

If we have $a\alpha^2 + b\alpha + c = 0$ and $2a\alpha + b = 0$, hence if $\alpha$ is a multiple root of the characteristic equation, then we look for a solution of the form $x_t = \gamma\, t^2\alpha^t$.

Substituting into the equation, it follows

$$a\gamma(t+2)^2\,\alpha^{t+2} + b\gamma(t+1)^2\,\alpha^{t+1} + c\gamma t^2\,\alpha^{t} = \beta\,\alpha^{t}\,.$$

Simplifying by $\alpha^t$, we get $\gamma\left[a(t+2)^2\,\alpha^2 + b(t+1)^2\,\alpha + ct^2\right] = \beta\,.$

In the left-hand side, the coefficient of $t^2$ is

$$\gamma\left[a\alpha^2 + b\alpha + c\right] = 0$$

and that of $t$ is

$$\gamma\left[4a\alpha^2 + 2b\alpha\right] = \gamma\,2\alpha(2a\alpha + b) = 0\,;$$

the term of degree zero is $\gamma\left[4a\alpha^2 + b\alpha\right] = \beta$, wherefrom

$$\gamma = \frac{\beta}{\alpha\,(4a\alpha + b)}\,.$$

The above arguments work for any $\alpha$, real or not.

## 2.9 The Samuelson – Hicks Model for Economic Growth

The Samuelson – Hicks model was described in **2.4.5** and we analyzed there the associated linear equation

$$Y_t - (v + c_1) Y_{t-1} + (v - c_2) Y_{t-2} = 0 .$$

We turn now to the affine equation

$$Y_t - (v + c_1) Y_{t-1} + (v - c_2) Y_{t-2} = \gamma_0 + A_t$$

and analyze the particular solutions corresponding to several possible forms of the autonomous investment $A_t$.

As we have seen in **2.4**, by denoting

$$v - c_2 = w, \; c_1 + c_2 = c = 1 - s ,$$

the equation becomes

$$Y_t - (1 + w - s) Y_{t-1} + w Y_{t-2} = \gamma_0 + A_t .$$

For the component $\gamma_0$, we obtain a constant component of the solution, $\hat{Y} = \dfrac{\gamma_0}{s}$.

Let us consider now an autonomous investment of the form $A_t = A(1 + r)^t$, $r > 0$, which means an exponential growth by rate $r$. According to the discussion in (2.7.), we get a response (solution) of the same form, if $1 + r$ is not a root of the characteristic equation; if the roots of the characteristic equation are smaller than 1 in modulus, then this condition is satisfied and we have a response of the form $\hat{Y}_t = \alpha (1 + r)^t$.

Substituting into the equation, we get

$$\alpha (1 + r)^t - \alpha(1 + w - s)(1 + r)^{t-1} + w\alpha (1 + r)^{t-2} = A (1 + r)^t ,$$

hence

$$\alpha = \frac{A(1+r)^2}{(1+r)^2 - (1+w-s)(1+r) + w}.$$

We obtain a general solution of the form

$$Y_t = D_1 q_1^t + D_2 q_2^t + \frac{\gamma_0}{s} +$$

$$+ \frac{A(1+r)^2}{(1+r)^2 - (1+w-s)(1+r) + w}(1+r)^t.$$

The constants $D_1$ and $D_2$ can be determined from the initial conditions

$$Y_0 = D_1 + D_2 + \frac{\gamma_0}{s} + \frac{A(1+r)^2}{(1+r)^2 - (1+w-s)(1+r) + w},$$

$$Y_1 = D_1 q_1 + D_2 q_2 + \frac{\gamma_0}{s} + \frac{A(1+r)^3}{(1+r)^2 - (1+w-s)(1+r) + w}.$$

Let us consider also the situation when $A_t = A\cos\theta t$, hence when the autonomous investments have an oscillating (cyclical) character.

We look for a response of the form $Y_t = \alpha\cos\theta t + \beta\sin\theta t$. Substituting into the equation, we get

$$\alpha\cos\theta t + \beta\sin\theta t - (1+w-s)\left[\alpha\cos(\theta t-\theta) + \beta\sin(\theta t-\theta)\right] +$$

$$+ w\left[\alpha\cos(\theta t-2\theta) + \beta\sin(\theta t-2\theta)\right] = A\cos\theta t;$$

$$\alpha\cos\theta t + \beta\sin\theta t -$$

$$-(1+w-s)\left[\alpha\left(\cos\theta t\cos\theta + \sin\theta t\sin\theta\right) +\right.$$

$$\left.+\beta\left(\sin\theta t\cos\theta - \cos\theta t\sin\theta\right)\right] +$$

$$+ w\left[\alpha\left(\cos\theta t\cos 2\theta + \sin\theta t\sin 2\theta\right) +\right.$$

$$\left.+\beta\left(\sin\theta t\cos 2\theta - \cos\theta t\sin 2\theta\right)\right] = A\cos\theta t.$$

Identifying the coefficients of $\cos\theta t$ and $\sin\theta t$, respectively, we obtain the system:

$$S_1\begin{cases}\alpha\left[1-(1+w-s)\cos\theta+w\cos 2\theta\right]+\\ \quad+\beta\left[(1+w-s)\sin\theta-w\sin 2\theta\right]=A\\ \alpha\left[-(1+w-s)\sin\theta+w\sin 2\theta\right]+\\ \quad+\beta\left[1-(1+w-s)\cos\theta+w\cos 2\theta\right]=0.\end{cases}$$

The determinant of this system is

$$\Delta_1=\left[1-(1+w-s)\cos\theta+w\cos 2\theta\right]^2+$$
$$+\left[(1+w-s)\sin\theta-w\sin 2\theta\right]^2.$$

It vanishes if and only if

$$S_2\begin{cases}1-(1+w-s)\cos\theta+w\cos 2\theta=0\ \text{and}\\ (1+w-s)\sin\theta=w\sin 2\theta.\end{cases}$$

The last equation gives

$$\sin\theta\left[(1+w-s)-2w\cos\theta\right]=0.$$

The case $\sin\theta=0$ leads to the trivial situations $\cos\theta=\pm1$, which reduce to the cases discussed earlier.

Assuming $\sin\theta\neq0$, we get $\cos\theta=\dfrac{1+w-s}{2w}$.

Substituting this value into the first equation of the above system and taking into account that $\cos 2\theta=2\cos^2\theta-1$, we get

$$1-\frac{(1+w-s)^2}{2w}+w\left[2\,\frac{(1+w-s)^2}{4w^2}-1\right]=0,$$

which reduces to $w=1$. Therefore,

$$\cos\theta=\frac{2-s}{2}=1-\frac{s}{2},$$

and the characteristic equation becomes $q^2-(2-s)q+1=0$; its roots are

$$q_{1,2} = 1 - \frac{s}{2} \pm \frac{\sqrt{s^2 - 4s}}{2} = 1 - \frac{s}{2} \pm i\frac{\sqrt{s(4-s)}}{2}.$$

We remark that the roots $q_{1,2}$ can be written as $\cos\theta \pm i\sin\theta$. Indeed, we saw that $\cos\theta = 1 - \frac{s}{2}$ and we have

$$\sin\theta = \sqrt{1 - \cos^2\theta} = \sqrt{1 - \left(1 - \frac{s}{2}\right)^2} = \sqrt{-\frac{s^2}{4} + s} = \frac{\sqrt{4s - s^2}}{2}.$$

We conclude that the determinant $\Delta_1$ of system $S_1$ is non-zero, if $\cos\theta + i\sin\theta$ is not a root of the characteristic equation; in this case, the coefficients $\alpha$ and $\beta$ can be uniquely determined from the system $S_1$.

Let us turn now to the case when $\cos\theta \pm i\sin\theta$ are the roots of the characteristic equation. We recognize here the phenomenon of *resonance*, discussed in **2.5**, in a continuous time framework. This is the case when the term in the right-hand side has the same period as the intrinsic oscillations of the system; this "forcing" term reinforces the intrinsic oscillations and the amplitude of the resulting oscillations will tend to infinity.

Indeed, we shall see that, in this case, we can choose a particular solution of the form: $\hat{Y}_t = t(\alpha\cos\theta t + \beta\sin\theta t)$.

Substituting into the equation

$$Y_t - (2 - s)Y_{t-1} + Y_{t-2} = A\cos\theta t,$$

we get

$$t(\alpha\cos\theta t + \beta\sin\theta t) -$$

$$-(2 - s)(t - 1)\left[\alpha\cos(\theta t - \theta) + \beta\sin(\theta t - \theta)\right] +$$

$$+(t - 2)\left[\alpha\cos(\theta t - 2\theta) + \beta\sin(\theta t - 2\theta)\right] = A\cos\theta t.$$

In the above identity, the coefficient of $t$ is

$$\alpha\left[\cos(\theta t - 2\theta) + \cos\theta t + (s-2)\cos(\theta t - \theta)\right] +$$
$$+\beta\left[\sin(\theta t - 2\theta) + \sin\theta t + (s-2)\sin(\theta t - \theta)\right] =$$
$$= \alpha\left[2\cos(\theta t - \theta)\cos\theta + (s-2)\cos(\theta t - \theta)\right] +$$
$$+\beta\left[2\sin(\theta t - \theta)\cos\theta + (s-2)\sin(\theta t - \theta)\right] =$$
$$= \left[2\cos\theta + (s-2)\right]\left[\alpha\cos(\theta t - \theta) + \beta\sin(\theta t - \theta)\right].$$

But

$$2\cos\theta + s - 2 = 2\left(1 - \frac{s}{2}\right) + s - 2 = 0,$$

hence the coefficient of $t$ vanishes. Identifying the terms of degree zero, we get

$$\left[\alpha\cos(\theta t - \theta) + \beta\sin(\theta t - \theta)\right](2 - s) -$$
$$- 2\left[\alpha\cos(\theta t - 2\theta) + \beta\sin(\theta t - 2\theta)\right] = A\cos\theta t,$$

or

$$\alpha\left(\cos\theta t\cos\theta + \sin\theta t\sin\theta\right)(2-s) +$$
$$+\beta\left(\sin\theta t\cos\theta - \cos\theta t\sin\theta\right)(2-s) -$$
$$-2\alpha\left(\cos\theta t\cos 2\theta + \sin\theta t\sin 2\theta\right) -$$
$$-2\beta\left(\sin\theta t\cos 2\theta - \cos\theta t\sin 2\theta\right) = A\cos\theta t.$$

Identifying the coefficients of $\cos\theta t$ and $\sin\theta t$, respectively, we get the system:

$$S_3\begin{cases}\alpha\left[(2-s)\cos\theta - 2\cos 2\theta\right] + \\ \quad +\beta\left[-(2-s)\sin\theta + 2\sin 2\theta\right] = A \\ \alpha\left[(2-s)\sin\theta - 2\sin 2\theta\right] + \\ \quad +\beta\left[(2-s)\cos\theta - 2\cos 2\theta\right] = 0.\end{cases}$$

The determinant of this system is

$$\Delta_3 = \left[(2-s)\cos\theta - 2\cos 2\theta\right]^2 + \left[(2-s)\sin\theta - 2\sin 2\theta\right]^2.$$

We shall show that this determinant cannot vanish. Indeed, for $\Delta_3$

to vanish we need to have simultaneously

$$S_4 \begin{cases} (2-s)\cos\theta - 2\cos 2\theta = 0 \text{ and} \\ (2-s)\sin\theta - 2\sin 2\theta = 0. \end{cases}$$

Taking into account that $\cos\theta = \dfrac{2-s}{2}$, the first equation above gives $\cos^2\theta = \cos 2\theta = \cos^2\theta - \sin^2\theta$, hence

$$\sin^2\theta = 0, \text{ or } \theta = k\pi.$$

Since we assumed $\theta \neq k\pi$, this cannot happen.

Thus the determinant $\Delta_3$ does not vanish for any $\theta \neq k\pi$, hence we can determine in a unique way the coefficients $\alpha$ and $\beta$ from system $S_3$, therefore the recurrence equation admits the solution

$$\hat{Y}_t = t(\alpha\cos\theta t + \beta\sin\theta t).$$

The general solution of the recurrence equation is

$$\hat{Y}_t = D_1\cos\theta t + D_2\sin\theta t + t(\alpha\cos\theta t + \beta\sin\theta t) =$$
$$= (t\alpha + D_1)\cos\theta t + (t\beta + D_2)\sin\theta t.$$

This is an oscillation with period $\dfrac{2\pi}{\theta}$ and amplitude

$$\sqrt{(t\alpha + D_1)^2 + (t\beta + D_2)^2}.$$

It can be seen that the amplitude tends to infinity, when $t \to \infty$.

## 2.10 Notes and References

The topics of this chapter are elementary and rather standard. As references, we mention Louis Brand [3] and Allen [1]. The economic models are essentially discussed in Allen [1], where more economic interpretations are given.

CHAPTER 3

# Linear Systems with Constant Coefficients

## 3.1 General Form of Solutions

A linear system of differential equations with constant coefficients is written as

$$x_i' = \sum_{j=1}^{n} a_{ij} x_j .$$

The approach is much easier if we introduce matrix notations.

We denote by $x$ the <u>*column − vector*</u> $\begin{pmatrix} x_1 \\ \vdots \\ x_n \end{pmatrix}$ and by $A$-the matrix with elements $a_{ij}$. Using these notations, the system of differential equations is written $x' = Ax$.

Starting form the experience of linear differential equations, we look for solutions of the form $x(t) = e^{rt}u$, where $u$ is a constant vector. We have $x'(t) = re^{rt}u$ and, substituting into the system, we get $re^{rt}u = Ae^{rt}u$. Simplifying by $e^{rt}$, we obtain $ru = Au$ and this relation shows that r is an eigenvalue of matrix $A$ and $u$–a corresponding eigenvector.

Since $u \neq 0$, we obtain the relation $det(rI - A) = 0$, which determines the eigenvalues. The polynomial $det(rI - A)$ is called *characteristic polynomial* of matrix $A$ and it is a *monic* polynomial ( the coefficient of $r^n$ is equal to 1).

Since $e^{rt}$ has a meaning for any complex number $r$, we shall not distinguish between real and non-real roots.

Let us assume, for the beginning, that the characteristic polynomial has distinct roots $r_1, r_2, \ldots, r_n$ and let $u_1, u_2, \ldots, u_n$ be corresponding eigenvectors.

**Proposition 3.1** *If the roots* $r_1, r_2, \ldots, r_n$ *are distinct, then* $u_1, u_2, \ldots, u_n$ *are linearly independent.*

***Proof*** We proceed by induction with respect to $n$.

If $n = 1$, then the property is obvious, linear independence meaning, in this case, $u_1 \neq 0$.

For $n = 2$, let us assume that $\alpha_1 u_1 + \alpha_2 u_2 = 0$. It follows that $A(\alpha_1 u_1 + \alpha_2 u_2) = 0$, hence $\alpha_1 A u_1 + \alpha_2 A u_2 = 0$.
Since $Au_1 = r_1 u_1$ and $Au_2 = r_2 u_2$, it follows $\alpha_1 r_1 u_1 + \alpha_2 r_2 u_2 = 0$.

From the first relation, we have $\alpha_1 r_1 u_1 + \alpha_2 r_1 u_2 = 0$ and, by subtraction, we get $\alpha_2 (r_2 - r_1) u_2 = 0$. Since $u_2 \neq 0$ and $r_1 \neq r_2$, it follows $\alpha_2 = 0$, hence $a_1 u_1 = 0$; since $u_1 \neq 0$, we deduce $\alpha_1 = 0$ and, therefore, $u_1$ and $u_2$ are linearly independent.

Let us assume that the property is true for $n = k$ and let us prove it for $n = k + 1$.

We assume $\alpha_1 u_1 + \alpha_2 u_2 + \ldots + \alpha_k u_k + \alpha_{k+1} u_{k+1} = 0$.

As above, we obtain

$$\alpha_1 r_1 u_1 + \alpha_2 r_2 u_2 + \ldots + \alpha_k r_k u_k + \alpha_{k+1} r_{k+1} u_{k+1} = 0.$$

From the first relation, we have

$$\alpha_1 r_{k+1} u_1 + \alpha_2 r_{k+1} u_2 + \ldots + \alpha_k r_{k+1} u_k + \alpha_{k+1} r_{k+1} u_{k+1} = 0$$ and,

by subtraction, we get

$$\alpha_1 (r_1 - r_{k+1}) u_1 + \alpha_2 (r_2 - r_{k+1}) u_2 + \ldots + \alpha_k (r_k - r_{k+1}) u_k = 0.$$

According to the induction assumption, $u_1, u_2, \ldots, u_k$ are linearly independent, hence

$$\alpha_1 (r_1 - r_{k+1}) = 0, \quad \alpha_2 (r_2 - r_{k+1}) = 0, \ldots, \alpha_k (r_k - r_{k+1}) = 0.$$

By assumption, the eigenvalues are distinct, hence

$$r_1 \neq r_{k+1}, r_2 \neq r_{k+1}, \ldots, r_k \neq r_{k+1},$$

and it follows that $\alpha_1 = 0, \alpha_2 = 0, \ldots, \alpha_k = 0$.

This implies $\alpha_{k+1} u_{k+1} = 0$ and, finally, $\alpha_{k+1} = 0$, what ends the proof.

From this proposition it follows that, if the eigenvalues are distinct, then there exists a basis of the space, consisting of eigenvectors.

If $r_1, r_2, \ldots, r_n$ are distinct roots of the characteristic polynomial and $u_1, u_2, \ldots, u_n$ are corresponding eigenvectors, then $e^{r_1 t} u_1$, $e^{r_2 t} u_2, \ldots, e^{r_n t} u_n$ are solutions of the system of differential equations.

It can be seen directly that, for any complex constants

$C_1, C_2, \ldots, C_n$, the function defined by

$$x(t) = C_1 e^{r_1 t} u_1 + C_2 e^{r_2 t} u_2 + \ldots + C_n e^{r_n t} u_n$$

is also a solution of the system of differential equations.

Since $u_1, u_2, \ldots, u_n$ form a basis, we can determine the constants $C_1, C_2, \ldots, C_n$ by means of the initial condition

$$x(0) = x_0,\ x_0 = C_1 u_1 + C_2 u_2 + \ldots + C_n u_n .$$

Thus, we have the following procedure for determining the solution of linear systems with constant coefficients.

1. We calculate $det(rI - A)$.

2. We calculate the roots of the equation $det(rI - A) = 0$.

3. For each root $r_k$, we determine vectors $u_k$ satisfying $(r_k I - A) u_k = 0$.

4. If we have distinct roots $r_1, r_2, \ldots, r_n$, then we determine the constants $C_1, C_2, \ldots, C_n$ from the relation

$$x_0 = C_1 u_1 + C_2 u_2 + \ldots + C_n u_n$$

and, with these values of the constants, the solution satisfying $x(0) = x_0$ can be written as

$$x(t) = C_1 e^{r_1 t} u_1 + C_2 e^{r_2 t} u_2 + \ldots + C_n e^{r_n t} u_n .$$

The question arises, what is the structure of solutions if the characteristic polynomial admits multiple roots?

Let us remark that the assumption of simple roots was used only to ensure the existence of a basis consisting of eigenvectors. Such a basis can exist also in the case of multiple roots; e.g. for the unit matrix all eigenvalues are equal to 1 and any vector is an eigenvector, therefore there exists a basis consisting of eigenvectors. But, in general, if the characteristic polynomial has multiple roots, we can not find a basis consisting of eigenvectors.

In what follows, we shall use a *theorem* that we do not prove here: given a matrix $A$, there exists $S$ (with complex elements), with $\det S \neq 0$ and such that $S^{-1}AS$ has the form

$$S^{-1}AS = \begin{pmatrix} J_1 & & 0 \\ & \ddots & \\ 0 & & J_s \end{pmatrix}, \text{ where}$$

$$J_k = \begin{pmatrix} r_k & 1 & 0 & \dots & 0 \\ 0 & r_k & 1 & & 0 \\ & & \ddots & \ddots & \\ 0 & \dots & 0 & r_k & 1 \\ 0 & \dots & 0 & 0 & r_k \end{pmatrix};$$

the matrices $J_k$ are called *Jordan cells*, and the above form is called *Jordan canonical form*.

If $\dim J_k = 1$, then we have $J_k = (r_k)$;

if $\dim J_k = 2$, then $J_k = \begin{pmatrix} r_k & 1 \\ 0 & r_k \end{pmatrix}$;

if $\dim J_k = 3$, then $J_k = \begin{pmatrix} r_k & 1 & 0 \\ 0 & r_k & 1 \\ 0 & 0 & r_k \end{pmatrix}$ a.s.o.

In order to see the significance of this result for the study of systems of differential equations, we consider the system

$x' = Ax$ and perform the change of variables $x = Sy$.

We have $Sy' = ASy$, $y' = S^{-1}ASy$, hence $y$ satisfies a system whose matrix has Jordan canonical form.

Taking into account that the Jordan canonical form is

"block-diagonal", the system satisfied by $y$ splits into $s$ independent systems whose matrices are Jordan cells.

In this way, the study of the system satisfied by $y$ reduces to the study of systems whose matrices are Jordan cells.

When the Jordan cell is of dimension 2, the system is written as $y_1' = ry_1 + y_2$; $y_2' = ry_2$ and we obtain

$$y_2(t) = e^{rt}\, y_2(0) \text{ and, then } y_1' = ry_1 + e^{rt}\, y_2(0).$$

Thus, we obtain for $y_1$ an affine first order differential equation; the representation formula obtained in Chapter 1 gives

$$y_1(t) = e^{rt} y_1(0) + \int_0^t e^{r(t-s)} e^{rs} y_2(0)\, ds, \text{ hence}$$

$$y_1(t) = e^{rt}\, y_1(0) + te^{rt}\, y_2(0) = e^{rt}\left[y_1(0) + ty_2(0)\right].$$

If the Jordan cell is of dimension 3, then the system is written as $y_1' = ry_1 + y_2$; $y_2' = ry_2 + y_3$; $y_3' = ry_3$; it follows, as above, that

$$y_3(t) = e^{rt}\, y_3(0),\ y_2(t) = e^{rt}\left[y_2(0) + ty_3(0)\right]$$

and then, from

$$y_1' = ry_1 + e^{rt}\left[y_2(0) + ty_3(0)\right],$$

we obtain

$$y_1(t) = e^{rt} y_1(0) + \int_0^t e^{r(t-s)} e^{rs}\left[y_2(0) + sy_3(0)\right] ds =$$

$$= e^{rt}\, y_1(0) + e^{rt}\left[ty_2(0) + \frac{t^2}{2}\, y_3(0)\right] =$$

$$= e^{rt}\left[y_1(0) + ty_2(0) + \frac{t^2}{2}\, y_3(0)\right].$$

It is easy to see that this structure propagates for Jordan cells of arbitrary dimension.

Taking into account the structure for $y$ and the fact that $x = Sy$, we deduce the following theorem.

**Theorem 3.2** *Any solution of the linear system* $x' = Ax$ *has the structure*

$$x(t) = \sum_{k=1}^{q} e^{r_k t}\, p_k(t),$$

*where* $r_1, r_2, \ldots, r_q$ *are the* <u>*distinct*</u> *roots of the characteristic polynomial* $\det(rI - A)$ *and* $p_k(t)$ *are polynomial functions whose degree is at most* $d_k - 1$, *where* $d_k$ *is the dimension of the largest Jordan cell having the eigenvalue* $r_k$ *on its diagonal.*

Since $d_k$ is at most equal to the multiplicity of the root $r_k$ in the characteristic polynomial, it follows that the degree of the polynomial $p_k$ is strictly smaller than the multiplicity of the root $r_k$.

For this reason, in order to determine the solution, we associate to any root of the characteristic polynomial an expression of the form $e^{r_k t}\, p_k(t)$, where $p_k(t)$ is a polynomial with <u>*vector – coefficients*</u>, of degree strictly smaller than the multiplicity of $r_k$, and we determine the coefficients so that the function $e^{r_k t}\, p_k(t)$ be a solution of the system.

## 3.2 Matrix Exponential

In **3.1** we found the structure of solutions of a system with constant coefficients. From this structure it follows that the solution is an analytic function, which admits the expansion

$$x(t) = x(0) + \frac{t}{1!}x'(0) + \frac{t^2}{2!}x''(0) + \ldots + \frac{t^k}{k!}x^{(k)}(0) + \ldots.$$

On the other hand, from $x'(t) = A\,x(t)$, we deduce, inductively, $x^{(k)}(t) = A^k\,x(t)$, hence $x^{(k)}(0) = A^k\,x(0)$.

Thus, we obtain the expansion

$$x(t) = x(0) + \frac{t}{1!}A\,x(0) + \frac{t^2}{2!}A^2\,x(0) + \ldots + \frac{t^k}{k!}A^k\,x(0) + \ldots.$$

By analogy with the situation in the case of real or complex arguments, we define the matrix exponential as

$$\exp A = e^A = I + \frac{A}{1!} + \frac{A^2}{2!} + \ldots + \frac{A^k}{k!} + \ldots.$$

A series whose terms are $(n \times n)$ matrices reduces, in fact, to $n^2$ numerical series.

The series defined above is absolutely convergent, in the sense that the series obtained by considering the <u>norms</u> of its terms is convergent.

There exist several norms that can be defined for a matrix, but all of them are equivalent from the viewpoint of convergence.

By means of the matrix exponential, the solution of the system of differential equations $x' = A\,x$ can be written in the form $x(t) = e^{At}\,x(0)$.

This formula leads to the following method of determining $e^{At}$: we choose $x\,(0)$ equal to the j-th column of the identity matrix and we obtain the j-th column of the matrix $e^{At}$ as the corresponding solution of the system.

We shall use the representation formula obtained above to answer the following question: what requirements should we impose on matrix $A$ so that, starting with a positive vector $x(0)$, the solution $x(t)$ be positive for any $t > 0$ ?

It can be seen immediately that if all elements of matrix $A$ are positive, then the elements of matrices $A^k$ are positive for any $k$, hence $e^{At}$ has all its elements positive for $t \geq 0$. Herefrom it follows that, if the elements of $x(0)$ are positive, then the elements of $x(t) = e^{At}\, x(0)$ will also be positive.

However, the above condition is not necessary, as we can illustrate by the simplest example: the equation $x' = -a\,x$, with $a > 0$ admits the solution $x(t) = e^{-at}\, x(0)$ and this solution is positive for any $t$, if $x(0) > 0$.

**Proposition 3.3** *If the non-diagonal elements of matrix A are positive, then for* $x(0)$ *with positive elements we obtain* $x(t)$ *with positive elements, for any* $t \geq 0$.

***Proof*** We consider $y(t) = e^{\alpha t}\, x(t)$; if the elements of $y(t)$ are positive, then the elements of $x(t)$ are also positive.

We also remark that $y(0) = x(0)$, hence if the elements of $x(0)$ are positive, the same is true for $y(0)$. We have

$$y'(t) = \alpha e^{\alpha t}\, x(t) + e^{\alpha t}\, x'(t) =$$

$$= \alpha y(t) + e^{\alpha t} A\, x(t) = (\alpha I + A)\, y(t),$$

hence $y$ is a solution of a linear system with constant coefficients, defined by the matrix $\alpha I + A$; this matrix differs from $A$ only by its diagonal elements and, for sufficiently large $\alpha > 0$, <u>all</u> its elements will be positive.

We deduce that, for a convenient $\alpha > 0$, $y(t)$ will be positive for $t \geq 0$, and so will be $x(t)$, if $x(0) > 0$.

This result is important since, in some economic applications, negative coordinates for $x$ are meaningless.

Let us apply the above proposition for the case of linear differential equations.

We begin with the simple situation of a second order linear differential equation $x'' + bx' + cx = 0$.

We associate the system $\begin{cases} x_1' = x_2 \\ x_2' = -cx_1 - bx_2 . \end{cases}$

It is obvious that, if $x$ is a solution of the equation, then the vector $\begin{pmatrix} x_1 \\ x_2 \end{pmatrix}$ is a solution of the system and, conversely, if $\begin{pmatrix} x_1 \\ x_2 \end{pmatrix}$ is a solution of the system, then $x_1$ is a solution of the equation.

The matrix of the above system is:

$$A = \begin{pmatrix} 0 & 1 \\ -c & -b \end{pmatrix};$$

its non-diagonal elements are 1 and $-c$.

Therefore, if $c < 0$ and if $x(0) \geq 0$, $x'(0) \geq 0$, then we obtain $x(t) \geq 0$, $x'(t) \geq 0$ for any $t$. Let us remark that, if $c < 0$, then the roots of the characteristic equation are necessarily real (and having opposite signs).

Let us consider now the third order linear equation

$x''' + bx'' + cx' + dx = 0$.

We associate the system $\begin{cases} x_1' = x_2 \\ x_2' = x_3 \\ x_3' = -dx_1 - cx_2 - bx_3 . \end{cases}$

Again, it is obvious that if $x$ is a solution of the equation, then $\begin{pmatrix} x_1 \\ x_2 \\ x_3 \end{pmatrix}$ is a solution of the system and, conversely, if $\begin{pmatrix} x_1 \\ x_2 \\ x_3 \end{pmatrix}$ is a solution of the system, then $x_1$ is a solution of the equation.

In this case, the matrix of the system is

$$A = \begin{pmatrix} 0 & 1 & 0 \\ 0 & 0 & 1 \\ -d & -c & -b \end{pmatrix};$$

its non-diagonal elements are 0, 1, $-\,d$ and $-\,c$.

We conclude that, if $d < 0$, $c < 0$ and if $x(0)$, $x'(0)$, $x''(0)$ are positive, then $x(t)$, $x'(t)$, $x''(t)$ will be positive for any $t \geq 0$.

It is clear that we can analyze in the same way the case of a linear equation of order $n$ ($n \in \mathbb{N}$).

## 3.3 Linear Affine Systems

We consider a system of the form $x' = A\,x + f(t)$.

We begin with the simplest situation, when $f$ is <u>*constant*</u>.

We look for a constant solution.

If $\hat{x}$ is such a solution, then its derivative vanishes, hence $\hat{x}$ satisfies the relation $A\,\hat{x} + f = 0$. We can find such a vector if the matrix $A$ is invertible (equivalently, if $det\, A \neq 0$). In this case, $\hat{x} = -\,A^{-1} f$.

Let us remark that $det\,(-\,A) = (-\,1)^n det\, A$, hence $det\, A \neq 0$ if $det\,(-\,A) \neq 0$, hence if $r = 0$ is not a root of the characteristic polynomial $det\,(r\, I - A)$.

Let us analyze what happens if 0 is a root of the characteristic polynomial; in this case, 0 occurs on the diagonal of one or several Jordan cells.

If such a Jordan cell is one-dimensional, then it leads to an equation of the form $x' = f$, hence $x(t) = tf + x(0)$, therefore $x$ is a polynomial of first degree.

If the cell is two-dimensional, we obtain a system of the form $\begin{cases} x_1' = x_2 + f_1 \\ x_2' = f_2 \end{cases}$, which leads to

$$x_2(t) = tf_2 + x_2(0),$$
$$x_1(t) = \frac{1}{2}t^2 f_2 + t\left[x_2(0) + f_1\right] + x_1(0),$$

and the solution is a polynomial of second degree.

It can be seen in the same way that if the cell is of dimension $k$, then the solution obtained is polynomial of degree $k$.

We conclude that, if $r = 0$ is a root of the characteristic polynomial, having multiplicity $k$, then the affine system with constant $f$ admits a polynomial solution of degree *at most* $k$.

The next case is that of a polynomial $f$:

$$f(t) = a_0 t^k + a_1 t^{k-1} + \ldots + a_{k-1} t + a_k .$$

We look for a polynomial solution *of the same degree*

$$x(t) = \xi_0 t^k + \xi_1 t^{k-1} + \ldots + \xi_{k-1} t + \xi_k .$$

Substituting into the equation, it follows

$$x'(t) = k\xi_0 t^{k-1} + (k-1)\xi_1 t^{k-2} + \ldots + \xi_{k-1} =$$
$$= A\left(\xi_0 t^k + \xi_1 t^{k-1} + \ldots + \xi_{k-1} t + \xi_k\right) +$$

$$+ a_0 t^k + a_1 t^{k-1} + \ldots + a_{k-1} t + a_k .$$

Identifying the coefficients, we obtain

$$A \xi_k + a_k = \xi_{k-1}, \ldots , A \xi_1 + a_1 = k\xi_0, A \xi_0 + a_0 = 0 .$$

We deduce that we shall be able to determine successively the coefficients $\xi_0, \xi_1, \ldots , \xi_k$ if $A$ is invertible, hence if $r = 0$ is not a root of the characteristic equation.

If $r = 0$ is a root of the characteristic equation, then the reasoning is the same as above ( for constant $f$). Thus, if 0 is present in a cell of dimension 1 and if $x' = f(t)$,where $f$ is polynomial of degree $k$, then $x$ is polynomial of degree $k + 1$.

If $\begin{cases} x_1' = x_2 + f_1(t) \\ x_2' = f_2(t) \end{cases}$ and $f_1$ , $f_2$ are polynomials of degree $k$, then $x_2$ is a polynomial of degree $k + 1$ and $x_1$ is a polynomial of degree $k + 2$.

Let us remark that, if $f$ is a polynomial of degree $k$, then it is possible that $f_1$ be of degree $k$ , while $f_2$ be of smaller degree; in this case, the solution will be a polynomial of degree $k + 1$.

Therefore, all that we can assert is that the degree of the polynomial solution is <u>*at most*</u> $k + 2$.

In general, we can assert that if $f$ is a polynomial of degree $k$, than the system admits a polynomial solution of degree <u>*at most*</u> $k + p$, where $p$ is the multiplicity of $r = 0$ as a root of the characteristic polynomial (if $r = 0$ is not a root of the characteristic polynomial, we consider that its multiplicity is zero).

We can now consider the general situation when $f(t) = e^{\alpha t} p(t)$, where $\alpha$ is an arbitrary complex number and $p$ is a polynomial.

We reduce this case to the above one, by taking $x(t) = e^{\alpha t} y(t)$. We have

$$x'(t) = \alpha e^{\alpha t} y(t) + e^{\alpha t} y'(t) = A e^{\alpha t} y(t) + e^{\alpha t} p(t).$$

Simplifying by $e^{\alpha t}$, we obtain $y'(t) = (A - \alpha I) y(t) + p(t)$.

We obtain for $y$ a polynomial solution whose degree exceeds the degree of $p$ with at most the multiplicity of $r = 0$ as a root of the characteristic polynomial for the matrix $A - \alpha I$.

The characteristic polynomial can be written as

$$det(rI - A + \alpha I) = det[(r + \alpha)I - A]$$

and $r = 0$ is its root if and only if $\alpha$ is a root of the characteristic polynomial for the matrix $A$.

Thus, we obtained the following result: if $f(t) = e^{\alpha t} p(t)$, then the system admits a solution of the form $x(t) = e^{\alpha t} q(t)$, where $q$ is a polynomial; the degree of $q$ can exceed the degree of $p$ with at most the multiplicity of $\alpha$ as a root of the characteristic polynomial for matrix $A$.

Let us analyze now the case when $f$ is an arbitrary continuous function.

We look for a solution of the form $x(t) = e^{At} y(t)$. We obtain

$$x'(t) = A e^{At} y(t) + e^{At} y'(t) =$$

$$= A\, x(t) + f(t) = A e^{At} y(t) + f(t), \text{ hence}$$

$$e^{At} y'(t) = f(t) \text{ and } y'(t) = e^{-At} f(t).$$

It follows that

$$y(t) = y(0) + \int_0^t e^{-As} f(s)\,ds$$

and, since $x(0) = y(0)$, we deduce

$$x(t) = e^{At} x(0) + e^{At} \int_0^t e^{-As} f(s)\,ds\,.$$

We used above the fact that

$\left(e^{At}\right)^{-1} = e^{-At}$, equivalent to $e^{At}\, e^{-At} = I$.

This property follows from the general fact that if $A$ and $B$ are such that $A\,B = B\,A$, then $e^{A+B} = e^A \cdot e^B$. The proof of this fact is based on a theorem concerning multiplication of absolutely convergent series. The same property allows to write $e^{At}\, e^{-As} = e^{A(t-s)}$ and we can write the general form of the solution as

$$x(t) = e^{At} x_0 + \int_0^t e^{A(t-s)} f(s)\,ds\,.$$

This formula is called the *variation of constants* formula (or variation of parameters formula). The name "variation of constants" was introduced by Lagrange, starting from the idea of replacing in the formula $x(t) = e^{At} x(0)$ (which gives the general solution of the linear system) the constant vector $x(0)$ by the variable vector $y(t)$.

## 3.4 Economic Models

**3.4.1** We discussed in **2.4.1** the Phillips model with multiplier and accelerator, reducing it to a second order affine differential equation.

Here, we shall approach it as a linear affine system.

We recall the assumptions of the model:

$$Y' = -\lambda(Y - Z), \quad Z = C + I + A,$$

$$C = cY, \quad I' = -k(I - vY').$$

We have

$$Y' = -\lambda(Y - cY - I - A) = -\lambda s Y + \lambda I + \lambda A$$

and

$$I' = -kI + kvY' = -kI + kv(-\lambda s Y + \lambda I + \lambda A) =$$

$$= k(v\lambda - 1)I - kv\lambda s Y + kv\lambda A.$$

Thus, the model is described by the affine system

$$\begin{cases} Y' = -\lambda s Y + \lambda I + \lambda A \\ I' = -kv\lambda s Y + k(v\lambda - 1)I + kv\lambda A. \end{cases}$$

We shall not pursue the analysis, since the system is equivalent to the second order equation which has been discussed in 2.4.1.

**3.4.2** We shall describe now a model of financial policy of a firm.

We denote by $(CF)(t)$ the cash-flow at time $t$, by $N(t)$-the level of debt, by $N'(t)$-the velocity of evolution of the debt, by $K(t)$-the capital, by $I(t)$-investment, by $D(t)$-dividends, by $R(t)$-gross income and by $\delta$-the interest rate. We have

$$(CF)(t) = R(t) - \delta N(t) = I(t) + D(t) - N'(t), \text{ hence}$$

$$N'(t) = I(t) + D(t) + \delta N(t) - R(t).$$

The evolution of capital is given by

$$K'(t) = I(t) - a K(t),$$

where $a$ is the coefficient of wear and tear of capital, which is assumed to be constant.

If we assume that $R$ is an affine function of $K$:

$$R(t) = \rho\, K(t) + \rho_0$$

and that investments obey a law of the type

$$I'(t) = -\,\sigma\left[I(t) - v\, K'(t)\right],$$

as in **3.4.1**, then we obtain the system

$$\begin{cases} K' = -\,a\,K + I \\ I' = -\,\sigma\, I + \sigma\, v\, K' = -\,\sigma\, I + \sigma\, v\,(-\,a\,K + I) = \\ \quad = a\, v\, \sigma\, K + \sigma\,(v-1)\, I \\ N' = -\,\rho\, K + I + \delta\, N - \rho_0 + D(t)\,. \end{cases}$$

Since the first two equations do not contain the variable $N$, they may be discussed independently, as in **3.4.1**, the evolution of debt being studied from the last equation, where the solution of the first system is used.

**3.4.3** We can also obtain a system of differential equations by considering models of price evolution.

On the basis of the law of supply and demand, we have $P' = \lambda\,(D - S)$; as before, we take $D = \alpha + a\,P$ and we modify the evolution law cf supply by admitting an "indirect " dynamic response to the signals given by the market; such a law is

$$S' = \mu\,(\beta + b\,P - S).$$

We obtain

$$\begin{cases} P' = \lambda\,(a\,P - S + \alpha) \\ S' = \mu\,(b\,P - S + \beta). \end{cases}$$

Assuming, as usually, that $a$, $\alpha$, $b$, $\beta$ are constant, the equilibrium corresponds to a constant solution $\hat{P}$, $\hat{S}$, given by the relations $a\hat{P} - \hat{S} = -\alpha$, $b\hat{P} - \hat{S} = -\beta$, hence $\hat{P} = \frac{\alpha - \beta}{b - a}$, as in the models studied before.

The matrix of the system is $\begin{pmatrix} \lambda a & -\lambda \\ \mu b & -\mu \end{pmatrix}$ and its characteristic polynomial is

$$\begin{vmatrix} r - \lambda a & \lambda \\ -\mu b & r + \mu \end{vmatrix} = r^2 + (\mu - \lambda a) r + \lambda \mu b .$$

If $\mu > \lambda a$, then the roots of the characteristic polynomial have strictly negative real parts; taking into account the structure of solutions, this means that the evolution of prices and of supply will approach the equilibrium.

**3.4.4** Other price evolution models which lead to systems of differential equations correspond to a market with several commodities, by taking into account that demand and supply for a given commodity depend on the prices of all commodities.

Denoting the commodities by the index $r$, we have the evolutions $P_r' = \lambda(D_r - S_r)$ and $D_r - S_r = C_r + \sum_{j=1}^{m} a_{rj} P_j$; we obtain an affine system of the form

$$P_r' = \lambda_r C_r + \sum_{j=1}^{m} \lambda_r a_{rj} P_j ; \; r = 1, 2, \ldots, m .$$

We shall not discuss in more detail such models, since the general theory is directly applied.

## 3.5 Leontieff – Type Models

**3.5.1** We consider a system of $n$ activities, with the additional *simplifying* assumption that each activity produces only one commodity and that each commodity is produced in only one activity.

The output of activity $r$ is denoted by $X_r$; it is used partly (e.g. as raw material) in other activities and partly as final product, for consumption.

We denote by $x_{rs}$ the part of $X_r$ which is used in activity $s$ and by $x_r$ the part allotted for consumption. The input-output relationships, determined by the *technical* conditions, are of the form $x_{rs} = a_{rs} X_s$; the coefficients $a_{rs}$ are called *input* coefficients; obviously $a_{rs} \geq 0$ and it is assumed that $a_{rr} = 0$ (no activity uses its own product for its own production).

The static, equilibrium model is written as

$$X_r = x_r + \sum_{s=1}^{n} x_{rs} = x_r + \sum_{s=1}^{n} a_{rs} X_s,$$

or, in matrix – form, $(I - A)X = x$.

Given a vector representing the final product and given the matrix $A$, we can determine the production structure that can provide the required final product:

$X = (I - A)^{-1} x$ (provided that $(I - A)^{-1}$ exists).

We shall not discuss in detail this *static* problem, but we introduce a *dynamic* element, by taking into account stocks; we denote by $S_{rs}$ the stock of the product of activity $r$, accumulated in activity $s$ (e.g. materials, equipment). We assume that

$S_{rs} = b_{rs} X_s$, where $b_{rs}$ are <u>*capital coefficients*</u>.

The matrix $B = \| b_{rs} \|$ describes the capital structure of the economy, while $A = \| a_{rs} \|$ describes the flow structure.

The conditions for the exact distribution of the product are now

$$X_r = x_r + \sum_{s=1}^{n} x_{rs} + \sum_{s=1}^{n} S'_{rs}\,,\ r = 1,2,\dots,n.$$

This means that the product goes in part to find consumers (first term), in part as current input for other activities (second term) and in part as inputs on capital account for other activities (third term).

Taking into account that $S'_{rs} = b_{rs} X'_s$, we have, in matrix notations, $X = x + AX + BX'$, or $BX' = (I - A)X - x$.

To obtain a system of differential equations of the type we discussed above, we have to assume that <u>*matrix B is invertible*</u>, what allows to write

$$X' = B^{-1}(I - A)X - B^{-1}x.$$

Such a model allows, in principle, to determine the evolution of production, starting from a production structure $X$ at time $t = 0$, from the technological structure described by matrices $A$ and $B$, and from an imposed evolution of the final product.

From the mathematical viewpoint, we are interested in the requirements that should be imposed on matrices $A$ and $B$ so that the model could work. In particular, it is desirable that for any vector $x$, such that $x_r \geq 0$, whatever $r$, we obtain $X_r \geq 0$, $\forall r$.

If we use the representation formula for the solutions of affine systems (the "variation of constants" formula), we obtain

$$X(t) = e^{B^{-1}(I-A)t}\, X(0) - \int_0^t e^{B^{-1}(I-A)(t-s)}\, B^{-1}\, x(s)\, ds\,.$$

From this formula, the following conditions follow:

**1)** all elements of $e^{B^{-1}(I-A)t}$ should be positive;

**2)** all elements of $-B^{-1}x$ should be positive, if all elements of $x$ are positive.

We saw that the first condition is satisfied if the <u>*non − diagonal*</u> elements of matrix $B^{-1}(I-A)$ are positive.

We shall not discuss, for the moment, the second condition, since it seems reasonable to accept only such structure of the final product which ensures that $X(t) \geq 0$ for any $t \geq 0$, if $X(0) \geq 0$.

**3.5.2** We shall describe now a dynamic model in which the technological matrix $A$ is again present and which starts from the assumption that the velocity of production evolution is proportional to profit.

We interpret $\sum_{s=1}^{n} a_{rs} X_s$ as being the <u>*demand*</u> for commodity $r$, and $X_r$ as representing <u>*supply*</u>. If $p_r$ is the price of commodity $r$, we have the evolution

$$p_r' = \delta_r \left( \sum_{s=1}^{n} a_{rs} X_s - X_r \right)$$

or, with matrix notations,

$$p' = \Delta(A - I)X \text{, where } \Delta = \begin{pmatrix} \delta_1 & & & \\ & \delta_2 & & 0 \\ 0 & & \ddots & \\ & & & \delta_n \end{pmatrix}.$$

For a given price system, the profit for a unit of commodity $r$ is $p_r - \sum_{s=1}^{n} a_{sr}\, p_s$ (the difference between price and cost).

Therefore, the assumption concerning the proportionality of the velocity of production evolution to profit is described by

$$X_r' = \gamma_r \left( p_r - \sum_{s=1}^{n} a_{sr}\, p_s \right)$$

or, with matrix notations,

$$X' = \Gamma\left(p - A^* p\right) \text{, where } \Gamma = \begin{pmatrix} \gamma_1 & & & \\ & \gamma_2 & & 0 \\ 0 & & \ddots & \\ & & & \gamma_n \end{pmatrix}.$$

We obtain the systems with variables $X$, $p$

$$\begin{cases} p' = \Delta(A - I)X \\ X' = \Gamma\left(I - A^*\right)p \end{cases},$$

where $\Gamma$ and $\Delta$ are diagonal matrices, whose diagonal elements are strictly positive.

To point out a property of the model, let us consider the quadratic form

$$V(X, p) = (\Gamma p, p) + (\Delta X, X),$$

where ( , ) means scalar product;

$$V(X, p) = \sum_{r=1}^{n} \gamma_r\, p_r^2 + \sum_{r=1}^{n} \delta_r\, X_r^2 .$$

We calculate the derivative of the function

$t \mapsto V\left(X(t),\, p(t)\right).$

Denoting $\tilde{V}(t) = V\left(X(t),\, p(t)\right)$, we have

$$\tilde{V}'(t) = \sum_{r=1}^{n} 2\gamma_r\, p_r(t)\, p_r'(t) + \sum_{r=1}^{n} 2\delta_r\, X_r(t)\, X_r'(t) =$$

$$= 2\sum_{r=1}^{n} \gamma_r\, \delta_r\, p_r(t) \left[\sum_{s=1}^{n} a_{rs}\, X_s(t) - X_r(t)\right] +$$

$$+ 2\sum_{r=1}^{n} \gamma_r\, \delta_r\, X_r(t) \left[p_r(t) - \sum_{s=1}^{n} a_{sr}\, p_s(t)\right] =$$

$$= 2\sum_{r=1}^{n}\sum_{s=1}^{n} \gamma_r\, \delta_r \left[a_{rs}\, p_r(t)\, X_s(t) - a_{sr}\, p_s(t)\, X_r(t)\right].$$

In the special case when $\gamma_r$ and $\delta_r$ do not depend on $r$, it follows $\tilde{V}'(t) = 0$, hence

$$\gamma \sum_{r=1}^{n} p_r^2(t) + \delta \sum_{r=1}^{n} X_r^2(t) = \gamma \sum_{r=1}^{n} p_r^2(0) + \delta \sum_{r=1}^{n} X_r^2(0),$$

indicating that the evolution is such that production and prices remain bounded.

## 3.6 Phase – Portrait for Second Order Linear Systems with Coastant Coefficients

We consider a second order system

$$\begin{cases} x_1' = a_{11}\, x_1 + a_{12}\, x_2 \\ x_2' = a_{21}\, x_1 + a_{22}\, x_2 \,. \end{cases}$$

A solution of this system is a pair of functions $\left(x_1(\cdot),\, x_2(\cdot)\right)$, which represents, in the plane $(x_1, x_2)$, a smooth path.

In what follows, we present a classification of systems, based on the geometric behaviour in the plane $(x_1, x_2)$ of the curves associated to the solutions. This classification was given by Poincaré, at the end of last century.

Let us remind that the structure of solutions depends on the roots of the characteristic polynomial $det(rI - A)$ which, in the case considered here, is a second degree polynomial.

**1)** We begin with the situation in which the characteristic polynomial has real, distinct roots, having *the same sign*.

We have the particular solutions $e^{r_1 t} u_1$ and $e^{r_2 t} u_2$; from the geometric viewpoint, these solutions represent the half-lines corresponding to the vectors $u_1$ and $u_2$, since, for any $t$, $e^{r_1 t} u_1$ is collinear with $u_1$ and $e^{r_2 t} u_2$ is collinear with $u_2$.

If $r_1$ and $r_2$ are negative, then we have

$$\lim_{t \to \infty} e^{r_1 t} u_1 = 0 \text{ and } \lim_{t \to \infty} e^{r_2 t} u_2 = 0,$$

hence, for $t \to \infty$, the point $(x_1(t), x_2(t))$ will converge towards the origin of the plane.

If $r_1$ and $r_2$ are positive, then we have

$$\lim_{t \to -\infty} e^{r_1 t} u_1 = 0 \text{ and } \lim_{t \to -\infty} e^{r_2 t} u_2 = 0$$

and the sense of motion is such that the point moves away from the origin.

Let us see now what can we say about the other trajectories. We know that any solution can be written in the form

$$x(t) = C_1 e^{r_1 t} u_1 + C_2 e^{r_2 t} u_2$$

hence, coordinate-wise,

$$\begin{cases} x_1(t) = C_1 e^{r_1 t} u_1^1 + C_2 e^{r_2 t} u_2^1 \\ x_2(t) = C_1 e^{r_1 t} u_1^2 + C_2 e^{r_2 t} u_2^2 . \end{cases}$$

Assuming that $r_1 < r_2 < 0$, we have

$$\frac{x_2(t)}{x_1(t)} = \frac{C_1 e^{r_1 t} u_1^2 + C_2 e^{r_2 t} u_2^2}{C_1 e^{r_1 t} u_1^1 + C_2 e^{r_2 t} u_2^1} = \frac{C_1 e^{t(r_1 - r_2)} u_1^2 + C_2 u_2^2}{C_1 e^{t(r_1 - r_2)} u_1^1 + C_2 u_2^1}.$$

Since $r_1 < r_2$, we have $\lim_{t \to \infty} e^{t(r_1 - r_2)} = 0$, hence $\lim_{t \to \infty} \frac{x_2(t)}{x_1(t)} = \frac{u_2^2}{u_2^1}$.

Let us remark that, in this case, $\lim_{t \to \infty} \frac{x_2(t)}{x_1(t)}$ represents the slope of the tangent to the curve at the point $x_1 = 0$, $x_2 = 0$, since $\frac{x_2(t)}{x_1(t)}$ is the slope of the chord linking the origin with the point $\left(x_1(t), x_2(t)\right)$ and for $t \to \infty$, the point $\left(x_1(t), x_2(t)\right)$ tends to the origin.

We deduce that, in the case when $r_1 < r_2 < 0$, any trajectory converges to the origin, being tangent at the origin to the vector $u_2$ corresponding to the greater eigenvalue (smaller in absolute value), except the half-line $x(t) = e^{r_1 t} u_1$.

In this case, we say that the origin is *a stable node* (sink).

We consider now the case $0 < r_1 < r_2$. We have

$$\lim_{t \to -\infty} \frac{x_2(t)}{x_1(t)} = \lim_{t \to -\infty} \frac{C_1 u_1^2 + C_2 e^{t(r_2 - r_1)} u_2^2}{C_1 u_1^1 + C_2 e^{t(r_2 - r_1)} u_2^1} = \frac{u_1^2}{u_1^1},$$

since $r_2 > r_1$ implies $\lim_{t \to -\infty} e^{(r_2 - r_1)t} = 0$.

We see that, in this case, the trajectory tends to the origin for $t \to -\infty$, being tangent to the vector $u_1$ corresponding to the smaller eigenvalue, except the half-line $x(t) = e^{r_2 t} u_2$.

In this case, we say that the origin is *an unstable node* (source).

We remark that in both cases discussed above, the trajectory reaches the origin being tangent to the eigenvector corresponding to the eigenvalue which is the smaller in absolute value.

**2)** Let us consider the case of equal real roots.

We distinguish two situations:

**2a)** There exist two linearly independent eigenvectors. In this case, any solution is written as

$$x(t) = e^{rt}\left(C_1 u_1 + C_2 u_2\right)$$

and we see that all trajectories are half-lines.

**2b)** If there are no linearly independent eigenvectors, then any solution is written as

$$x(t) = e^{rt}\left(t u + v\right)$$

or, coordinate-wise,

$$\begin{cases} x_1(t) = e^{rt}\left(t u^1 + v^1\right) \\ x_2(t) = e^{rt}\left(t u^2 + v^2\right) \end{cases}; \quad \frac{x_2(t)}{x_1(t)} = \frac{t u^2 + v^2}{t u^1 + v^1}$$

and for any $t \to \infty$, if $r < 0$, or for any $t \to -\infty$, if $r > 0$, all trajectories tend to the origin, tangent to the vector $u$, except for the trajectories corresponding to the half-lines defined by $v$; the vector $u$ is the eigenvector associated to $r$.

In these cases we also say that the origin is a node.

**3)** We consider the case when the roots of the characteristic polynomial are real, having opposite signs: $r_1 < 0$, $r_2 > 0$.

In this case we also have trajectories that are half-lines: on the two trajectories associated to root $r_1$, the point $\left(x_1(t), x_2(t)\right)$ tends to the origin for $t \to \infty$, while on those corresponding to $r_2$

the point tends to the origin for $t \to -\infty$.

Let us remark that $x(t) = C_1 e^{r_1 t} u_1 + C_2 e^{r_2 t} u_2$, hence

$$x(t) - C_2 e^{r_2 t} u_2 = C_1 e^{r_1 t} u_1,$$

therefore $\lim_{t \to \infty} \left[x(t) - C_2 e^{r_2 t} u_2\right] = 0$, what shows that the line defined by $u_2$ is asymptotic to the trajectory, as $t \to \infty$.

In the same way,

$$x(t) - C_1 e^{r_1 t} u_1 = C_2 e^{r_2 t} u_2$$

and $\lim_{t \to -\infty} \left[x(t) - C_1 e^{r_1 t} u_1\right] = 0$, hence the line defined by $u_1$ is asymptotic to the trajectory, as $t \to -\infty$.

In this case, the origin is called a *saddle point*.

**4)** We consider now the case when $r_1$ and $r_2$ are purely imaginary.

If $r_1 = i\nu, r_2 = -i\nu$, then the eigenvectors are complex conjugate: $u_1 = v + iw, r_2 = v - iw$.

The solution can be written as

$$x(t) = C_1 e^{i\nu t}(v + iw) + C_2 e^{-i\nu t}(v - iw).$$

If $x$ is *real*, we have to take $C_1 = \alpha + i\beta, \quad C_2 = \alpha - i\beta$. Then

$$\begin{aligned} x(t) &= (\alpha + i\beta)e^{i\nu t}(v + iw) + (\alpha - i\beta)e^{-i\nu t}(v - iw) = \\ &= 2\,\mathrm{Re}\left[(\alpha + i\beta)e^{i\nu t}(v + iw)\right] = \\ &= 2\,\mathrm{Re}\left[(\alpha + i\beta)(\cos\nu t + i\sin\nu t)(v + iw)\right] = \\ &= 2\,\mathrm{Re}\left\{(\cos\nu t + i\sin\nu t)\left[\alpha v - \beta w + i(\beta v + \alpha w)\right]\right\} = \\ &= 2\left[(\alpha v - \beta w)\cos\nu t - (\beta v + \alpha w)\sin\nu t\right]. \end{aligned}$$

Thus, we can write

$$\begin{cases} x_1(t) = A\cos\nu t + B\sin\nu t \\ x_2(t) = C\cos\nu t + D\sin\nu t \end{cases};$$

herefrom we obtain (solving a system of linear equations!),

$$\cos\nu t = a\,x_1(t) + b\,x_2(t);\ \sin\nu t = c\,x_1(t) + d\,x_2(t)$$

whence

$$[a\,x_1(t) + b\,x_2(t)]^2 + [c\,x_1(t) + d\,x_2(t)]^2 = 1.$$

We remark that $\alpha v - \beta w = \frac{1}{2}x(0)$, hence $\alpha$ and $\beta$ can be determined from the initial values $x_1(0)$, $x_2(0)$; in their turn, $A$, $B$, $C$, $D$ will depend on $x_1(0)$ and $x_2(0)$ and so will $a$, $b$, $c$, $d$, since

$$\begin{pmatrix} a & b \\ c & d \end{pmatrix} = \begin{pmatrix} A & B \\ C & D \end{pmatrix}^{-1}.$$

All trajectories are ellypses, having the same axes. In this case, the origin is called <u>*centre*</u>.

**5)** The real parts, as well as the imaginary parts of $r_1$ and $r_2$ are non-zero: $r_1 = \mu + i\nu$, $r_2 = \mu - i\nu$, $\mu \neq 0$, $\nu \neq 0$.

In this case, performing the same kind of transformations as above, we get

$$\begin{cases} x_1(t) = e^{\mu t}(A\cos\nu t + B\sin\nu t) \\ x_2(t) = e^{\mu t}(C\cos\nu t + D\sin\nu t). \end{cases}$$

The trajectories are spyrrals that tend to the origin if $\mu < 0$ or move away from the origin if $\mu > 0$ (for increasing $t$).

In the first situation, we call the origin a <u>*stable focus*</u>; in the second, an <u>*unstable focus*</u>.

## 3.7 Notes and References

Most of the facts concerning systems of differential equations can be found in almost all textbooks. A proof of the reduction to Jordan canonical form, which uses an idea of I.G. Petrovsky and was developed by I.M. Gelfand, can be found in the book "Differential Equations" [7] by A. Halanay.

For the theory, as well as for interesting examples, we refer also to the book [12] by Hirsch and Smale.

Most of the economic models are adapted from R.G.D. Allen [1]. The model in **3.5.2** is taken from [4].

CHAPTER 4

# General Theory of Nonlinear Systems. Stability

## 4.1 Existence and Uniqueness Theorem for the Initial Value Problem

Linear systems with constant coefficients correspond to the simplest types of models.

The general case is that of systems of the form

$$\begin{cases} x_1' = f_1(t, x_1, x_2, \dots, x_n) \\ x_2' = f_2(t, x_1, x_2, \dots, x_n) \\ \vdots \\ x_n' = f_n(t, x_1, x_2, \dots, x_n) \,. \end{cases}$$

In this case too, the vector notation simplifies the exposition.

We denote by $x$ the vector

$$x = \begin{pmatrix} x_1 \\ \vdots \\ x_n \end{pmatrix}$$

and by $f$ the function having the components $f_1, f_2, \dots, f_n$.

In this way, a system of differential equations is defined by a function

$$f : I \times G \subset \mathrm{R} \times \mathrm{R}^n \to \mathrm{R}^n$$

and can be written as $x' = f(t,x)$; $I$ is an open interval and $G$ is an open set in $\mathrm{R}^n$.

We call *solution* of such a system a function

$x : J \subset I \to G$ such that

**1)** $x$ is differentiable;

**2)** $x'(t) = f(t, x(t))$, for any $t \in J$.

We call *Cauchy problem* or *initial value problem* the problem of determining a solution of the system defined by $f$, satisfying $x(t_0) = x_0$.

In connection with this problem, it is necessary to answer the following questions:

**a)** Does such a solution exist?

**b)** If such a solution exists, is it unique?

**c)** If the solution exists and is unique, in what way does it depend on the data?

**d)** If the solution exists, it is unique and depends continuously upon data, how can we compute it (approximately)?

The answer to the first three questions is given by the fundamental theory of differential equations; the fourth question is answered by the numerical methods.

The following theorem answers the first two questions.

**Theorem 4.1 (Existence and uniqueness theorem)**

*Let* $f : I \times G \subset \mathrm{R} \times \mathrm{R}^n \to \mathrm{R}^n$ *be of class* $\mathscr{C}^1$ *(it admits partial derivatives of first order and these partial derivatives are continuous).*

*Then for any* $t_0 \in I$ *and* $x_0 \in G$ *there exists an interval* $I_{(t_0,x_0)} \subset I$ *and a function* $x : I_{(t_0,x_0)} \to G$*, differentiable, such that* $x'(t) = f(t,x(t))$, $x(t_0) = x_0$.

*If* $x : I_x \to G$, $y : I_y \to G$ *satisfy* $x(t_0) = y(t_0)$, *(hence* $t_0 \in I_x \cap I_y$*), if x and y are differentiable and if* $x'(t) = f(t,x(t))$, $y'(t) = f(t,y(t))$, *then* $x(t) = y(t)$ *for all* $t \in I_x \cap I_y$.

The proof of this theorem requires some preliminaries.

**A.** Given $t_0 \in I$, $x_0 \in G$, since $I$ and $G$ are open sets, there exist $a > 0$ and $b > 0$ such that

$$\left\{ t ; |t - t_0| \le a \right\} = I_a \subset I,$$

$$\left\{ x ; |x - x_0| \le b \right\} = S_b \subset G.$$

We denoted by $|x - x_0|$ the norm of the vector $x - x_0$, representing the distance between $x_0$ and $x$ in the space $\mathrm{R}^n$.

The sets $I_a$ and $S_b$ are compact (bounded and closed) and so is their product $I_a \times S_b$. The function $f$ is continuous on the compact set $I_a \times S_b$, hence it is bounded on this set; therefore, there exists $M > 0$ such that $|f(t,x)| \le M$ for all $t \in I_a$, $x \in S_b$; we denoted by $|f|$ the norm of the vector $f$.

We choose $\alpha = \min\left(a, \dfrac{b}{M}\right)$ and we define

$$I_{(t_0,x_0)} = \left\{ t ; |t - t_0| \le \alpha \right\}.$$

The interval $I_{(t_0, x_0)}$ depends on $t_0$ and $x_0$, since $\alpha$ depends on $t_0$ and $x_0$; indeed, $a$ depends on $t_0$, $b$ depends on $x_0$, $M$ depends on $t_0$ and on $x_0$.

**B.** Since $f$ is of class $\mathscr{C}^1$, we can apply a *general Lagrange theorem*:

$$f(t, x_2) - f(t, x_1) = \int_0^1 \frac{\partial f}{\partial x}(t, x_1 + s(x_2 - x_1))(x_2 - x_1)\, ds.$$

Here $\frac{\partial f}{\partial x}$ denotes the matrix with elements $\frac{\partial f^i}{\partial x^j}$; the formula is obtained directly from the Leibniz – Newton theorem.

We denote $g(s) = f(t, x_1 + s(x_2 - x_1))$, $s \in [0, 1]$, where $t$, $x_1$, $x_2$ are *fixed*; if $G$ is *convex* and $x_1 \in G$, $x_2 \in G$, then the point $x_1 + s(x_2 - x_1)$ belongs to $G$ for all $s \in [0, 1]$.

The function $g$ is differentiable, since $f$ is of class $\mathscr{C}^1$; on the basis of the chain rule, we have

$$g'(s) = \frac{\partial f}{\partial x}(t, x_1 + s(x_2 - x_1))(x_2 - x_1).$$

The Leibniz–Newton theorem gives $g(1) - g(0) = \int_0^1 g'(s)\, ds$.

Taking into account that $g(1) = f(t, x_2)$ and $g(0) = f(t, x_1)$, the Lagrange formula is proved.

From this formula, it follows

$$|f(t, x_2) - f(t, x_1)| \le \max_{0 \le s \le 1} \left| \frac{\partial f}{\partial x}(t, x_1 + s(x_2 - x_1)) \right| |x_2 - x_1|.$$

We denote $L = \max_{(t,x) \in I_a \times S_b} \left| \frac{\partial f}{\partial x}(t, x) \right|$ (the maximum exists, since $\frac{\partial f}{\partial x}$ is continuous and $I_a \times S_b$ is compact).

We deduce that $\left|f(t,x_2) - f(t,x_1)\right| \le L\left|x_2 - x_1\right|$,

for any $(t,x_1) \in I_a \times S_b$, $(t,x_2) \in I_a \times S_b$.

The property obtained is expressed by saying that $f$ is Lipschitz continuous with respect to the second argument, with Lipschitz constant not depending upon the first argument.

**C.** We define the sequence of *successive approximations* taking $x_0(t) = x_0$ for all $t$,

$$x_1(t) = x_0 + \int_{t_0}^{t} f(s, x_0(s))\, ds \text{ and, in general,}$$

$$x_{k+1}(t) = x_0 + \int_{t_0}^{t} f(s, x_k(s))\, ds .$$

We show that, for any $t \in I_{(t_0,x_0)}$, $x_k(t) \in S_b$.

This means that the sequence may be defined (the point $(s, x_k(s))$ belongs to the domain of $f$). The proof will be performed by induction. We have

$$\left| x_1(t) - x_0 \right| = \left| \int_{t_0}^{t} f(s, x_0(s))\, ds \right| \le M \left| t - t_0 \right| \le M\alpha \le b,$$

hence $x_1(t) \in S_b$ for any $t \in I_{(t_0,x_0)}$.

We assume that $x_k(t) \in S_b$ for any $t \in I_{(t_0,x_0)}$ and we shall prove that the property is preserved for $x_{k+1}$. According to the definition, we have

$$\left| x_{k+1}(t) - x_0 \right| = \left| \int_{t_0}^{t} f(s, x_k(s))\, ds \right| \le M \left| t - t_0 \right| \le M\alpha \le b,$$

hence $x_{k+1}(t) \in S_b$ for any $t \in I_{(t_0,x_0)}$.

**D.** We prove that the sequence of successive approximations is uniformly convergent on $I_{(t_0,x_0)}$.

We have $|x_1(t) - x_0(t)| = |x_1(t) - x_0| \le M|t - t_0|$.

Since $x_2(t) = x_0 + \int_{t_0}^{t} f(s, x_1(s))\,ds$, we deduce

$$x_2(t) - x_1(t) = \int_{t_0}^{t} f(s, x_1(s))\,ds - \int_{t_0}^{t} f(s, x_0(s))\,ds =$$

$$= \int_{t_0}^{t} \left[f(s, x_1(s)) - f(s, x_0(s))\right] ds,$$

hence

$$|x_2(t) - x_1(t)| \le \left| \int_{t_0}^{t} |f(s, x_1(s)) - f(s, x_0(s))|\,ds \right| \le$$

$$\le \left| \int_{t_0}^{t} L\,|x_1(s) - x_0(s)|\,ds \right| \le \left| \int_{t_0}^{t} L M\,|s - t_0|\,ds \right| \le$$

$$\le \frac{L M (t - t_0)^2}{2}.$$

Further,

$$|x_3(t) - x_2(t)| = \left| \int_{t_0}^{t} \left[ f(s, x_2(s)) - f(s, x_1(s)) \right] ds \right| \le$$

$$\le \left| \int_{t_0}^{t} |f(s, x_2(s)) - f(s, x_1(s))|\,ds \right| \le$$

$$\le \left| \int_{t_0}^{t} L\,|x_2(s) - x_1(s)|\,ds \right| \le \left| \int_{t_0}^{t} \frac{L^2 M (s - t_0)^2}{2}\,ds \right| =$$

$$= \frac{M L^2 |t - t_0|^3}{2 \cdot 3}.$$

The estimates obtained suggest that we have, in general,

$$|x_k(t) - x_{k-1}(t)| \le \frac{M L^{k-1} |t - t_0|^k}{k!}.$$

We check this estimate by induction. We have

$$|x_{k+1}(t) - x_k(t)| = \left| \int_{t_0}^{t} \left[ f(s, x_k(s)) - f(s, x_{k-1}(s)) \right] ds \right| \le$$

$$\leq \left| \int_{t_0}^{t} \left| f(s, x_k(s)) - f(s, x_{k-1}(s)) \right| ds \right| \leq$$

$$\leq \left| \int_{t_0}^{t} L \left| x_k(s) - x_{k-1}(s) \right| ds \right| \leq \left| \int_{t_0}^{t} \frac{M L^k \left| s - t_0 \right|^k}{k!} ds \right| =$$

$$= \frac{M L^k \left| t - t_0 \right|^{k+1}}{k!(k+1)}$$ and the estimate is validated.

**E.** From the above estimates, we deduce

$$\left| x_k(t) - x_{k-1}(t) \right| \leq \frac{M L^{k-1} \left| t - t_0 \right|^k}{k!} \leq \frac{M L^{k-1} \alpha^k}{k!},$$

hence the series

$\sum_{k \geq 1} \left( x_k(t) - x_{k-1}(t) \right)$ is <u>uniformly convergent</u> on $I_{(t_0, x_0)}$,

therefore the sequence $(x_k)_{k \geq 0}$ is uniformly convergent on $I_{(t_0, x_0)}$.

We define $x(t) = \lim_{k \to \infty} x_k(t)$. Since the functions $x_k$ are continuous (they are even differentiable, with continuous derivatives) it follows that $x$ is a continuous function. Further, we can write

$$\left| f(t, x(t)) - f(t, x_k(t)) \right| \leq L \left| x(t) - x_k(t) \right|;$$

herefrom it follows that $\lim_{k \to \infty} f(t, x_k(t)) = f(t, x(t))$, the convergence being uniform on $I_{(t_0, x_0)}$.

Therefore, we can take the limit in the relation defining $x_{k+1}$ and we get $x(t) = \int_{t_0}^{t} f(s, x(s)) ds$. This equality shows that $x$ is differentiable and $x'(t) = f(t, x(t))$.

**F.** In order to prove uniqueness, we shall show that $y(t) = x(t)$, for $t \in J$, where $J$ is an arbitrary compact interval contained in $I_x \cap I_y$.

From $x'(t) = f(t, x(t))$ and $y'(t) = f(t, y(t))$ we deduce

$$x(t) = x(t_0) + \int_{t_0}^{t} f(s, x(s))\, ds,$$

$$y(t) = y(t_0) + \int_{t_0}^{t} f(s, y(s))\, ds,$$

$$x(t) - y(t) = \int_{t_0}^{t} \left[f(s, x(s)) - f(s, y(s))\right] ds.$$

Using the Lagrange formula, we can write

$$|x(t) - y(t)| \le \tilde{L} \left| \int_{t_0}^{t} |x(s) - y(s)|\, ds \right|.$$

We denote $\mu = \max_{t \in J} |x(t) - y(t)|$. On the basis of the above estimate, we have $|x(t) - y(t)| \le \tilde{L} \mu |t - t_0|$.

Further, we deduce

$$|x(t) - y(t)| \le \tilde{L} \left| \int_{t_0}^{t} \tilde{L} \mu |s - t_0|\, ds \right| = \frac{\mu \tilde{L}^2 |t - t_0|^2}{2}.$$

We can proceed in the same way and obtain, inductively, the estimate

$$|x(t) - y(t)| \le \frac{\mu \tilde{L}^k |t - t_0|^k}{k!},$$

hence

$$|x(t) - y(t)| \le \frac{\mu \left(\tilde{L} l\right)^k}{k!},$$

where $l$ is the length of interval $J$.

For $k \to \infty$, we have $\dfrac{\left(\tilde{L} l\right)^k}{k!} \to 0$, hence $|x(t) - y(t)| \le 0$ for any $t \in J$, which ends the proof.

Let us make the important remark that, in the general case

of nonlinear systems, the interval $I_{(t_0,x_0)}$ is usually smaller than the interval $I$ on which $f$ is defined.

***Example*** Let us consider the scalar equation $x' = x^2 + 1$.

Here the function $f$ is defined on R×R, $f(t,x) = x^2 + 1$, hence $I$ = R, $G$ = R. The equation can be solved by the procedure described for equations with separated variables. If $x$ is a solution of the Cauchy problem (which exists, according to the above general theorem, and is unique) then we have

$$x'(t) = x^2(t) + 1,\ x(t_0) = x_0,\ \frac{x'(t)}{x^2(t)+1} = 1,$$

$$\int_{t_0}^{t} \frac{x'(s)}{x^2(s)+1} ds = t - t_0;$$

further, performing a change of variables, we have

$$\int_{x_0}^{x(t)} \frac{dx}{x^2+1} = t - t_0,$$

hence

$$\arctan x(t) - \arctan x_0 = t - t_0,$$

$$\arctan x(t) = \arctan x_0 + t - t_0$$

and, finally,

$$x(t) = \tan\left[\arctan x_0 + t - t_0\right].$$

If, for instance, $t_0 = 0$, $x_0 = 0$, we get $x(t) = \tan t$ and the solution is defined only for $t \in \left(-\frac{\pi}{2}, \frac{\pi}{2}\right)$.

Let us also remark that, if $f$ is not of class $\mathscr{C}^1$, then the uniqueness property can be lost.

***Example*** We consider the equation $x' = \sqrt[3]{x^2}$.

Again $I$ = R, $G$ = R; the function $f$ is defined by

$f(t,x) = \sqrt[3]{x^2}$ and is continuous; it is also differentiable, except for the point $x = 0$.

Let us consider the Cauchy problem defined by the condition $x(0) = 0$. Obviously, this problem admits the solution $x(t) \equiv 0$.

We show that it also admits other solutions.

The function defined by $x(t) = \dfrac{t^3}{27}$ is such a solution; indeed, we have $x(0) = 0$ and $x'(t) = \dfrac{t^2}{9} = \sqrt[3]{x^2(t)}$.

Let us examine the procedure of successive approximations in the special case of linear systems with constant coefficients.

If the system is written as $x' = A x$, then we have $I = \mathrm{R}$, $G = \mathrm{R}^n$, $f(t,x) = A x$. It follows

$$x_1(t) = x_0 + \int_{t_0}^{t} A x_0 \, ds = x_0 + (t - t_0)A x_0 = [I + (t - t_0)A] x_0 ,$$

$$x_2(t) = x_0 + \int_{t_0}^{t} A x_1(s) \, ds = x_0 + \int_{t_0}^{t} [I + (s - t_0)A] x_0 \, ds =$$

$$= x_0 + (t - t_0)A x_0 + \frac{(t - t_0)^2}{2!} A^2 x_0 =$$

$$= \left( I + \frac{(t - t_0)A}{1!} + \frac{[(t - t_0)A]^2}{2!} \right) x_0 .$$

Proceeding in the same way, we find out that

$$x_k(t) = \left[ I + \frac{(t - t_0)A}{1!} + \frac{[(t - t_0)A]^2}{2!} + \ldots + \frac{(t - t_0)^k A^k}{k!} \right] x_0 ,$$

hence the sequence $x_k(t)$ is the sequence of partial sums for the series $e^{A(t-t_0)} x_0$.

Thus, the general theorem we proved provides the representation $x(t) = e^{A(t-t_0)} x(t_0)$.

In this case, the solution obtained is defined for all $t \in \mathrm{R}$.

This property is a consequence of linearity. In fact, we have a general result, that will be stated here without proof:

*If* $f : I \times \mathrm{R}^n \to \mathrm{R}^n$ *and for any compact set* $J \subset I$ *there exist* $L_1$, $L_0$ *such that, whatever* $t \in J$, *we have*

$$|f(t,x)| \le L_1 |x| + L_0,\ x \in \mathrm{R}^n,$$

*then the solution of the initial value problem is defined on the whole interval* $I$ (the sequence of successive approximations can be defined for all $t \in I$ and it can be proved that it converges uniformly on any compact interval $J \subset I$).

Let us end this section by the remark that the functions $x_k$ obtained in the process of successive approximations are continuous functions with respect to $(t_0,\ x_0)$ and, since the sequence is uniformly convergent, it follows that the solution $x$ will be a continuous function of $(t_0,\ x_0)$, what gives an answer to the third question put at the beginning.

In what follows, we shall state the continuity property with respect to $x_0$ in a global form (while the argument based on the process of successive approximations worked only in a neighbourhood of $t_0$).

**Theorem 4.2** *Let* $f : I \times G \subset \mathrm{R} \times \mathrm{R}^n \to \mathrm{R}^n$ *be of class* $\mathscr{C}^1$.

*Let* $t_0 \in I$, $x_0 \in G$; *let* $x(\cdot, t_0, x_0)$ *be the unique solution of the initial value problem defined by* $(f, t_0, x_0)$.

*Let* $J$ *be a compact interval containing* $t_0$ *and being contained in the domain of the solution* $x(\cdot, t_0, x_0)$.

*Then, for any* $\varepsilon > 0$ *there exists* $\delta > 0$, *depending on* $J$ *and on* $(t_0, x_0)$, *such that* $|\tilde{x}_0 - x_0| < \delta$ *implies that* $x(\cdot, t_0, \tilde{x}_0)$ *is defined on* $J$ *and* $|x(t, t_0, \tilde{x}_0) - x(t, t_0, x_0)| < \varepsilon$ *for all* $t \in J$.

From the practical viewpoint, the difficulties related to this property result from the possibility that, for an interval $J$ which is of interest, the value of $\delta$ be *too small*.

## 4.2 Equilibria. Stability. Continuous Time

We consider $f: G \subset \mathbf{R}^n \to \mathbf{R}^n$ defining the system of differential equations $x' = f(x)$.

We call *equilibrium* a constant solution of the system, hence a point $\hat{x}$ satisfying $f(\hat{x}) = 0$.

**Definition 4.3** *The equilibrium* $\hat{x}$ *is called stable if for any* $\varepsilon > 0$ *there exists* $\delta(\varepsilon) > 0$ *such that the solutions for which* $|x(0) - \hat{x}| < \delta(\varepsilon)$ *satisfy* $|x(t) - \hat{x}| < \varepsilon$ *for any* $t \geq 0$.

*The equilibrium is called asymptotically stable if it is stable and, moreover, there exists* $\delta_0 > 0$, *such that the solutions for which* $|x(0) - \hat{x}| < \delta_0$ *satisfy* $\lim_{t \to \infty} |x(t) - \hat{x}| = 0$.

*The equilibrium is called exponentially stable if there exist* $\delta_0 > 0$, $\alpha > 0$, $\beta > 0$ *such that the solutions for which* $|x(0) - \hat{x}| < \delta_0$ *satisfy* $|x(t) - \hat{x}| < \beta e^{-\alpha t}$ *for any* $t > 0$.

It follows from these formal definitions that an equilibrium is stable if small initial deviations from this point do not have significant effects, in the sense that the trajectory stays in the neighbourhood of the equilibrium point, asymptotic stability meaning, moreover, that the initial deviations are damped; in the case of exponential stability, the character of this damping is specified.

Let us analyze these properties in the simple case of an affine system of the form $x' = A\,x + b$, where $A$ is a constant matrix and $b$ a constant vector.

The equation defining the equilibrium is $A\,\hat{x} + b = 0$; it has a (unique) solution if $A$ is invertible, namely $\hat{x} = -A^{-1}\,b$.

Let us denote $y(t) = x(t) - \hat{x}$, $x$ being an arbitrary solution of the system. We have:

$$y'(t) = x'(t) = A\,x(t) + b = A\left(\hat{x} + y(t)\right) + b =$$

$$= A\,\hat{x} + A\,y(t) + b = A\,y(t).$$

We deduce that

$$y(t) = e^{At}\,y(0) = e^{At}\left[\,x(0) - \hat{x}\,\right],$$

hence

$$x(t) - \hat{x} = e^{At}\left[\,x(0) - \hat{x}\,\right].$$

From the general results concerning the structure of solutions of linear systems with constant coefficients, we know that

$$e^{At} = \sum_{k=1}^{s} e^{r_k t}\,P_k(t),$$

where $r_k$ are the roots of the characteristic polynomial and $P_k(t)$ are polynomial functions with matrix coefficients; the degree of $P_k$ is strictly smaller than the dimension of the largest Jordan cell having $r_k$ on its diagonal.

From this structure we deduce that $e^{At}$ is bounded for any $t \geq 0$ if and only if the real parts of all roots $r_k$ are strictly negative or zero and purely imaginary roots can only occur in Jordan cells of dimension 1.

If the above property holds, then

$$|x(t) - \hat{x}| \leq |e^{At}||x(0) - \hat{x}| \leq M|x(0) - \hat{x}|,\ t \geq 0,$$

hence for any $\varepsilon > 0$ we can choose $\delta(\varepsilon) = \dfrac{\varepsilon}{M}$ and

$$|x(0) - \hat{x}| < \delta(\varepsilon) \text{ implies } |x(t) - \hat{x}| \leq \varepsilon, \text{ for all } t \geq 0.$$

In this case, the equilibrium is <u>*stable*</u>.

If the real parts of all roots $r_k$ are strictly negative, then there exists $\alpha > 0$ such that $\operatorname{Re} r_k < -\alpha$ for all $k$. Then, we can write

$$e^{At} = e^{-\alpha t}\sum_{k=1}^{s} e^{(\alpha + r_k)t} P_k(t).$$

Since $\operatorname{Re}(\alpha + r_k) < 0$, we have $\lim_{t \to \infty} e^{(\alpha + r_k)t} P_k(t) = 0$, hence the function $t \mapsto e^{(\alpha + r_k)t} P_k(t)$ is bounded; therefore, there exists $\beta > 0$ such that $|e^{At}| < \beta e^{-\alpha t}$; we deduce that

$$|x(t) - \hat{x}| \leq \beta e^{-\alpha t}|x(0) - \hat{x}|$$

and the equilibrium is <u>*exponentially stable*</u> .

Let us remark that the matrix $A$ and the matrix $A^*$ obtained by <u>*transposition*</u>, have the same characteristic polynomial, hence if matrix $A$ has eigenvalues with strictly negative real parts, then the same property is valid for matrix $A^*$.

Herefrom it follows that the integral

$$V = \int_0^\infty e^{A^* t} \cdot e^{At}\, dt \text{ is } \underline{\text{convergent}}.$$

We have

$$A^* V + VA = \int_0^\infty \left[ A^* e^{A^* t} \cdot e^{At} + e^{A^* t} \cdot e^{At} A \right] dt =$$

$$= \int_0^\infty \frac{d}{dt} \left[ e^{A^* t} \cdot e^{At} \right] dt = -I.$$

We used the fact that $\lim\limits_{t \to \infty} e^{At} = \lim\limits_{t \to \infty} e^{A^* t} = 0$ if the eigenvalues of $A$ have strictly negative real parts.

From the definition of matrix $V$, we have for any vector $x$

$$x^* V x = \int_0^\infty x^* e^{A^* t} \cdot e^{At} x\, dt = \int_0^\infty \left| e^{At} x \right|^2 dt \ge 0.$$

If we had $x^* V x = 0$, then it would follow $e^{At} x = 0$ for all $t$ and, for $t \to 0$, we deduce $x = 0$. This means that the *quadratic form* associated to matrix $V$ is *positive definite*, hence there exists $\mu > 0$ such that $x^* V x \ge \mu |x|^2$ for any $x$.

We shall use these properties of the function $x \mapsto x^* V x$ in order to prove the following result, known as the ***theorem on stability by the first approximation***.

**Theorem 4.4** *Let $f : G \subset \mathbb{R}^n \to \mathbb{R}^n$ be of class $\mathscr{C}^1$ and let $\hat{x}$ be an equilibrium for the system of differential equations $x' = f(x)$.*

*Let $A = \frac{\partial f}{\partial x}(\hat{x})$. If the eigenvalues of matrix $A$ have strictly negative real parts, then the equilibrium $\hat{x}$ is exponentially stable.*

***Proof*** **a)** Since $f$ is of class $\mathscr{C}^1$, using a Lagrange formula (for $y$ such that $y + \hat{x} \in G$) we deduce

$$f(\hat{x} + y) - f(\hat{x}) = \int_0^1 \frac{\partial f}{\partial x}(\hat{x} + sy)\, y\, ds =$$

$$= \int_0^1 \left[\frac{\partial f}{\partial x}(\hat{x} + sy) - \frac{\partial f}{\partial x}(\hat{x})\right] y\, ds + \frac{\partial f}{\partial x}(\hat{x})\, y = A\, y + g(y),$$

where

$$g(y) = \int_0^1 \left[\frac{\partial f}{\partial x}(\hat{x} + sy) - \frac{\partial f}{\partial x}(\hat{x})\right] y\, dy$$

has the property that $\lim\limits_{|y| \to 0} \dfrac{|g(y)|}{|y|} = 0$.

**b)** Let $x$ be an arbitrary solution of the system defined by $f$ and $y(t) = x(t) - \hat{x}$. We have

$$y'(t) = x'(t) = f(x(t)) = f(\hat{x} + y(t)) = A\, y(t) + g(y(t)).$$

Since, by assumption, $A$ has eigenvalues with strictly negative real parts, we can associate to it the matrix $V$ defined above, satisfying

$$\mu |x|^2 \le x^* V x \le M |x|^2 \text{ and } A^* V + VA = -I.$$

Let us define $\widetilde{V}(t) = y^*(t)\, V\, y(t)$.

The function $\widetilde{V}$ is differentiable and we have

$$\widetilde{V}'(t) = \left[\frac{d}{dt} y^*(t)\right] V\, y(t) + y^*(t)\, V \frac{d}{dt} y(t) =$$

$$= \left[y^*(t)\, A^* + g^*(y(t))\right] V\, y(t) + y^*(t)\, V \left[A\, y(t) + g(y(t))\right] =$$

$$= y^*(t)\left[A^*\, V + V\, A\right] y(t) + g^*(y(t))\, V\, y(t) + y^*(t)\, V\, g(y(t)) =$$

$$= -\, y^*(t)\, y(t) + g^*(y(t))\, V\, y(t) + y^*(t)\, V\, g(y(t)).$$

Since $\lim\limits_{|y| \to 0} \dfrac{|g(y)|}{|y|} = 0$ there exists $\delta$ such that $|y| < 2\delta \sqrt{\dfrac{M}{\mu}}$ implies $\dfrac{|g(y)|}{|y|} < \dfrac{1}{4M}$, hence $|g(y)| < \dfrac{1}{4M} |y|$.

We assume now that the solution $x$ satisfies $|x(0) - \hat{x}| < \delta$.

From continuity of $x$, there exists a time interval such that

$$|x(t) - \hat{x}| < 2\delta\sqrt{\frac{M}{\mu}} \text{ for } t \text{ in this interval;}$$

it follows $|y(t)| < 2\delta\sqrt{\frac{M}{\mu}}$, hence

$$|g(y(t))| < \frac{1}{4M}|y(t)|.$$

In this interval, we have

$$\tilde{V}'(t) \leq -|y(t)|^2 + 2\frac{1}{4M}M|y(t)|^2 = -\frac{1}{2}|y(t)|^2.$$

But

$$|y(t)|^2 \geq \frac{1}{M}\tilde{V}(t), \quad -|y(t)|^2 \leq -\frac{1}{M}\tilde{V}(t),$$

hence

$$\tilde{V}'(t) \leq -\frac{1}{2M}\tilde{V}(t), \quad \tilde{V}'(t) + \frac{1}{2M}\tilde{V}(t) \leq 0, \quad \frac{d}{dt}\left[e^{\frac{t}{2M}}\tilde{V}(t)\right] \leq 0.$$

We deduce that the function $t \mapsto e^{\frac{t}{2M}}\tilde{V}(t)$ is decreasing and therefore

$$e^{\frac{t}{2M}}\tilde{V}(t) \leq \tilde{V}(0) \leq M|y(0)|^2 \leq M|x(0) - \hat{x}|^2 \text{ for } t \geq 0;$$

it follows

$$\mu|y(t)|^2 \leq \tilde{V}(t) \leq e^{-\frac{1}{2M}t}M|x(0) - \hat{x}|^2,$$

$$|y(t)| \leq \sqrt{\frac{M}{\mu}}e^{-\frac{1}{4M}t}|x(0) - \hat{x}| \leq \delta\sqrt{\frac{M}{\mu}} < 2\delta\sqrt{\frac{M}{\mu}}.$$

We deduce that our estimates are valid for the whole half-line $t \geq 0$ and

$$|x(t) - \hat{x}| \leq \beta e^{-\alpha t}|x(0) - \hat{x}|, \quad \beta = \sqrt{\frac{M}{\mu}}; \quad \alpha = \frac{1}{4M}$$

what ends the proof.

## 4.3 Stability. Discrete Time

Consider the discrete-time system $x_{t+1} = A\,x_t$, where $A$ is a constant $n \times n$ matrix. The solution is obviously written as

$$x_t = A^t\, x_0 .$$

To understand the behaviour of the system, perform a change of variables $y_t = S\,x_t$, with invertible $S$. We shall have

$$y_{t+1} = S\,x_{t+1} = S\,A\,x_t = S\,A\,S^{-1} y_t = B\,y_t .$$

Choose $S$ such that $B$ be in Jordan normal form:

$$B = \begin{pmatrix} J_1 & & 0 \\ & \ddots & \\ 0 & & J_r \end{pmatrix},$$

where $J_k$ is a Jordan cell:

$$J_k = \begin{pmatrix} \lambda_k & 1 & & 0 \\ & \ddots & \ddots & \\ 0 & & \lambda_k & 1 \end{pmatrix} = \lambda_k I + N .$$

$$N = \begin{pmatrix} 0 & 1 & & 0 \\ & \ddots & \ddots & \\ 0 & & 0 & 1 \end{pmatrix}$$

is a nillpotent matrix: $N^d = 0$, where $d$ is the dimension of the cell $J_k$.

From $B = S\,A\,S^{-1}$ it follows that $A^t = S^{-1} B^t\, S$. We have further

$$B = \begin{pmatrix} J_1^t & & & \\ & J_2^t & & \\ & & \ddots & \\ & & & J_r^t \end{pmatrix};$$

$$J_k^t = (\lambda_k I + N)^t = \sum_{j=0}^{t} \binom{j}{t} \lambda_k^{t-j} N^j = \sum_{j=0}^{d-1} \binom{j}{t} \lambda_k^{t-j} N^j .$$

Let us remark that in this sum the term for $j = 0$ is a diagonal matrix, the next one is non-zero only on the first parallel to the diagonal and so on.

We deduce that $\left\| J_k^t \right\| \le p_k(t) \left| \lambda_k \right|^t$, where $p_k$ is a polynomial.

Herefrom it follows that $\left\| A^t \right\| \le \sum_{k=1}^{r} q_k(t) \left| \lambda_k \right|^t$.

The case of interest is the one when all eigenvalues of matrix $A$ are located inside the unit disk: $\left| \lambda_k \right| < 1$. In this case, let $\mu$ be such that $\max_k \left| \lambda_k \right| < \mu < 1$.

We have $\left\| A^t \right\| \le q(t) \left( \dfrac{\max \left| \lambda_k \right|}{\mu} \right)^t \mu^t$, where $q$ is a polynomial.

Since $\lim_{t\to\infty} q(t)\alpha^t = 0$, for $0 < \alpha < 1$, we deduce that there exists $\beta \ge 1$ such that

$$q(t) \left( \frac{\max \left| \lambda_k \right|}{\mu} \right)^t \le \beta \text{ and } \left\| A^t \right\| \le \beta \mu^t, \ 0 < \mu < 1 .$$

We have obtained the following result

**Theorem 4.5** *If the eigenvalues of matrix A are located inside the unit disk, then there exist* $\beta \ge 1$, $0 < \mu < 1$, *such that for every solution of the linear discrete-time system defined by A we have* $\left\| x_t \right\| \le \beta \mu^t \left\| x_0 \right\|$.

We call such behaviour again *exponential stability*.

Let us now state a lemma which will play an important role in the sequel.

**Lemma 4.6** *Let $A$ be a matrix with all eigenvalues located inside the unit disk and let $C > 0$ ($C$ is a symmetric matrix with strictly positive eigenvalues). Then the equation*

$$A^*PA - P + C = 0$$

*has a solution $P > 0$, defined by* $P = \sum_{t=0}^{\infty} (A^*)^t C (A)^t$.

***Proof*** The absolute convergence of the series defining $P$ follows from the estimate $\| A^t \| \le \beta \mu^t$. We have, further,

$$A^*PA = \sum_{t=0}^{\infty} (A^*)^{t+1} C (A)^{t+1} = \sum_{t=1}^{\infty} (A^*)^t C (A)^t = P - C.$$

To show that $P > 0$, we have to prove that, for every $x \ne 0$, $x^*Px > 0$.

We have $x^*Px = \sum_{t=0}^{\infty} x^* (A^*)^t C A^t x \ge 0$, since $C > 0$.

If $x^*Px = 0$, then $x^* (A^*)^t C A^t x = 0$ for any $t$ and for $t = 0$ we have $x^*Cx = 0$, hence $x = 0$.

**Remark 4.7** We see from the definition of $P$ that

$$x^*Px \ge x^*Cx \ge \lambda \| x \|^2,$$

where $\lambda$ is the smallest eigenvalue of $C$. Let us remark also that

$$x^*Px \le \sum_{t=0}^{\infty} \Lambda \| A^t x \|^2,$$

where $\Lambda$ is the largest eigenvalue of matrix $C$.

It follows that $x^*Px \le \Lambda \beta^2 \sum_{t=0}^{\infty} \mu^{2t} \| x \|^2 = \dfrac{\Lambda \beta^2}{1 - \mu^2} \| x \|^2$.

We are now in position to study the stability problem for

nonlinear discrete-time systems.

For the system $x_{t+1} = f(x_t)$, an <u>equilibrium</u> will be a point $\hat{x}$ such that $f(\hat{x}) = \hat{x}$ (a fixed point for $f$); $\hat{x}$ defines a <u>*constant*</u> solution of the system.

**Definition 4.8**

*The equilibrium $\hat{x}$ is called <u>exponentially stable</u> if there exist $\delta > 0$, $\beta \geq 1$, $0 < \mu < 1$, such that for any solution $x_t$ with $\| x_0 - \hat{x} \| \leq \delta$ it follows that $\| x_t - \hat{x} \| \leq \beta \mu^t \| x_0 - \hat{x} \|$.*

Our aim is to prove a theorem ***on stability by the first approximation***, that is to infer exponential stability of the equilibrium from the exponential stability of the linear system obtained by linearization around the equilibrium.

**Theorem 4.9** *Let $f$ be $\mathscr{C}^1$. Assume that the eigenvalues of the matrix $A = \frac{\partial f}{\partial x}(\hat{x})$ are located inside the unit disk.*

*Then the equilibrium $\hat{x}$ of the system $x_{t+1} = f(x_t)$ is exponentially stable.*

***Proof*** **a)** Let $y_t = x_t - \hat{x}$. We have

$$y_{t+1} = x_{t+1} - \hat{x} = f(x_t) - f(\hat{x}) = f(\hat{x} + y_t) - f(\hat{x}) =$$

$$= \left[ \int_0^1 \left( \frac{\partial f}{\partial x}(\hat{x} + \alpha\, y_t) \right) d\alpha \right] \cdot y_t =$$

$$= \frac{\partial f}{\partial x}(\hat{x})\, y_t + \left[ \int_0^1 \left( \frac{\partial f}{\partial x}(\hat{x} + \alpha\, y_t) - \frac{\partial f}{\partial x}(\hat{x}) \right) d\alpha \right] \cdot y_t .$$

Denote $g(y) = \left[ \int_0^1 \left( \frac{\partial f}{\partial x}(\hat{x} + \alpha\, y) - \frac{\partial f}{\partial x}(\hat{x}) \right) d\alpha \right] \cdot y$. We shall have $\| g(y) \| \leq \left[ \int_0^1 \left\| \left( \frac{\partial f}{\partial x}(\hat{x} + \alpha\, y) - \frac{\partial f}{\partial x}(\hat{x}) \right) \right\| d\alpha \right] \cdot \| y \|$.

Since $\dfrac{\partial f}{\partial x}$ is continuous, it is uniformly continuous in the ball $\| y \| \le 1$; let $\omega(r)$ be the corresponding modulus of continuity.

We deduce $\| g(y) \| \le \left[ \int_0^1 \omega\left( \alpha \| y \| \right) d\alpha \right] \cdot \| y \| \le \omega\left( \| y \| \right) \cdot \| y \|$.

Since $\lim_{r \to 0} \omega(r) = 0$, it follows that for any $\gamma > 0$ there exists $\eta(\gamma) > 0$ such that if $\| y \| \le \eta(\gamma)$ then $\| g(y) \| \le \gamma \| y \|$.

**b)** We have seen that $y_{t+1} = A y_t + g(y_t)$, $A = \dfrac{\partial f}{\partial x}(\hat{x})$, and $g$ having the property above.

By assumption, the eigenvalues of matrix $A$ are located inside the unit disk, hence, according to Lemma 4.6, there exists $P > 0$ satisfying $A^* P A - P = - I$.

Let $y_t$ be an arbitrary solution and denote $V_t = y_t^* P y_t$.

We have

$$V_{t+1} = y_{t+1}^* P y_{t+1} = \left[ y_t^* A^* + g^*(y_t) \right] P \left[ A y_t + g(y_t) \right] =$$

$$= y_t^* A^* P A y_t + y_t^* A^* P g(y_t) + g^*(y_t) P A y_t + g^*(y_t) P g(y_t).$$

We have, further,

$$V_{t+1} - V_t = y_t^* \left[ A^* P A - P \right] y_t + y_t^* A^* P g(y_t) +$$

$$+ g^*(y_t) P A y_t + g^*(y_t) P g(y_t) =$$

$$= - \| y_t \|^2 + y_t^* A^* P g(y_t) + g^*(y_t) P A y_t + g^*(y_t) P g(y_t).$$

Herefrom, we obtain

$$V_{t+1} - V_t \le - \| y_t \|^2 + 2 \| y_t \| \| g(y_t) \| \| P \| \| A \| + \| P \| \| g(y_t) \|^2.$$

Let now $\gamma$ be such that $\gamma^2 \| P \| + 2 \gamma \| P \| \| A \| \le \dfrac{1}{2}$.

Let $\| y_0 \| < \dfrac{\eta(\gamma)}{\beta}$, $\beta = 2 \sqrt{\Lambda}$ (see Remark 4.7).

In any case, $\| y_0 \| < \eta(\gamma)$. As long as $\| y_t \| < \eta(\gamma)$, we have $V_{t+1} - V_t \le -\frac{1}{2}\| y_t \|^2$.

On the other hand, we have

$$\Lambda \| y_t \|^2 \ge V_t \ge \| y_t \|^2, \quad \| y_t \|^2 \ge \frac{1}{\Lambda} V_t, \quad -\frac{1}{2}\| y_t \|^2 \le -\frac{1}{2\Lambda} V_t .$$

We deduce $V_{t+1} \le \left(1 - \frac{1}{2\Lambda}\right) V_t$.

Denote $\mu = \sqrt{1 - \frac{1}{2\Lambda}}$; $\Lambda \ge 1$ implies $0 < \mu < 1$.

From $V_{t+1} \le \mu^2 V_t$, it follows that $V_t \le \mu^{2t} V_0$.

Taking into account the inequalities satisfied by $V_t$, we deduce $\| y_t \|^2 \le \Lambda \mu^{2t} \| y_0 \|^2$, that is $\| y_t \| \le \frac{\beta}{2} \mu^t \| y_0 \|$.

Since, for all $t$, we have $\mu^t < 1$, it follows that, for all $t$,

$$\| y_t \| \le \frac{\beta}{2} \| y_0 \| \le \frac{\beta}{2} \frac{\eta(\gamma)}{\beta} \le \frac{\eta(\gamma)}{2},$$

and the estimates above result valid for all $t$.

With $\beta$, $\mu$ as above and $\delta = \frac{\eta(\gamma)}{\beta}$, the exponential stability of the equilibrium is thus obtained.

## 4.4 Discrete – Time Logistic Equation

We consider the equation $x_{t+1} = \lambda x_t (1 - x_t)$, $\lambda > 0$, which can be considered as a discrete-time analogue of the model of limited growth.

It is used mainly in the models of population evolution, but it can also be interpreted as a model of economic growth. At the same time, it is very instructive in pointing out some specific aspects that occur in connection with discrete-time systems.

**A.** We begin by determining the equilibria. These are given by $x = \lambda x(1-x)$, hence they are $\hat{x} = 0$ and $\tilde{x} = \dfrac{\lambda - 1}{\lambda} = 1 - \dfrac{1}{\lambda}$.

**B.** We analyze the stability properties of these equilibria, by using the *linear approximations*.

For $\hat{x}$, the linear (first) approximation is $x_{t+1} = \lambda x_t$ and we have stability for $0 < \lambda < 1$.

For $\tilde{x}$, we define $y_t = x_t - \tilde{x}$ and we get

$$y_{t+1} = x_{t+1} - \tilde{x} = \lambda x_t (1 - x_t) - \tilde{x} =$$

$$= \lambda(\tilde{x} + y_t)(1 - \tilde{x} - y_t) - \tilde{x} = (1 - \lambda \tilde{x}) y_t - \lambda y_t^2 =$$

$$= \left[1 - \lambda\left(1 - \frac{1}{\lambda}\right)\right] y_t - \lambda y_t^2 = (2 - \lambda) y_t - \lambda y_t^2 .$$

The condition for stability is $-1 < 2 - \lambda < 1$, hence $1 < \lambda < 3$.

If we accept only evolutions with $x_t > 0$, then we can take into account the equilibrium $\tilde{x}$ only if $\lambda > 1$.

**C.** We look for periodic solutions with period 2, *i.e.* such that $x_{t+2} = x_t$ for all $t$. We have

$$x_{t+2} = \lambda x_{t+1}(1 - x_{t+1}) = \lambda\{\lambda x_t (1 - x_t)[1 - \lambda x_t (1 - x_t)]\} =$$

$$= \lambda^2 x_t (1 - x_t)[1 - \lambda x_t (1 - x_t)].$$

If $x_{t+2} = x_t \neq 0$ then $\lambda^2 (1 - x_t)[1 - \lambda x_t (1 - x_t)] = 1$.

We deduce that the periodic solutions with period 2 correspond to the roots of the third degree equation

$$\lambda^3 x^3 - 2\lambda^3 x^2 + \lambda^2 (1 + \lambda) x + 1 - \lambda^2 = 0 .$$

The equilibrium $\tilde{x} = 1 - \dfrac{1}{\lambda}$ is necessarily one of the roots of this equation.

If we denote $\xi = \lambda x$, the equation becomes

$$\xi^3 - 2\lambda\xi^2 + \lambda(1+\lambda)\xi + 1 - \lambda^2 = 0$$

and admits the root $\xi = \lambda - 1$; the other two roots are

$$\frac{1}{2}\left[\lambda + 1 \pm \sqrt{(\lambda+1)(\lambda-3)}\right],$$

hence the periodic solutions that are not equilibria are given by

$$\frac{\sqrt{1+\lambda}}{2\lambda}\left[\sqrt{1+\lambda} \pm \sqrt{\lambda-3}\right].$$

They are real and positive for $\lambda \geq 3$.

The stability of periodic solutions with period 2 is given by the stability of the corresponding equilibria of the equation

$$y_{t+1} = \lambda^2 y_t (1 - y_t)\left[1 - \lambda y_t (1 - y_t)\right].$$

We shall discuss stability by determining the linear approximations for the equilibria.

We denote

$f(y) = \lambda^2 y(1-y)\left[1 - \lambda y(1-y)\right]$ and $z_t = y_t - \hat{y}$; we have

$$z_{t+1} = y_{t+1} - \hat{y} = f(y_t) - \hat{y} = f(\hat{y} + z_t) - \hat{y} =$$

$$= f(\hat{y}) + f'(\hat{y})z_t + \ldots - \hat{y}.$$

But $f(\hat{y}) = \hat{y}$ and the linear approximation is $z_{t+1} = f'(\hat{y})z_t$.

We have

$$f'(y) = \lambda^2(1-y)\left[1 - \lambda y(1-y)\right] - \lambda^2 y\left[1 - \lambda y(1-y)\right] +$$

$$+ \lambda^2 y(1-y)\left[\lambda y - \lambda(1-y)\right] =$$

$$= \lambda^2(1-2y)\left[1 - 2\lambda y(1-y)\right]$$

We have to evaluate this derivative at $\frac{\sqrt{1+\lambda}}{2\lambda}\left[\sqrt{1+\lambda} \pm \sqrt{\lambda-3}\right]$.

These points satisfy

$$\lambda^3 \hat{y}^3 - 2\lambda^3 \hat{y}^2 + \lambda^2 (1+\lambda)\hat{y} + 1 - \lambda^2 = 0$$

hence

$$f'(\hat{y}) = -2\lambda^3 \hat{y}^2 + 2\lambda^2 (1+\lambda)\hat{y} + 4 - 3\lambda^2;$$

on the other hand, the points of interest also satisfy

$$\lambda^2 \hat{y}^2 - \lambda(1+\lambda)\hat{y} + \lambda + 1 = 0,$$

hence

$$f'(\hat{y}) = -2\lambda\left[\lambda(1+\lambda)\hat{y} - (1+\lambda)\right] + 2\lambda^2 (1+\lambda)\hat{y} + 4 - 3\lambda^2 =$$

$$= 2\lambda(1+\lambda) + 4 - 3\lambda^2 = -\lambda^2 + 2\lambda + 4.$$

The stability condition is $|f'(\hat{y})| < 1$, what leads to the inequalities

$$-1 < -\lambda^2 + 2\lambda + 4 < 1, \text{ i.e.}$$

$$-\lambda^2 + 2\lambda + 3 < 0, \; -\lambda^2 + 2\lambda + 5 > 0.$$

The first inequality is satisfied for $\lambda < -1$ and $\lambda > 3$, while the second is satisfied for $1 - \sqrt{6} < \lambda < 1 + \sqrt{6}$.

As we assumed $\lambda > 3$, for stability we have the additional condition $\lambda < 1 + \sqrt{6}$.

Let us summarize the informations we got. If we are interested only in evolutions with $x_t > 0$ then, for $0 < \lambda < 1$, the only equilibrium is $\hat{x} = 0$ and this equilibrium is an attractor.

For $\lambda > 1$, the equilibrium $\hat{x}$ is repulsive; we also have an equilibrium at $\tilde{x} = 1 - \frac{1}{\lambda}$, which is also repulsive for $\lambda > 3$ and attractive for $1 < \lambda < 3$.

If $\lambda > 3$, we also have two periodic solutions with period 2, corresponding to the values $\frac{\sqrt{1+\lambda}}{2\lambda}\left[\sqrt{1+\lambda} \pm \sqrt{\lambda-3}\right]$.

These solutions are attractive for $3 < \lambda < \lambda_1 = 1 + \sqrt{6}$.

We can proceed further and find out that, for $\lambda_1 < \lambda < \lambda_2$, there exists an attractive cycle with period 4 and, in general, we can obtain values $\lambda_k$ such that, for $\lambda_k < \lambda < \lambda_{k+1}$, there exists an attractive cycle with period $2^{k+1}$.

These values converge to a limit $\tilde{\lambda}$, beyond which regular behaviour ceases.

## 4.5 Stable Polynomials

We saw in which manner the condition that the real parts of the roots of the characteristic polynomial be strictly negative occurs in the study of stability of linear continuous time systems. The question occurs in a natural way, how can we recognize, without determining the roots, whether all roots have strictly negative real parts.

This problem was of concern for engineers and mathematicians even since the last century and it was completely solved more than 100 years ago (Hurwitz, Math. Ann., 1895). It is easy to obtain particular results; we shall discuss here the cases $n = 2$ and $n = 3$.

We begin with a second degree polynomial of the form $r^2 + br + c$.

If $c \le 0$, then the roots are real and of opposite signs (possibly one of them is zero), hence it is not possible that both of them be negative; if $b \le 0$, then the sum of the roots is positive (possibly zero), and again it is not possible that both roots had strictly negative real parts.

Thus, the conditions $b > 0$, $c > 0$ are *necessary*. These conditions are also sufficient. Indeed, if $c > 0$, then the roots are real, having the same sign, or complex conjugate; if they are complex conjugate, then their sum, $-b$, is equal to the sum of the real parts, hence the real parts are negative; if the roots are real, having the same sign and their sum being negative, then they are negative.

Let us study now the third degree polynomials of the form $r^3 + br^2 + cr + d$.

If $r_1, r_2, r_3$ are the roots, then we have the relations:

$$r_1 + r_2 + r_3 = -b,$$

$$r_1 r_2 + r_2 r_3 + r_3 r_1 = c,$$

$$r_1 r_2 r_3 = -d.$$

If $r_1$ is real and $r_2, r_3$ are complex conjugate, then $r_2 r_3 > 0$ and, if $r_1 < 0$, then $d > 0$. If all the roots are real and negative, then again $r_1 r_2 r_3 < 0$, hence $d > 0$.

If $r_1 < 0$ and $r_2, r_3$ are complex conjugate with strictly negative real parts or if they are real and strictly negative, then $r_1 + r_2 + r_3 < 0$, hence $b > 0$.

Further, we have

$$bc - d = b\left[r_1(r_2 + r_3) + r_2 r_3\right] + r_1 r_2 r_3 =$$

$$= b r_1 (r_2 + r_3) + (b + r_1) r_2 r_3 = b r_1 (r_2 + r_3) - (r_2 + r_3) r_2 r_3 =$$

$$= (r_2 + r_3)(b r_1 - r_2 r_3).$$

We have $b r_1 < 0$, $r_2 r_3 > 0$, $r_2 + r_3 < 0$, hence $b c - d > 0$.

Therefore, if the roots of the third degree polynomial have strictly negative real parts, then $b > 0$, $d > 0$, $b c - d > 0$.

Let us assume, conversely, that the above conditions are satisfied.

If all roots are real, then they are necessarily negative, since from $b > 0$, $c > 0$, $d > 0$ it follows that $r^3 + b r^2 + c r + d > 0$ for all $r \geq 0$. If only one root is real then, from $d > 0$ it follows that this root is negative.

Since $b c - d > 0$ implies $(r_2 + r_3)(b r_1 - r_2 r_3) > 0$ and since $b r_1 < 0$, $r_2 r_3 > 0$, it follows that $r_2 + r_3 < 0$, hence the roots $r_2$, $r_3$ have strictly negative real parts.

We shall not discuss the case of polynomials of higher degree, but we shall discuss the analogous problems that occur in the study of discrete models.

As we saw, in this case the stability condition requires that the roots be strictly smaller than 1 in modulus.

We can reduce this problem to the previous one, by performing a change of variables.

Let us define $z = \dfrac{1 + r}{1 - r}$ and assume that $r = \mu + i \nu$;

we have

$$|z|^2 = \frac{|1 + \mu + i \nu|^2}{|1 - \mu - i \nu|^2} = \frac{(1 + \mu)^2 + \nu^2}{(1 - \mu)^2 + \nu^2}$$

and we have $|z| < 1$ if and only if $(1+\mu)^2 + \nu^2 < (1-\mu)^2 + \nu^2$, hence if and only if $\mu < 0$.

Given the second degree polynomial $P(z) = z^2 + bz + c$, we associate the polynomial obtained by substituting $z = \frac{1+r}{1-r}$, hence

$$\left(\frac{1+r}{1-r}\right)^2 + b\frac{1+r}{1-r} + c =$$

$$= \left[(1+r)^2 + b\left(1-r^2\right) + c\left(1-r\right)^2\right]\frac{1}{(1-r)^2}.$$

The roots of the polynomial $P(z)$ will correspond to the roots of the polynomial

$$Q(r) = (1+r)^2 + b\left(1-r^2\right) + c(1-r)^2 =$$

$$= (1+c-b)r^2 + 2(1-c)r + 1 + b + c.$$

The roots of $P(z)$ will be strictly smaller than 1 in modulus if and only if the roots of $Q(r)$ will have strictly negative real parts.

For this property to hold, it is necessary and sufficient that all coefficients of $Q(r)$ had the same sign, hence that

**(1)** $(1+c-b)(1+c+b) > 0$,

**(2)** $(1+c-b)(1-c) > 0$.

The first condition can be written as $(1+c)^2 - b^2 > 0$, or $|b| < |1+c|$.

If the roots of $P(z)$ are smaller than 1 in modulus, then $|c| < 1$, hence $1+c > 0$, $1-c > 0$.

If $|b| < 1 + c$, then condition **(2)** is automatically satisfied and, thus, the <u>*necessary and sufficient*</u> conditions are $|c| < 1$, $|b| < 1 + c$.

It is possible to analyze in the same way the case of polynomials of third degree.

## 4.6 Some Properties of Matrices that occur in Economic Models

For the Leontieff type models, analyzed in **3.5**, the equilibrium corresponds to $(I - A)X = x$, hence to $X = (I - A)^{-1}x$.

Such an equilibrium has all its components $X_j \geq 0$ for any final consumption vector $x$ with positive components, if all elements of $(I - A)^{-1}$ are positive. On the other hand, all non-diagonal elements of $(I - A)$ are negative. We are thus led to look for conditions that should be satisfied by a matrix with negative non-diagonal elements, in order that all elements of its inverse be positive.

**Theorem 4.10** *Let $D$ be an $n \times n$ matrix such that $d_{ij} \leq 0$ for $i \neq j$. Then all elements of $D^{-1}$ are positive if and only if one of the following <u>equivalent</u> conditions holds:*

**(I)** *There exist $c_i > 0$, $i = 1,\dots,n$ such that the system $\sum_{j=1}^{n} d_{ij} x_j = c_i$ $(i = 1,\dots,n)$ had a solution with $x_j \geq 0$ for all $j$.*

**(II)** *For any $c_i \geq 0$, $i = 1,\dots,n$, the above system admits a* *solution with* $x_j \geq 0$, $j = 1,\dots,n$.

**(III)** *The principal diagonal minors of matrix D are strictly positive.*

**(IV)** *Any principal minor is positive.*

***Proof* A.** We prove first the equivalence of conditions **(I)**, **(II)**, **(III)**, **(IV)**.

We prove the implication **(I)→(III)** by induction with respect to $n$:

For $n = 1$, the equation is $d_{11} x_1 = c_1$ and if there exists $c_1 > 0$ for which $x_1 \geq 0$, then $d_{11} > 0$, i.e. **(III)** holds.

We assume the implication **(I)→(III)** valid for dimension $n-1$. We consider $c_1,\dots,c_n$ with $c_i > 0$ for all $i$ and let $x_1,\dots,x_n$ with $x_j \geq 0$ for all $j$ be the solution of the system

$$\sum_{j=1}^{n} d_{ij} x_j = c_i, \ i = 1,\dots,n.$$

We write the first equation in the form

$$d_{11} x_1 = c_1 - \sum_{j=2}^{n} d_{1j} x_j.$$

We have $c_1 > 0$, $d_{ij} \leq 0$, $x_j \geq 0$; according to the argument we made for $n = 1$, we deduce $d_{11} > 0$. We eliminate $x_1$ from the other equations and we obtain

$$\sum_{j=2}^{n} d^{*}_{ij} x_j = c_i^{*}, \ i = 2,\dots,n,$$

where

$$d_{ij}^{*} = d_{ij} - \frac{d_{i1} d_{1j}}{d_{11}} \leq 0 \text{ for } i \neq j, i > 1, j > 1$$

and $c_i^* = c_i - \dfrac{d_{i1}\, c_1}{d_{11}} > 0,\ i > 1.$

Using the induction assumption, we conclude that the principal diagonal minors of matrix $\left(d_{ij}^*\right)_{i,j=2}^{n}$ are strictly positive.

Since a principal diagonal minor of $D$ is obtained by multiplying by $d_{11}$ the principal diagonal minor of matrix $\left(d_{ij}^*\right)_{i,j=2}^{n}$ whose order is one unit smaller, we conclude that **(III)** holds.

We also prove by induction the implication **(III)→(II)**.

For $n = 1$, if $d_{11} > 0$ and $c_{11} \geq 0$, then $x_j \geq 0$, hence the implication is valid.

Assume the implication **(III)→(II)** valid for $n - 1$. We have again $d_{11} > 0$, we can eliminate $x_1$ and obtain a system of dimension $n - 1$ whose principal diagonal minors are strictly positive, hence, according to the induction assumption, it follows $x_j \geq 0,\ j = 2,\ldots,n$ for $c_i \geq 0$; from the first equation we also obtain $x_1 \geq 0$, the implication being thus proved.

We obviously have **(II)→(I)** and **(IV)→(III)**.

Therefore, we have **(I)→(III)→(II)→(I)**, hence the first conditions are equivalent.

On the other hand, **(II)→(IV)**, since we always can transform a system of principal minors into a system of principal diagonal minors, by renumbering the equations and the variables, what does not alter property **(II)**. Herefrom, taking into account that **(II)→(III)**, we also have **(II)→(IV)**, what gives the cycle **(II)→(IV)→(III)→(II)**.

**B.** After the equivalences were established, it is easy to prove the statement of the theorem.

If the elements of $D^{-1}$ are positive, then **(II)** obviously holds. If **(II)** holds, then **(III)** holds, hence $D$ is invertible, the solution of the system $Dx = c$ is unique: $x = D^{-1}c$ and since, according to **(II)**, $x_j \geq 0$ for $c_i \geq 0$, we deduce that all elements of $D^{-1}$ are positive, what ends the proof.

**Proposition 4.11** *Let $A$ be a square matrix with positive elements. We define*

$$M(A) = \{\rho \,/\, \rho I - A \text{ admits an inverse with positive elements}\}.$$

*Then there exists $\lambda \geq 0$ such that $M(A) = (\lambda, +\infty)$.*

***Proof*** Let $x$ be a vector with all its coordinates strictly positive; we choose $\rho$ sufficiently large so that

$$\rho x_j > \sum_k a_{jk} x_k, \text{ for all } j;$$

for such $\rho$, all coordinates of the vector $(\rho I - A)x$ will be strictly positive; denoting this vector by $c$, we see that the matrix $\rho I - A$ satisfies condition **(II)** of the above theorem, hence all elements of $(\rho I - A)^{-1}$ are positive, what indicates that $M(A) \neq \Phi$.

Further, if $\rho \in M(A)$ and $\sigma \geq \rho$, then $\sigma$ also belongs to $M(A)$; indeed, if $\rho \in M(A)$, then there exist $x_i \geq 0$, $i = 1, \ldots, n$ such that all elements of $(\rho I - A)x$ are positive (property **(II)** of Theorem 4.10) and hence all elements of

$$(\sigma I - A)x = (\rho I - A)x + (\sigma - \rho)x$$

are also positive.

From Theorem 4.10 it follows that the elements of $(\sigma I - A)^{-1}$ are positive, hence $\sigma \in M(A)$.

We define $\lambda = \inf M(A)$; we prove that $\lambda \notin M(A)$. If $\lambda$ belonged to $M(A)$, then there would exist a vector $x$ with positive elements, such that all coordinates of $(\lambda I - A)x$ would be strictly positive (property **(I)** of Theorem 4.10), hence

$$\lambda x_i > \sum_j a_{ij} x_j, \quad i = 1,\dots,n;$$

the inequalities being strict, they would be preserved for a value $\rho < \lambda$, sufficiently close to $\lambda$. Such $\rho$ would belong to $M(A)$, what contradicts the definition of $\lambda$ as the infimum.

**Proposition 4.12** *For* $\lambda = \inf M(A)$ *there exists* $x \neq 0$ *with positive coordinates, such that* $Ax = \lambda x$.

***Proof*** Let $c$ be a vector having all its coordinates strictly positive; for $\rho \in M(A)$, we denote $y(\rho) = (\rho I - A)^{-1} c$.

Consider $\sigma \geq \rho$, hence $\sigma \in M(A)$. We have

$$(\rho I - A)[y(\rho) - y(\sigma)] = c - (\rho I - A)\, y(\sigma) =$$

$$= c - (\rho - \sigma)\, y(\sigma) - (\sigma I - A)\, y(\sigma) = (\sigma - \rho)\, y(\sigma),$$

hence

$$y(\rho) - y(\sigma) = (\sigma - \rho)(\rho I - A)^{-1} y(\sigma),$$

therefore

$$y_j(\rho) \geq y_j(\sigma), \quad j = 1,\dots,n, \text{ if } \sigma \geq \rho.$$

We deduce that the function $\rho \mapsto y_j(\rho)$ is *decreasing*.

We choose a decreasing sequence $\rho_\nu$, such that $\lim\limits_{\nu\to\infty} \rho_\nu = \lambda$ (such a sequence can be defined according to the definition of the infimum). It follows that the sequence $y_j(\rho_\nu)$ is increasing, hence $z_\nu = \sum_{j=1}^{n} y_j(\rho_\nu)$ is an increasing sequence.

If this sequence were bounded, then it would follow that $y_j(\rho_\nu)$ is bounded and, since it is increasing, it would have a positive limit $\hat{y}_j$, $j = 1,\dots,n$.

From $(\rho_\nu I - A)\, y(\rho_\nu) = c$ we deduce that, for $\nu \to \infty$, $(\lambda I - A)\,\hat{y} = c$, hence $\lambda \in M(A)$. We saw that this is not possible, hence $z_\nu$ is <u>*unbounded*</u> and tends to infinity.

We define $x_j^\nu = \dfrac{y_j(\rho_\nu)}{z_\nu}$; we have $x_j^\nu \geq 0$, $\sum_{j=1}^{n} x_j^\nu = 1$, hence the sequence $x_j^\nu$ is bounded and we can take a convergent subsequence; let $\hat{x}_j$ be its limit. We have

$$(\rho_\nu I - A)\, y(\rho_\nu) = c,$$

$$\left(\rho_{\nu_l} I - A\right) \frac{y(\rho_{\nu_l})}{z_{\nu_l}} = \frac{c}{z_{\nu_l}},$$

hence

$$\left(\rho_{\nu_l} I - A\right) x^{\nu_l} = \frac{c}{z_{\nu_l}}$$

For $l \to \infty$, it follows $(\lambda I - A)\,\hat{x} = 0$. Since $\sum_{j=1}^{n} \hat{x}_j = 1$, it follows $\hat{x} \neq 0$.

Proposition 4.12 shows that $\lambda = \inf M(A)$ is an eigenvalue of matrix $A$ and that there exists a corresponding eigenvector with positive coordinates. We remark that $\lambda$ is the largest real eigenvalue; indeed, if $\rho > \lambda$, then $\rho \in M(A)$, $\rho I - A$ is invertible, hence $\rho$ cannot be an eigenvalue.

**Proposition 4.13** *If there exists* $y \neq 0$ *with positive coordinates, such that* $\sum_{j=1}^{n} a_{ij} y_j \geq \mu y_i$, $i = 1, \dots, n$, *then* $\lambda \geq \mu$.

***Proof*** If $\mu > \lambda$, then $\mu \in M(A)$, hence all elements of $(\mu I - A)^{-1}$ are positive. Since, by assumption, $(\mu I - A) y$ has negative elements, it would follow that $y$ had negative elements, a contradiction.

**Proposition 4.14** *If* $\mu$ *is an eigenvalue of* $A$, *then* $|\mu| \leq \lambda$.

***Proof*** Let $\mu$ be an eigenvalue of $A$ and $u$ a corresponding eigenvector. We have $\sum_{j=1}^{n} a_{ij} u_j = \mu u_i$, hence

$$\sum_{j=1}^{n} a_{ij} |u_j| \geq |\mu| |u_i|$$

and, from the above proposition, it follows $|\mu| \leq \lambda$.

Proposition 4.11 – Proposition 4.14. give the <u>*Perron – Frobenius theorem*</u>, which states that for a matrix $A$ with positive elements the eigenvalue which is largest in modulus is real, positive and there exists a corresponding eigenvector with positive elements.

We denote $\lambda(A) = \inf M(A)$.

**Theorem 4.15** *The following properties hold:*

**a)** $\lambda(A) = \lambda(A^*)$;

**b)** $\lambda(\alpha A) = \alpha \lambda(A)$ *for* $\alpha \geq 0$;

**c)** $\lambda\left(A^k\right) = \left[\lambda\left(A\right)\right]^k$ *for any positive integer* $k$;

**d)** $\lambda\left(A\right) \geq \lambda\left(B\right)$ *if* $A - B$ *has positive elements*;

**e)** $\lambda\left(A\right) \geq \lambda\left(C\right)$ *for any principal submatrix* $C$ *of* $A$;

**f)** $\lambda\left(A\right) = 0$ *if and only if* $A$ *is nillpotent (there exists* $k$ *such that* $A^k = 0$*).*

***Proof***

**a)** We have $M\left(A\right) = M\left(A^*\right)$, hence $\lambda\left(A^*\right) = \lambda\left(A\right)$.

**b)** We have $A\,x = \lambda\,x$, hence $\alpha\,A\,x = \alpha\,\lambda\,x$, therefore $\lambda\,\alpha$ is an eigenvalue for $\alpha\,A$ and $x$ is also an eigenvector for $\alpha\,A$.

**c)** From $A\,x = \lambda\,x$ it follows $A^k\,x = \lambda^k\,x$; we see that $\lambda^k$ is indeed an eigenvalue for $A^k$ and it is associated to the same eigenvector $x$.

If $\mu$ is an eigenvalue for $A^k$, then $\mu = \rho^k$, where $\rho$ is an eigenvalue for $A$, hence $\left|\mu\right| = \left|\rho\right|^k \leq \lambda^k$ and the conclusion follows.

**d)** If $\rho \in M(A)$, then there exists a vector $x$ with positive coordinates, such that $\left(\rho\,I - A\right)x$ had strictly positive coordinates, hence

$$\left(\rho\,I - B\right)x = \left(\rho\,I - A\right)x + \left(A - B\right)x$$

has strictly positive coordinates. It follows that $\rho \in M(B)$, therefore $M(A) \subset M(B)$, what implies $\lambda\left(A\right) \geq \lambda\left(B\right)$.

**e)** The eigenvalues do not change if we perform a suitable renumbering of rows and columns, hence we may assume that $C$ is a diagonal principal submatrix; we complete it with zeros and we apply the conclusion of **d)**. Completing with zeros does not modify the largest eigenvalue.

**f)** If $A = 0$ then we obviously have $\lambda(A) = 0$. If $A^k = 0$, it follows $\lambda(A^k) = 0$, hence $[\lambda(A)]^k = 0$, therefore $\lambda(A) = 0$.

We still have to prove that if $\lambda(A) = 0$, then $A$ is nillpotent. If $\lambda(A) = 0$, then we have $\lambda(A^k) = 0$ for all $k$.

Let $x$ be a vector with strictly positive coordinates and let $\theta > 0$ be such that $\theta x_i \geq \sum_{j=1}^{n} a_{ij} x_j$ for all $i$. It follows

$$\theta \sum_{k=1}^{n} a_{ki} x_i \geq \sum_{k=1}^{n} a_{ki} \sum_{j=1}^{n} a_{ij} x_j,$$

hence $\theta (A x)_i \geq (A^2 x)_i$ for all $i$.

Repeating this reasoning, it follows $(\theta A^s x)_i \geq (A^{s+1} x)_i$.

We define

$N_s = \{i \,/\, \textit{the i-th coordinate of vector } A^s x \textit{ is stictly positive}\}$.

Taking into account the inequality established, we have

$$N_0 \supset N_1 \supset \ldots \supset N_s \supset N_{s+1} \supset \ldots.$$

We show that $N_n = \Phi$; denote $\mu_s = \min\limits_{i \in N_s} \dfrac{(A^{s+1} x)_i}{(A^s x)_i}$; $\mu_s$ can be defined for any $s$ for which $N_s \neq \Phi$. We have

$$\mu_s \leq \frac{(A^{s+1} x)_i}{(A^s x)_i} \text{ for all } i \in N_s,$$

hence

$$\mu_s \cdot (A^s x)_i \leq (A^{s+1} x)_i \text{ for all } i \in N_s, \text{ and}$$

$$\mu_s \cdot (A^s x)_i = (A^{s+1} x)_i \text{ for all } i \notin N_s.$$

From Proposition 4.12 it follows that $\mu_s \leq \lambda(A)$ and, since

$\lambda(A) = 0$, it follows $\mu_s = 0$, hence $\left(A^{s+1}x\right)_i = 0$ for at least one $i \in N_s$, what shows that $N_{s+1}$ is strictly included in $N_s$. Therefore, as long as $N_s$ is *not empty*, the inclusion is strict and, after $n$ steps, we reach necessarily the empty set.

Therefore, $\lambda(A) = 0$ implies $N_n = \Phi$, hence $A^n x = 0$ and since all elements of $A^n$ are positive and all elements of $x$ are strictly positive, it follows $A^n = 0$, what ends the proof.

**Theorem 4.16** *A matrix $D$ with $d_{ij} \le 0$ for $i \ne j$ has an inverse $D^{-1}$ with all elements positive if an only if the eigenvalues of matrix $-D$ have strictly negative real parts.*

***Proof*** We choose $\rho > 0$ sufficiently large in order that all elements of $\rho I - D$ be positive. We denote $A = \rho I - D$.

If $\mu$ is an eigenvalue of $D$, then there exists $x \ne 0$ such that $Dx = \mu x$, hence $Ax = \rho x - Dx = (\rho - \mu)x$ and $\rho - \mu$ is an eigenvalue for $A$; we deduce that $\rho - \operatorname{Re}\mu$ is the real part of an eigenvalue of $A$, hence smaller than the modulus of $\rho - \mu$, therefore $\rho - \operatorname{Re}\mu \le \lambda(A)$; it follows $\operatorname{Re}\mu \ge \rho - \lambda(A)$ and $\rho - \lambda(A)$ is an eigenvalue for $D$.

Let $\hat{x}$ be a vector with positive elements, such that $A\hat{x} = \lambda(A)\hat{x}$; we have $D\hat{x} = [\rho - \lambda(A)]\hat{x}$.

But $\rho I - A = D$ has an inverse with positive elements if and only if $\rho > \lambda(A) = \inf M(A)$, hence $D$ has an inverse with positive elements if an only if, for any eigenvalue $\mu$ of $D$, we have $\operatorname{Re}\mu \ge \rho - \lambda(A) > 0$. Since $\mu$ is an eigenvalue of $D$ if an only if

$-\mu$ is an eigenvalue of $-D$, we deduce that $D^{-1}$ has positive elements if and only if the eigenvalues of $-D$ have strictly negative real parts, what ends the proof.

Returning to the Leontieff type models, we deduce that, for any given vector of final consumption, there exists an equilibrium production vector if an only if the eigenvalues of $A - I$ have strictly negative real parts, which means that the equilibrium of the system $X' = (A - I)X + x$ is exponentially stable.

## 4.7 Notes and References

The existence and uniqueness theorem is a standard result and may be found in any text on differential equations, hence it is difficult to give a specific reference.

The theorem on stability by the first approximation may also be found in most of advanced books on differential equations and again we shall not give specific references.

For the discrete - time case, we shall refer to the book of La Salle [19].

For the material in **4.6**, a reference can be Nikaido [23].

CHAPTER 5

# Numerical Solution of Differential Equations

## 5.1 Euler Method

We begin with the description of the simplest numerical method of solving differential equations. Although this method is not efficient and it is not used currently, it allows to understand the basic ideas of numerical procedures. At the same time, it is the starting point for more efficient methods. Although the results are valid for systems of differential equations of any order, we consider only the case of scalar equations, in order to avoid certain unessential difficulties.

Let $f : I \times J \subset \mathrm{R} \times \mathrm{R} \rightarrow \mathrm{R}$ be a smooth function; in what follows , we consider the class $\mathscr{C}^2$, but in this chapter we shall also deal with greater degree of smoothness.

Assume $t_o \in I$, $x_o \in J$; from the fundamental existence and uniqueness theorem follows the existence and uniqueness of the solution of the equation $x' = f(t,x)$, satisfying $x(t_0) = x_0$. This solution is defined on an interval $I_{(t_0, x_0)} \subset I$. We intend to describe a procedure allowing to compute approximately the values of this solution.

We consider an interval $[t_0, t_0 + T] \subset I_{(t_0, x_0)}$ and we choose a step size $\tau = \frac{T}{N}$, obtained by dividing the interval $[t_0, t_0 + T]$ into $N$ equal parts. We denote by $t_k$ the point $t_0 + k\tau$.

The numerical procedure will provide approximate values of the solution at the points $t_k$.

The computational procedure will be as follows: starting from the value $x_0$, we build the sequence $x_k$ given recurrently by

$$x_{k+1} = x_k + \tau f(t_k, x_k).$$

**Theorem 5.1** *There exists a constant* $\nu$, *independent of* $\tau$, *such that* $\|x(t_k) - x_k\| \le \nu\tau$, *for all* $k = 1, 2, \dots, N$.

***Proof*** We denote $\varepsilon_k = x(t_k) - x_k$. We have

$$\varepsilon_{k+1} = x(t_{k+1}) - x_{k+1} = x(t_k + \tau) - x_k - \tau f(t_k, x_k).$$

But $x(t_k + \tau) = x(t_k) + \tau x'(t_k) + \frac{\tau^2}{2} x''(\theta_k)$ (according to the Taylor formula). Since $x$ is a solution of the differential equation, we have $x'(t_k) = f(t_k, x(t_k))$. Hence, we can write:

$$\varepsilon_{k+1} = x(t_k) + \tau f(t_k, x(t_k)) + \frac{\tau^2}{2} x''(\theta_k) - x_k - \tau f(t_k, x_k) =$$

$$= \varepsilon_k + \tau\left[f(t_k, x(t_k)) - f(t_k, x_k)\right] + \frac{\tau^2}{2} x''(\theta_k).$$

Using the Lagrange formula, we write:

$$f(t_k, x(t_k)) - f(t_k, x_k) = \frac{\partial f}{\partial x}(t_k, \xi_k)\left(x(t_k) - x_k\right).$$

Thus, we obtain the relation:

$$\varepsilon_{k+1} = \varepsilon_k + \tau \frac{\partial f}{\partial x}(t_k, \xi_k)\varepsilon_k + \frac{\tau^2}{2}x''(\theta_k).$$

We denote by $\mu$ the maximum of $\left\|\frac{\partial f}{\partial x}(t, x)\right\|$, for $t \in [t_0, t_0 + T]$ and for $x$ in a compact containing $x_0$. We also denote by $2M$ the maximum of $\|x''(t)\|$, for $t \in [t_0, t_0 + T]$. With these notations, we have the estimate

$$\|\varepsilon_{k+1}\| \le \|\varepsilon_k\|(1 + \tau\mu) + \tau^2 M.$$

Let us denote by $\{\eta_k\}$ the sequence defined by

$$\eta_{k+1} = \eta_k(1 + \tau\mu) + \tau^2 M.$$

We have

$$\|\varepsilon_{k+1}\| - \eta_{k+1} \le \left(\|\varepsilon_k\| - \eta_k\right)(1 + \tau\mu)$$

hence, if we take $\eta_0 = \|\varepsilon_0\| = 0$, it follows by induction that

$$\|\varepsilon_k\| - \eta_k \le 0, \text{ or } \|\varepsilon_k\| \le \eta_k.$$

But $\eta_k$ satisfies a first-order affine recurrence; we denote by $\hat{\eta}$ the constant solution given by:

$$\hat{\eta} = -\frac{\tau M}{\mu}$$

and we obtain the general solution of the form

$$\eta_k = -\frac{\tau M}{\mu} + c(1 + \tau\mu)^k;$$

since $\eta_0 = 0$, it follows

$$c = \frac{\tau M}{\mu}, \quad \eta_k = \frac{\tau M}{\mu}\left[(1 + \tau\mu)^k - 1\right] \le \frac{\tau M}{\mu}(1 + \tau\mu)^k.$$

Recalling now that $\tau = \dfrac{T}{N}$, we can write:

$$(1+\tau\mu)^k = \left(1+\frac{\mu T}{N}\right)^k \le \left(1+\frac{\mu T}{N}\right)^N \le e^{\mu T}.$$

We obtained thus the estimate $\eta_k \le \dfrac{\tau M}{\mu} e^{\mu T}$, hence

$$\|\varepsilon_k\| = \|x(t_k) - x_k\| \le \tau \frac{M}{\mu} e^{\mu T} = \nu\tau$$

and this ends the proof.

The theorem shows that the sequence $x_k$ approximates uniformly the values $x(t_k)$, the error being of the order of the step size $\tau$; if the step size is sufficiently small, then the error can be made arbitrarily small. At the same time, let us remark that

$$\nu = \frac{M}{\mu} e^{\mu T},$$

so that its value increases exponentially with respect to the length of the time interval and with respect to the constant $\mu$ which depends on the derivative $\dfrac{\partial f}{\partial x}$.

The Euler method, the simplest one from the conceptual viewpoint, has a reduced efficiency, since the error is of the order of $\tau$.

## 5.2 Richardson Extrapolation

We indicate now a simple procedure by which, starting with the Euler scheme, we can obtain an error of the order of $\tau^2$.

We try to obtain for the sequence $x_k$ given by the Euler scheme a representation of the form:

$$x_k = x(t_k) + \tau\, v(t_k) + \tau^2\, w_k,$$

where the function $v$ is independent of $\tau$ and the sequence $w_k$ is bounded with respect to $\tau$. We have

$$w_k = \frac{1}{\tau^2}\left[x_k - x(t_k) - \tau\, v(t_k)\right]$$

and we try to find the function $v$ so that the sequence $w_k$ would remain bounded with respect to $\tau$. We have

$$w_{k+1} = \frac{1}{\tau^2}\left[x_{k+1} - x(t_{k+1}) - \tau\, v(t_{k+1})\right] =$$

$$= \frac{1}{\tau^2}\left[x_k + \tau f(t_k, x_k) - x(t_k + \tau) - \tau\, v(t_k + \tau)\right].$$

Further, we have

$$x(t_k + \tau) = x(t_k) + \tau\, x'(t_k) + \frac{\tau^2}{2} x''(t_k) + \frac{\tau^3}{3!} x'''(\theta_k) =$$

$$= x(t_k) + \tau f\big(t_k, x(t_k)\big) + \frac{\tau^2}{2} x''(t_k) + \frac{\tau^3}{3!} x'''(\theta_k).$$

We also have

$$v(t_k + \tau) = v(t_k) + \tau\, v'(t_k) + \frac{\tau^2}{2} v''(\hat{\theta}_k).$$

We deduce

$$w_{k+1} = \frac{1}{\tau^2}\Big[\, x_k + \tau f(t_k, x_k) - x(t_k) - \tau f\big(t_k, x(t_k)\big) -$$

$$- \frac{\tau^2}{2} x''(t_k) - \frac{\tau^3}{3!} x'''(\theta_k) - \tau^2\, v'(t_k) - \frac{\tau^3}{2} v''(\hat{\theta}_k) - \tau\, v(t_k)\Big] =$$

$$= \frac{1}{\tau^2}\Big[\, \tau^2 w_k + \tau\big(f(t_k, x_k) - f\big(t_k, x(t_k)\big)\big) -$$

$$- \frac{\tau^2}{2} x''(t_k) - \frac{\tau^3}{3!} x'''(\theta_k) - \tau^2\, v'(t_k) - \frac{\tau^3}{2} v''(\hat{\theta}_k)\Big].$$

But

$$f(t_k, x_k) - f\big(t_k, x(t_k)\big) = \frac{\partial f}{\partial x}(t_k, \xi_k)\big(x_k - x(t_k)\big) =$$

$$= \frac{\partial f}{\partial x}(t_k, x(t_k))\left(\tau v(t_k) + \tau^2 w_k\right) +$$

$$+ \left[\frac{\partial f}{\partial x}(t_k, \xi_k) - \frac{\partial f}{\partial x}(t_k, x(t_k))\right](x_k - x(t_k)).$$

Hence, we can write;

$$w_{k+1} = \frac{1}{\tau^2}\left\{\tau^2 w_k + \tau^2 \frac{\partial f}{\partial x}(t_k, x(t_k))\, v(t_k) + \tau^3 \frac{\partial f}{\partial x}(t_k, x(t_k))\, w_k + \right.$$

$$+ \tau\left[\frac{\partial f}{\partial x}(t_k, \xi_k) - \frac{\partial f}{\partial x}(t_k, x(t_k))\right](x_k - x(t_k)) -$$

$$\left. - \tau^2 v'(t_k) - \frac{\tau^2}{2} x''(t_k) - \frac{\tau^3}{3!} x'''(\theta_k) - \frac{\tau^3}{2} v''(\hat{\theta}_k)\right\}.$$

We choose now $v$ as the solution of the problem:

$$v'(t) = \frac{\partial f}{\partial x}(t, x(t))\, v(t) - \frac{1}{2} x''(t), \quad v(t_0) = 0,$$

and we obtain

$$w_{k+1} = w_k\left[1 + \tau \frac{\partial f}{\partial x}(t_k, x(t_k))\right] - \frac{\tau}{3!} x'''(\theta_k) - \frac{\tau}{2} v''(\hat{\theta}_k) +$$

$$+ \frac{1}{\tau}\left[\frac{\partial f}{\partial x}(t_k, \xi_k)) - \frac{\partial f}{\partial x}(t_k, x(t_k))\right](x_k - x(t_k)).$$

We also have

$$|x_k - x(t_k)| \le \nu\tau,$$

$$\left|\frac{\partial f}{\partial x}(t_k, \xi_k) - \frac{\partial f}{\partial x}(t_k, x(t_k))\right| \le \mu_2 |x_k - x(t_k)| \le \mu_2 \nu\tau$$

and we deduce:

$$|w_{k+1}| \le |w_k|(1 + \tau\mu) + \tau \hat{M}.$$

This inequality is of the same type as that obtained for $\varepsilon_k$ and it provides $|w_k| \le \dfrac{\hat{M}}{\mu} e^{\mu T}$; here $\hat{M}$ is an upper bound for $\left|\dfrac{\partial^2 f}{\partial x^2}(t, x)\right|$ for $t \in [\, t_0, t_0 + T\,]$ and $x$ in a compact containing $x_0$.

Thus we obtained for $w_k$ the desired estimate independent of $\tau$.

In what follows, we shall show how this estimate will allow to design a procedure providing an error of the order of $\tau^2$.

Since we shall use the step size $\tau$ and also the step size $\frac{\tau}{2}$, we shall denote the corresponding sequences by $x_k^\tau$ and $x_k^{\tau/2}$, respectively. We have

$$x_k^\tau = x(t_k) + \tau\, v(t_k) + \tau^2\, w_k^\tau;$$

$$x_{2k}^{\tau/2} = x(t_k) + \frac{\tau}{2} v(t_k) + \frac{\tau^2}{4} w_{2k}^{\tau/2}.$$

Herefrom

$$2\, x_{2k}^{\tau/2} = 2\, x(t_k) + \tau v(t_k) + \frac{\tau^2}{2} w_{2k}^{\tau/2}.$$

We deduce, by subtraction

$$2\, x_{2k}^{\tau/2} - x_k^\tau = x(t_k) + \frac{\tau^2}{2} w_{2k}^{\tau/2} - \tau^2\, w_k^\tau.$$

If we denote $\bar{x}_k = 2\, x_{2k}^{\tau/2} - x_k^\tau$ then we have

$$|\, x(t_k) - \bar{x}_k \,| \le \frac{3\tau^2}{2} \frac{\hat{M}}{\mu} e^{\mu T}.$$

We conclude that the sequence $\bar{x}_k$ approximates $x(t_k)$ with an *error of the order of* $\tau^2$.

In this estimate, the maximum of $|\, x'' \,|$ is replaced by the maximum of $\left| \frac{\partial^2 f}{\partial x^2} \right|$.

## 5.3 Predictor – Corrector Methods

The Euler scheme uses a single value of the function $f$ in order to compute the value $x_{k+1}$. When performing Richardson extrapolation, we have to compute two Euler procedures, one for the step size $\tau$ and another for $\frac{\tau}{2}$, what increases the amount of computation required. This is a general phenomenon: if we wish the procedure to provide a better precision, we have to increase the amount of computation.

We describe now a new scheme, which uses several values of the function $f$. We begin with a somehow intuitive reasoning.

Since $x$ is a solution of the differential equation, we have $x'(t) = f(t, x(t))$. We deduce

$x(t+\tau) - x(t) = \tau\, x'(\theta)$, where $t < \theta < t + \tau$; it follows

$x(t+\tau) - x(t) = \tau\, f(\theta, x(\theta))$.

In the Euler scheme, we approximated $f(\theta, x(\theta))$ by $f(t, x(t))$, but we could also have approximated by $f(t+\tau, x(t+\tau))$. Since we have no other information on $\theta$, we can hope that the arithmetic mean

$$\frac{f(t, x(t)) + f(t+\tau, x(t+\tau))}{2}$$

could give a better approximation.

Since we do not know $x(t+\tau)$ we replace it by the approximation given by the Euler scheme, which provides a first *prediction* of the value $x(t+\tau)$. Writing these relations at the point $t_k$, we obtain the <u>*predictor – corrector scheme*</u>:

$$\bar{x}_{k+1} = x_k + \tau f(t_k, x_k);$$

$$x_{k+1} = x_k + \frac{\tau}{2}\left[f(t_k, x_k) + f\left(t_{k+1}, \bar{x}_{k+1}\right)\right].$$

The scheme obtained in this way belongs also to the class of the frequently used schemes of the Runge-Kutta type; we remark that, for *one step* of the scheme we have to compute *two* values of the function $f$. This is the price we pay for the greater precision we hope for.

Let us try to obtain again an estimate of the error. We denote again $\varepsilon_k = x(t_k) - x_k$. We have

$$\varepsilon_{k+1} = x(t_k + \tau) - x_{k+1} = x(t_k) + \tau x'(t_k) +$$

$$+ \frac{\tau^2}{2} x''(\theta_k) - x_k - \frac{\tau}{2}\left[f(t_k, x_k) + f\left(t_k + \tau, \bar{x}_{k+1}\right)\right] =$$

$$= \varepsilon_k + \tau f\left(t_k, x(t_k)\right) - \frac{\tau}{2} f(t_k, x_k) - \frac{\tau}{2} f\left(t_k + \tau, \bar{x}_{k+1}\right) + \frac{\tau^2}{2} x''(\theta_k).$$

Further, we have

$$f\left(t_k + \tau, \bar{x}_{k+1}\right) = f\left(t_k + \tau, x(t_k + \tau)\right) + f\left(t_k + \tau, \bar{x}_{k+1}\right) -$$

$$- f\left(t_k + \tau, x(t_k + \tau)\right).$$

We deduce that

$$\varepsilon_{k+1} = \varepsilon_k + \frac{\tau}{2}\left[f\left(t_k, x(t_k)\right) - f\left(t_k, x_k\right)\right] +$$

$$+ \frac{\tau}{2}\left[f\left(t_k + \tau, x(t_k + \tau)\right) - f\left(t_k + \tau, \bar{x}_{k+1}\right)\right] +$$

$$+ \frac{\tau}{2}\left[f\left(t_k, x(t_k)\right) - f\left(t_k + \tau, x(t_k + \tau)\right)\right] + \frac{\tau^2}{2} x''(\theta_k).$$

On the other hand

$$f\left(t_k + \tau, x(t_k + \tau)\right) - f\left(t_k, x(t_k)\right) = x'(t_k + \tau) - x'(t_k) = \tau x''\left(\tilde{\theta}_k\right),$$

hence

$$\varepsilon_{k+1} = \varepsilon_k\left[1+\frac{\tau}{2}\frac{\partial f}{\partial x}(t_k,\xi_k)\right]+\frac{\tau^2}{2}\Big(x''(\theta_k)-x''(\tilde{\theta}_k)\Big)+$$

$$+\frac{\tau}{2}\Big[f\big(t_k+\tau,\,x(t_k+\tau)\big)-f(t_k+\tau,\,\bar{x}_{k+1})\Big]=$$

$$=\varepsilon_k\left[1+\frac{\tau}{2}\frac{\partial f}{\partial x}(t_k,\xi_k)\right]+\frac{\tau}{2}\frac{\partial f}{\partial x}(t_k+\tau,\,\tilde{\xi}_{k+1})\,\bar{\varepsilon}_{k+1}+$$

$$+\frac{\tau^2}{2}\Big(x''(\theta_k)-x''(\tilde{\theta}_k)\Big),$$

where we denoted $\bar{\varepsilon}_k = x(t_k)-\bar{x}_k$. Thus we have

$$\bar{\varepsilon}_{k+1} = x(t_k+\tau)-\bar{x}_{k+1} =$$

$$= x(t_k)+\tau f\big(t_k,\,x(t_k)\big)+\frac{\tau^2}{2}x''(\theta_k)-x_k-\tau f(t_k,\,x_k)=$$

$$=\varepsilon_k\left[1+\tau\frac{\partial f}{\partial x}(t_k,\xi_k)\right]+\frac{\tau^2}{2}x''(\theta_k).$$

Substituting $\bar{\varepsilon}_{k+1}$ in the estimate for $\varepsilon_{k+1}$, we obtain

$$\varepsilon_{k+1} = \varepsilon_k\left[1+\frac{\tau}{2}\frac{\partial f}{\partial x}(t_k,\xi_k)\right]+$$

$$+\frac{\tau}{2}\frac{\partial f}{\partial x}(t_k+\tau,\,\tilde{\xi}_{k+1})\,\varepsilon_k\left[1+\tau\frac{\partial f}{\partial x}(t_k,\xi_k)\right]+\frac{\tau^3}{4}\frac{\partial f}{\partial x}(t_k+\tau,\,\tilde{\xi}_{k+1})+$$

$$+\frac{\tau^2}{2}\Big(x''(\theta_k)-x''(\tilde{\theta}_k)\Big).$$

Finally we remark that

$$\left|x''(\theta_k)-x''(\tilde{\theta}_k)\right| = \left|x'''\big(\hat{\theta}_k\big)\right|\left|\theta_k-\tilde{\theta}_k\right| \le \tau\hat{M}$$

and we obtain

$$\left|\varepsilon_{k+1}\right| \le \left|\varepsilon_k\right|\Big(1+\tau\hat{M}\Big)+\tau^3\tilde{M}.$$

This inequality will lead, as in the case of the Euler scheme, to the explicit estimate $\left|\varepsilon_k\right| \le \tau^2\dfrac{\hat{M}}{\mu}e^{\hat{\mu}T}$, therefore the scheme gives an error of the order of $\tau^2$.

## 5.4 Numerical Quadrature

Let us consider $f:[a, b] \to \mathrm{R}$ ; we assume that $f$ is continuous, hence integrable. We denote $x(t) = \int_a^t f(s)\,ds$ and we obviously have $x(b) = \int_a^b f(s)\,ds$.

We can also write $x'(t) = f(t)$, $x(a) = 0$ and it follows that $x$ is a solution of a simple differential equation, in which $f$ does not depend on the second argument. We shall use the above described schemes for this particular case.

**A.** ***The Euler scheme*** **becomes:**

$$x_{k+1} = x_k + \tau f(t_k)$$

and corresponds to *the* ***rectangles rule*** (the area bounded by the graph of function $f$, the $Ox$ axis and the lines $t = a$, $t = b$ is approximated by a sum of areas of rectangles).

The approximation error will be of the order of $\tau$. In our simple situation we have

$\varepsilon_{k+1} = \varepsilon_k + \frac{\tau^2}{2} x''(\theta_k) = \varepsilon_k + \frac{\tau^2}{2} f'(\theta_k)$ and the estimate

$|\varepsilon_k| \le \frac{\tau}{2} T M_1$ follows directly, where $M_1 = \max\limits_{t\in[a,\, b]} |f'(t)|$.

It can be seen that, in order to estimate the error, we had to assume that $f$ is differentiable, with continuous derivative.

If $\tau = \frac{b-a}{N}$, then the integral is approximated by $x_N$.

## B. *The Euler scheme, with Richardson extrapolation*

We have $x(t_k)$ approximated by $\bar{x}_k = 2\,x_{2k}^{\tau/2} - x_k^{\tau}$, where

$$x_k^{\tau} = \tau\left[f(a) + f(t_1) + \ldots + f(t_{k-1})\right],$$

$$x_{2k}^{\tau/2} = \frac{\tau}{2}\left[f(a) + f(\hat{t}_1) + f(t_1) + \ldots + f(\hat{t}_{k-1}) + f(t_{k-1})\right],$$

$$\hat{t}_1 = a + \frac{\tau}{2},\ \hat{t}_2 = t_1 + \frac{\tau}{2} \quad \text{a.s.o.}$$

Hence, we have

$$\bar{x}_k = \tau\left[f(\hat{t}_1) + \ldots + f(\hat{t}_{k-1})\right],$$

corresponding also to an approximation by a sum of rectangles, but differently chosen.

## C. *The predictor – corrector scheme*

We have

$$\bar{x}_{k+1} = x_k + \tau f(t_k),$$

$$x_{k+1} = x_k + \frac{\tau}{2}\left[f(t_k) + f(t_{k+1})\right].$$

It can be noticed that the sequence $\bar{x}_k$ is no longer needed. The area is approximated, in this case, by the sum of areas of certain trapezia; for this reason, the corresponding quadrature formula is called the ***trapezoidal rule***.

It provides, according to the general theory, an error of the order of $\tau^2$. The study of the error can be made directly in this case too.

## 5.5 Adams Type Methods

If $x$ is a solution of the differential equation, then we have

$$x'(t) = f(t, x(t))$$

hence, integrating on the interval $(t_0, t)$, it follows

$$x(t) = x(t_0) + \int_{t_0}^{t} f(s, x(s))\, ds.$$

Herefrom, for $t_k = t_0 + k\tau$, we deduce

$$x(t_{k+1}) = x(t_k) + \int_{t_k}^{t_{k+1}} f(s, x(s))\, ds.$$

In this formula, we shall approximate the integrant by using interpolation.

The simplest situation is when we replace $f(s, x(s))$ by a first degree polynomial. For such interpolation, we have to use the values at $s = t_k$ and $s = t_{k+1}$; we replace $x(t_k)$ by the approximate value $x_k$ given by the scheme and $x(t_{k+1})$ by a value obtained through prediction, e.g. by an Euler scheme. Hence, we approximate $f(s, x(s))$ by

$$f(t_k, x_k) + \frac{s - t_k}{\tau}\left[f(t_{k+1}, \bar{x}_{k+1}) - f(t_k, x_k)\right].$$

We obtain the scheme:

$$\bar{x}_{k+1} = x_k + \tau f(t_k, x_k),$$

$$x_{k+1} = x_k + \int_{t_k}^{t_{k+1}} \left\{f(t_k, x_k) + \frac{s - t_k}{\tau}\left[f(t_{k+1}, \bar{x}_{k+1}) - f(t_k, x_k)\right]\right\} ds =$$

$$= x_k + \tau f(t_k, x_k) + \frac{\tau}{2}\left[f(t_{k+1}, \bar{x}_{k+1}) - f(t_k, x_k)\right] =$$

$$= x_k + \frac{\tau}{2}\left[f(t_k, x_k) + f(t_{k+1}, \bar{x}_{k+1})\right].$$

Thus, we rediscovered the predictor – corrector scheme discussed above, but we included this scheme in a general approach. For instance, we can approximate $f(s, x(s))$ by a second degree polynomial, which uses the values $x_{k-1}, x_k, \bar{x}_{k+1}$. For $\bar{x}_{k+1}$ we can also use the values $x_{k-1}$ and $x_k$, e.g. in the form:

$$\bar{x}_{k+1} = x_k + \frac{\tau}{2}\left[f(t_{k-1}, x_{k-1}) + f(t_k, x_k)\right].$$

Then

$$f(s, x(s)) \simeq f(t_{k-1}, x_{k-1}) \frac{(s - t_k)(s - t_{k+1})}{(t_{k-1} - t_k)(t_{k-1} - t_{k+1})} +$$

$$+ f(t_k, x_k) \frac{(s - t_{k-1})(s - t_{k+1})}{(t_k - t_{k-1})(t_k - t_{k+1})} +$$

$$+ f(t_{k+1}, \bar{x}_{k+1}) \frac{(s - t_{k-1})(s - t_k)}{(t_{k+1} - t_{k-1})(t_{k+1} - t_k)} =$$

$$= \frac{1}{\tau^2}\left[f(t_{k-1}, x_{k-1}) \frac{(s - t_k)(s - t_{k+1})}{2} - \right.$$

$$\left. - f(t_k, x_k)(s - t_{k-1})(s - t_{k+1}) + f(t_{k+1}, \bar{x}_{k+1}) \frac{(s - t_{k-1})(s - t_k)}{2}\right].$$

Substituting into the integral, the corresponding formulae are obtained, It can be seen that $x_{k+1}$ is obtained by using the values $x_{k-1}$ and $x_k$.

A formula of this type has the general structure

$$x_{k+1} = f_k(x_k, x_{k-1}, \tau).$$

In order to study the properties of this formula, we define

$$\varepsilon_k = x(t_k) - x_k$$

and we seek for $\varepsilon_k$ a recurrence of the form

$$\varepsilon_{k+1} + b\varepsilon_k + c\varepsilon_{k-1} = \varphi_k(\tau).$$

To obtain convenient estimates of the error $\varepsilon_k$, it is necessary, on the one hand, that $|\varphi_k(\tau)| \le c\,\tau^p$, $p \ge 2$ and, on the other hand, that the equation $\varepsilon_{k+1} + b\varepsilon_k + c\varepsilon_{k-1} = 0$ be stable, in the sense that the roots of the equation $\lambda^2 + b\lambda + c = 0$ be strictly smaller than 1 in modulus.

This stability property is important for the proof of convergence, but also in order to be able to control the error due to uncertainties in the initial conditions.

## 5.6 Stiff Systems

The procedures described in the previous sections work in a satisfactory way in most usual situations. However, there exist certain situations, some of them seemingly simple, when we should have to work with too small step size in order to obtain significant informations.

Let us consider, for instance, the very simple equation $x' = -10^2 x$, with $x(0) = 1$. The exact solution of this equation is $x(t) = e^{-10^2 t}$; it decreases very rapidly to zero.

The Euler scheme for the above equation is

$$x_{k+1} = x_k - \tau\, 10^2 x_k = \left(1 - \tau\, 10^2\right) x_k$$

and leads to the sequence $x_k = \left(1 - \tau\, 10^2\right)^k$.

For $\tau = \dfrac{1}{50}$, we have $x_k = (-1)^k$, without any connection to the exact solution; for $\tau = \frac{1}{200}$, we have $x_k = \left(\frac{1}{2}\right)^k$, which converges to zero, while the exact solution gives

$$x(t_k) = x\left(\frac{k}{200}\right) = e^{-\frac{k}{200}\cdot 100} = e^{-\frac{k}{2}} = \left(\frac{1}{\sqrt{e}}\right)^k .$$

Let us analyze the behaviour of <u>*the implicit Euler scheme*</u> $x_{k+1} = x_k + \tau f(t_{k+1}, x_{k+1})$ in this case. This scheme gives

$$x_{k+1} = x_k - 100\,\tau\, x_{k+1}\,, \text{ hence } (1 + 100\,\tau)\, x_{k+1} = x_k\,,$$

$$x_{k+1} = \frac{1}{1 + 100\,\tau}\, x_k\,,\; x_k = \left(\frac{1}{1 + 100\,\tau}\right)^k .$$

For any $\tau$, the convergence to zero is preserved.

For instance, for $\tau = \frac{1}{50}$, we have $x_k = \left(\frac{1}{3}\right)^k$ and for the exact solution we have

$$x(t_k) = x\left(\frac{k}{50}\right) = e^{-\frac{k}{50}\cdot 100} = e^{-2k} = \left(\frac{1}{e^2}\right)^k .$$

Qualitatively everything is all right, but not quantitatively.

For $\tau = \frac{1}{100}$, we have $x_k = \left(\frac{1}{2}\right)^k$ and the exact solution gives

$$x(t_k) = x\left(\frac{k}{100}\right) = e^{-\frac{k}{100}\cdot 100} = e^{-k} = \left(\frac{1}{e}\right)^k ;$$

the approximation is much better.

Finally, for $\tau = \frac{1}{200}$ we have $x_k = \left(\frac{1}{1 + 1/2}\right)^k = \left(\frac{2}{3}\right)^k$ and the exact solution gives $x(t_k) = x\left(\frac{k}{200}\right) = \left(\frac{1}{\sqrt{e}}\right)^k$, the approximation being sufficiently good.

We see that also for the smallest value considered, $\tau = \frac{1}{200}$, the implicit scheme leads to better results than the explicit one. The phenomenon is general.

In all cases of <u>*stiffness*</u>, when we have fast variations, the use of implicit schemes is recommended.

## 5.7 Some Applications of Differential Equations in Numerical Analysis and Optimization

### 5.7.1 Implicit Functions

We saw in the previous sections the numerical solution of the initial value (Cauchy) problem. The procedures are iterative, easy to implement. At present, there exist efficient codes and the numerical solution of differential equations and of systems of differential equations is a routine application. For this reason, there is a natural tendency to reduce the solution of some other problems to the solution of an initial value problem. One of such problems is that of determining an implicit function.

We shall discuss the case of scalar functions, but the method is also applicable in the case of systems of equations.

The theorem of implicit functions is stated in the following way:

*Let* $F : I \times J \subset \mathrm{R} \times \mathrm{R} \to \mathrm{R}$ *be a function of class* $\mathscr{C}^1$. *We consider a point* $(t_0, x_0) \in I \times J$ *such that* $F(t_0, x_0) = 0$. *We assume that* $\dfrac{\partial F}{\partial x}(t_0, x_0) \neq 0$. *In these conditions, there exists a function* $x(\cdot)$ *defined in a neighborhood of* $t_0$, *taking values in a neighborhood of* $x_0$, *such that:* $x(t_0) = x_0$, $F(t, x(t)) \equiv 0$, $x$ *is differentiable and:*

$$x'(t) = -\frac{\frac{\partial F}{\partial t}(t, x(t))}{\frac{\partial F}{\partial x}(t, x(t))}.$$

If we denote $f(t, x) = -\frac{\frac{\partial F}{\partial t}(t, x(t))}{\frac{\partial F}{\partial x}(t, x(t))}$, then it can be seen that the function $x(\cdot)$ is a solution of the initial value problem:

$$x'(t) = f(t, x),\ x(t_0) = x_0.$$

Conversely, let $x$ be a solution of the above Cauchy problem; such a solution exists and it is unique, if $f$ is of class $\mathscr{C}^1$, hence $F$ is of class $\mathscr{C}^2$.

For this solution, we have:

$$\frac{d}{dt}F(t, x(t)) = \frac{\partial F}{\partial t}(t, x(t)) + \frac{\partial F}{\partial x}(t, x(t)) \cdot x'(t) =$$
$$= \frac{\partial F}{\partial t}(t, x(t)) + \frac{\partial F}{\partial x}(t, x(t)) \cdot f(t, x).$$

Taking into account how we defined the function $f$, it follows that: $\frac{d}{dt}F(t, x(t)) = 0$, hence $t \mapsto F(t, x(t))$ is constant and, since $F(t_0, x_0) = 0$, we infer $F(t, x(t)) \equiv 0$.

In this way, for the functions $F$ of class $\mathscr{C}^2$, the fundamental existence and uniqueness theorem for the solution of the Cauchy problem provides a simple proof for the theorem of implicit functions. Moreover, any numerical procedure for the solution of the Cauchy problem provides a numerical procedure for determining the implicit function.

Thus, the Euler scheme leads to

$$x_{k+1} = x_k - \tau \frac{\dfrac{\partial F}{\partial t}(t_k, x_k)}{\dfrac{\partial F}{\partial x}(t_k, x_k)},$$

and the predictor-corrector scheme leads to

$$\bar{x}_{k+1} = x_k - \tau \frac{\dfrac{\partial F}{\partial t}(t_k, x_k)}{\dfrac{\partial F}{\partial x}(t_k, x_k)},$$

$$x_{k+1} = x_k - \frac{\tau}{2}\left[\frac{\dfrac{\partial F}{\partial t}(t_k, x_k)}{\dfrac{\partial F}{\partial x}(t_k, x_k)} + \frac{\dfrac{\partial F}{\partial t}(t_{k+1}, \bar{x}_{k+1})}{\dfrac{\partial F}{\partial x}(t_{k+1}, \bar{x}_{k+1})}\right].$$

### 5.7.2 Nonlinear Equations

We shall describe now an application of differential equations to the solution of nonlinear equations.

Let us consider $f: G \subset \mathrm{R}^n \to \mathrm{R}^n$ of class $\mathscr{C}^1$. We assume that the Jacobian matrix $\dfrac{\partial f}{\partial x}(x)$ is <u>invertible</u> for any $x \in G$. We look for a procedure to obtain a solution of the equation $f(x) = 0$.

i) We choose $x_0 \in G$, we denote $f_0 = f(x_0)$ and we associate the <u>*system of differential equations*</u>:

$$x' = -\left[\frac{\partial f}{\partial x}(x)\right]^{-1} \cdot f_0 \text{ and the initial condition } x(0) = x_0.$$

ii) We assume that the solution of the problem is defined on the interval $0 \le t \le 1$. Then $f(x(1)) = 0$.

Let us prove this assertion. We consider $y(t) = x\left(1 - e^{-t}\right)$;

we have $y(0) = x(0) = x_0$, $\lim_{t \to \infty} y(t) = x(1)$.

The function $y$ is differentiable and

$$y'(t) = x'\left(1 - e^{-t}\right)e^{-t} =$$

$$= -\left[\frac{\partial f}{\partial x}\left(x\left(1 - e^{-t}\right)\right)\right]^{-1} \cdot f_0\, e^{-t} = -\left[\frac{\partial f}{\partial x}(y(t))\right]^{-1} \cdot f_0\, e^{-t}.$$

We deduce that $\frac{\partial f}{\partial x}(y(t))\, y'(t) = -f_0\, e^{-t}$, which is equivalent to

$$\frac{d}{dt}[f(y(t))] = -f_0\, e^{-t}.$$

Integrating from 0 to $t$, we get

$$f(y(t)) - f(y(0)) = -f_0 \int_0^t e^{-s}\, ds = \left(e^{-t} - 1\right) f_0.$$

We have $f(y(0)) = f(x(0)) = f(x_0) = f_0$ and it follows $f(y(t)) = f_0\, e^{-t}$, hence $\lim_{t \to \infty} f(y(t)) = \lim_{t \to \infty} f_0\, e^{-t} = 0$. But

$$\lim_{t \to \infty} f(y(t)) = f\left(\lim_{t \to \infty} y(t)\right) = f(x(1))$$

and therefore $f(x(1)) = 0$.

In this way, any numerical procedure for the solution of the system of differential equations leads to a procedure for solving nonlinear systems of the form $f(x) = 0$.

Difficulties occur from the necessity to compute at each step the inverse $\left[\frac{\partial f}{\partial x}(x)\right]^{-1}$.

Let us illustrate the general theory by a simple example.

Let us consider $f(x) = x^3 + 3x + 1$; the equation $f(x) = 0$ has only one real root and this must be negative. Since $f(0) = 1$ and $f(-1) = -3$, the root is situated between $-1$ and 0.

We have $f'(x) = 3x^2 + 3 = 3\left(x^2 + 1\right)$, $f_0 = f(-1) = -3$ and the associated differential equation will be

$$x' = -\frac{1}{3(x^2+1)}(-3) = \frac{1}{x^2+1},$$

with initial condition $x(0) = -1$.

This is an equation with separated variables; integrating, we obtain

$$\frac{1}{3}x^3(t) + x(t) - \frac{1}{3}x^3(0) - x(0) = t,$$

which gives, for $t = 1$, $\frac{1}{3}x^3(1) + x(1) + \frac{1}{3} + 1 = 1$; in this way, we checked that $x(1)$ satisfies the equation $x^3(1) + 3x(1) + 1 = 0$.

We shall be able to determine the root approximately, by using any numerical procedure for solving the problem:

$$x' = \frac{1}{x^2+1}, \quad x(0) = -1.$$

The function $x \mapsto \frac{1}{x^2+1}$ is bounded, the solution exists globally and the procedure works.

### 5.7.3 Free Optimization

Let us consider the function $f : G \subset \mathrm{R}^n \to \mathrm{R}$.

A local minimum for $f$ is a point $\hat{x} \in G$ such that $f(x) \geq f(\hat{x})$ for all $x$ in a neighborhood of $\hat{x}$; if the inequality is valid for all $x \in G$, then $\hat{x}$ is called a global minimum.

It is known that if $\hat{x}$ is a local minimum and $f$ is of class $\mathscr{C}^1$, then $\frac{\partial f}{\partial x_i}(\hat{x}) = 0$ for all $i$. In this way, to find a local minimum, we have to find $\hat{x}$ such that $\frac{\partial f}{\partial x_i}(\hat{x}) = 0$.

A sufficient condition for a point satisfying the above conditions to be a minimum point is that the quadratic form defined by the Hessian matrix

$$\frac{\partial^2 f}{\partial x_i\,\partial x_j}(\hat{x})$$

be positive definite. This follows from the Taylor formula:

$$f(x) - f(\hat{x}) = \sum_{i=1}^{n} \frac{\partial f}{\partial x_i}(\hat{x})(x_i - \hat{x}_i) +$$

$$+\frac{1}{2}\sum_{i=1}^{n}\sum_{j=1}^{n} \frac{\partial^2 f}{\partial x_i\,\partial x_j}(\hat{x})(x_i - \hat{x}_i)(x_j - \hat{x}_j) + \ldots.$$

If $\frac{\partial f}{\partial x_i}(\hat{x}) = 0$ for all $i$ and the quadratic form defined by $\frac{\partial^2 f}{\partial x_i\,\partial x_j}(\hat{x})$ is positive definite, then it can be seen that $f(x) - f(\hat{x}) \geq 0$ for $x$ in a neighborhood of $\hat{x}$ (for which the missing terms in the Taylor formula do not influence the sign).

Let us consider these conditions satisfied and associate the system of differential equations:

$$x_i' = -\frac{\partial f}{\partial x_i}(x),\ i = 1,\ldots,n.$$

For this system, the point $\hat{x}$ is an <u>*equilibrium*</u>, since $\frac{\partial f}{\partial x_i}(\hat{x}) = 0$.

If we denote $y_i = x_i - \hat{x}_i$, then we have

$$y_i' = x_i' = -\frac{\partial f}{\partial x_i}(y + \hat{x}) = -\sum_{j=1}^{n} \frac{\partial^2 f}{\partial x_i\,\partial x_j}(\hat{x})\, y_j + \ldots$$

hence the linear approximation has the matrix $-\frac{\partial^2 f}{\partial x_i\,\partial x_j}(\hat{x})$; since the quadratic form is positive definite, the eigenvalues of the matrix $\frac{\partial^2 f}{\partial x_i \, \partial x_j}(\hat{x})$ are strictly positive, hence the eigenvalues of the matrix of the linear approximation are strictly negative.

It follows that the equilibrium is exponentially stable and hence, there exists $\delta_0 > 0$ such that $|x(0) - \hat{x}| < \delta_0$ implies $\lim_{t \to \infty} x(t) = \hat{x}$, $\lim_{t \to \infty} |x(t) - \hat{x}| = 0$ exponentially.

Therefore, if the starting point is close to the minimum and if we apply a numerical procedure for solving the system of differential equations, then we obtain a numerical procedure for finding the minimum.

For instance, the Euler scheme leads to the algorithm:

$$x_i^{k+1} = x_i^k - \tau \frac{\partial f}{\partial x_i}\left(x^k\right)$$

called gradient algorithm.

Let us remark that, in order to be close to the minimum, we have to integrate on large intervals, what increases the error.

Let us point out an important property of the equation system of equation associated to the function $f$. Let $x(\cdot)$ be an arbitrary solution of the system; let us denote $\bar{f}(t) = f(x(t))$.

We have

$$\bar{f}'(t) = \sum_{i=1}^{n} \frac{\partial f}{\partial x_i}(x(t))\, x_i'(t) = -\sum_{i=1}^{n} \left[\frac{\partial f}{\partial x_i}(x(t))\right]^2 \le 0,$$

hence the function $\bar{f}$ is *decreasing*.

In this way, moving along solutions of the system of differential equations leads to values of the function decreasing to the minimum.

### 5.7.4 Linear Programming

We consider a linear programming problem of the form

$$\min\left\{ c^* x \,/\, x \geq 0, Ax = b \right\}.$$

In what follows, we admit that the rows of matrix $A$ are linearly independent and that there exists a vector $d$ such that all its coordinates are positive and $Ad = b$.

We associate the system of differential equations:

$$x' = -D(x)\left[I - D(x)\, A^*\left(AD^2(x)A^*\right)^{-1} AD(x)\right] D(x)\, c$$

where we denoted by $D(x)$ the diagonal matrix having on the diagonal the coordinates of the vector $x$.

We check some properties of this system.

**a)** If all coordinates of $x(0)$ are strictly positive, then all coordinates of $x(t)$ are strictly positive.

Indeed, the system is written coordinate-wise in the form

$$\frac{d}{dt} x^i = -x^i \gamma^i(x)$$

hence, for any solution, we have:

$$\frac{d}{dt} x^i + x^i \gamma^i(x) = 0,$$

$$e^{\int_0^t \gamma^i(x(s))\,ds} \frac{d}{dt} x^i(t) + x^i(t) \frac{d}{dt} e^{\int_0^t \gamma^i(x(s))\,ds} = 0.$$

Therefore

$$\frac{d}{dt}\left[ e^{\int_0^t \gamma^i(x(s))\,ds} x^i(t) \right] = 0,$$

what implies that

$$e^{\int_0^t \gamma^i(x(s))\,ds} x^i(t) = x^i(0),$$

$$x^i(t) = e^{-\int_0^t \gamma^i(x(s))\,ds}\, x^i(0)$$

and hence, for $x^i(0) > 0$ it follows $x^i(t) > 0$ for all $t$.

**b)** The function $t \mapsto Ax(t)$ is constant for any solution $x$ of the system.

We denote

$$F(x) = -D(x)\left[I - D(x)\,A^*\left(AD^2(x)A^*\right)^{-1}AD(x)\right]D(x)\,c$$

and we remark that:

$$AF(x) = -AD(x)\left[I - D(x)\,A^*\left(AD^2(x)A^*\right)^{-1}AD(x)\right]D(x)\,c =$$

$$= -\left[AD(x) - AD^2(x)\,A^*\left(A\,D^2(x)\,A^*\right)^{-1}A\,D(x)\right]D(x)\,c = 0\,.$$

It follows that

$$\frac{d}{dt}Ax(t) = Ax'(t) = AF(x(t)) = 0$$

and hence the function $t \mapsto Ax(t)$ is constant.

From properties **a)** and **b)** it follows that if $x(0)$ is such that $x^i(0) > 0$ for all $i$ and $Ax(0) = b$, then $Ax(t) = b$, $x^i(t) > 0$ for any $t$ for which the solution is defined.

In others words, if we start with an admissible program, then the solution of the system of differential equations will give an admissible program for any $t$.

Let us remark that, moreover, for any solution we have

$$\frac{d}{dt}c^*x(t) \le 0\,,$$

what shows that, along the solution of the system, the value of the cost decreases for $t \ge 0$. Indeed, we have

$$\frac{d}{dt}c^*x(t) = c^*x'(t) =$$

$$= -c^*D(x)\left[I - D(x)\,A^*\left(AD^2(x)A^*\right)^{-1}AD(x)\right]D(x)\,c \le 0\,,$$

since

$$I - D(x)\,A^*\left(AD^2(x)A^*\right)^{-1}AD(x) \ge 0$$

(we shall not prove here this fact).

From the above considerations, we derive the following algorithm for linear programming problems.

We start with $d$ satisfying $Ad = b$ and $d^i > 0$ for all $i$.

We put $x_0 = d$ and define $x_k$ by recurrence:

$$x_{k+1} = x_k - \tau_k\,D(x_k)\left[I - D(x_k)\,A^*\left(AD^2(x_k)A^*\right)^{-1}AD(x_k)\right]D(x_k)\,c\,.$$

Coordinate-wise, we have

$$x_{k+1}^i = x_k^i - \tau_k\,x_k^i\,\gamma^i(x_k)\,,$$

hence

$$x_{k+1}^i = \left[1 - \tau_k\ \gamma^i(x_k)\right]x_k^i$$

and if we choose $\tau_k < \dfrac{1}{\gamma^i(x_k)}$, then $x_k^i > 0$ implies $x_{k+1}^i > 0$.

Further, from $x_{k+1} = x_k - \tau_k\,F(x_k)$ it follows

$$Ax_{k+1} = Ax_k - \tau_k AF(x_k) = Ax_k\,,$$

hence $Ax_k = b$ implies $Ax_{k+1} = b$ and the iterations starting from admissible programs will provide admissible programs.

Moreover, as above, $c^*x_{k+1} \le c^*x_k$, hence at each iteration we obtain a smaller value of the cost.

These ideas lie on the basis of certain classes of algorithms for solving linear programming problems, called "algorithms with interior points", since we approach the optimal solution not by testing the vertices, but evolving in the interior of the admissible domain.

## 5.8 Notes and References

This chapter is meant to describe the philosophy of numerical methods to treat the initial value problems for differential equations and to show some applications to some mathematical problems apparently far apart from the domain of differential equations.

There are several books specially devoted to numerical methods, some of them with excellent codes (Shampine and Gordon (1975)). We mention also [9], where Richardson extrapolation is extensively treated. Some elements concerning numerical methods can be found also in the book of Louis Brand [3].

The application of differential equations to the implicit functions problem is known as Davidenko – Keller procedure. The application to solving nonlinear equations seems to be known, but it is difficult to give specific references.

The application of differential equations to linear programming is more recent and it is related to so–called interior point methods (see [20]).

The applications to free non–linear optimization are standard.

CHAPTER 6

# Control Systems. Stabilization of Linear Systems

## 6.1 Stabilization Problem. Stabilization by Linear State Feed-Back

The mathematical models for evolution processes that can be influenced through decisions are the dynamic control systems, defined by functions $f : I \times G \times U \subset \mathrm{R} \times \mathrm{R}^n \times \mathrm{R}^m \to \mathrm{R}^n$; such system is written as $x' = f(t, x, u)$.

The vector $u$ represents the control parameters, whose choice enables the decision-maker to influence the evolution.

In what follows, we shall deal only with the case when $f$ does not depend explicitly on the first argument, hence to the case of systems of the form $x' = f(x, u)$.

By choosing certain constant values of the control parameters, we can influence the properties of the equilibrium.

More precisely, if $f$ is of class $\mathscr{C}^1$, if there exists a pair $(\hat{x}, \hat{u})$

such that $f(\hat{x},\hat{u}) = 0$ and if the matrix $\dfrac{\partial f}{\partial x}(\hat{x},\hat{u})$ is invertible, then a fundamental theorem of Mathematical Analysis, the implicit functions theorem ensures existence of a function $\tilde{x}$, defined in a neighbourhood of $\hat{u}$, such that $\tilde{x}(\hat{u}) = \hat{x}$ and $f(\tilde{x}(u),u) = 0$; the function $\tilde{x}$ defines equilibrium points associated to constant values of the control and the appropriate choice of such values allows the equilibrium to have certain desired properties.

Let us assume that such a choice had been made and, for the sake of simplicity, assume that it corresponds to the pair $(\hat{x},\hat{u})$. Therefore, if we have a control with fixed value $\hat{u}$, then the system admits an equilibrium at $\hat{x}$. However, in general it is not the only equilibrium: moreover, we are not sure that an evolution starting in the neighbourhood of the equilibrium will not leave that neighbourhood, unless the equilibrium is stable.

If the required stability properties are not satisfied, then a control correction is necessary. Such a correction can be achieved by taking into account the informations concerning the deviation of the solution from the equilibrium. When the control is defined on the basis of such informations, it is called *error feed-back control*.

Thus, let us see how does the system behave when we choose a control $u$ of the form $u = \hat{u} + F(x - \hat{x})$, where $F$ is a constant $m \times n$ matrix. For such a choice of the control, the evolution is described by the system

$$x' = f(x, u + F(x - \hat{x})) = \tilde{f}(x).$$

This system still has the equilibrium $\hat{x}$:

$$\tilde{f}(\hat{x}) = f(\hat{x},\hat{u}) = 0 .$$

In order to analyze the stability of this equilibrium, we refer to the theorem on stability by the linear approximation and write

$$\frac{\partial \tilde{f}}{\partial x}(\hat{x}) = \frac{\partial f}{\partial x}(\hat{x},\hat{u}) + \frac{\partial f}{\partial u}(\hat{x},\hat{u})\,F .$$

If we denote $A = \frac{\partial f}{\partial x}(\hat{x},\hat{u})$, $B = \frac{\partial f}{\partial u}(\hat{x},\hat{u})$, it follows that $F$ has to be chosen such that the matrix $A + BF$ had eigenvalues with strictly negative real parts.

**Theorem 6.1** *Let $A$ be an $n\times n$ matrix and $B$ an $n\times m$ matrix. We assume that* $\operatorname{rank}\left(B\ AB \ldots A^{n-1}B\right) = n$.

*Then there exists $F$ such that the eigenvalues of the matrix $A + BF$ had strictly negative real parts.*

***Proof*** **A.** We show first that, if the condition of the statement is satisfied, then the matrix $\int_0^T e^{-At}BB^* e^{-A^* t}\,dt$ is invertible.

Let us denote this matrix by $W_T$ and calculate

$$x^* W_T\, x = \int_0^T x^* e^{-At}BB^* e^{-A^* t}x\,dt = \int_0^T \left|x^* e^{-At}B\right|^2 dt .$$

If $x^* W_T\, x = 0$, then $x^* e^{-At}B \equiv 0$ and, by differentiating successively, we also obtain

$$x^* e^{-At}AB \equiv 0,\ x^* e^{-At}A^2B \equiv 0, \ldots,\ x^* e^{-At}A^{n-1}B \equiv 0 .$$

For $t = 0$ it follows

$$x^* B = 0,\ x^* AB = 0, \ldots,\ x^* A^{n-1}B = 0 ;$$

the rank condition in the statement implies $x^* = 0$, what implies that $W_T$ is invertible.

**B.** We choose $\alpha > 0$ such that the eigenvalues of matrix $\alpha I + A$ had strictly positive real parts and we consider the matrix

$$W_T^\alpha = \int_0^T e^{-(\alpha I + A)t} BB^* e^{-(\alpha I + A^*)t} dt =$$
$$= \int_0^T e^{-2\alpha t} e^{-At} BB^* e^{-A^* t} dt.$$

Since, for any $x$, we have

$$x^* W_T^\alpha x = \int_0^T e^{-2\alpha t} x^* e^{-At} BB^* e^{-A^* t} x \, dt \geq$$
$$\geq e^{-2\alpha T} \int_0^T x^* e^{-At} BB^* e^{-A^* t} x \, dt = e^{-2\alpha T} x^* W_T x,$$

we deduce that $x^* W_T^\alpha x \geq 0$ and $x^* W_T^\alpha x = 0$ only for $x = 0$, hence $W_T^\alpha$ is also invertible.

Since $\alpha I + A$ has eigenvalues with strictly positive real parts, it follows that the eigenvalues of $-(\alpha I + A)$ have strictly negative real parts, hence there exists

$$W^\alpha = \lim_{T \to \infty} \int_0^T e^{-(\alpha I + A)t} BB^* e^{-(\alpha I + A^*)t} dt = \lim_{T \to \infty} W_T^\alpha.$$

The function $T \mapsto x^* W_T^\alpha x$ is increasing, hence $x^* W^\alpha x \geq 0$ and $x^* W^\alpha x = 0$ only if $x = 0$, what shows that $W^\alpha$ is invertible. We have

$$W^\alpha = \int_0^\infty e^{-(\alpha I + A)t} BB^* e^{-(\alpha I + A^*)t} dt,$$

hence

$$(\alpha I + A) W^\alpha + W^\alpha (\alpha I + A^*) =$$
$$= \int_0^\infty - \frac{d}{dt} \left[ e^{-(\alpha I + A)t} BB^* e^{-(\alpha I + A^*)t} \right] dt = BB^*.$$

Therefore

$$(\alpha I + A) W^{\alpha} + W^{\alpha} (\alpha I + A^*) - BB^* = 0$$

$$\alpha I + A - BB^* \left(W^{\alpha}\right)^{-1} = - W^{\alpha} (\alpha I + A^*) \left(W^{\alpha}\right)^{-1}.$$

The matrix $- W^{\alpha} (\alpha I + A^*) \left(W^{\alpha}\right)^{-1}$ has the same eigenvalues as $-\left(\alpha I + A^*\right)$, which has the same eigenvalues as $-(\alpha I + A)$, hence the matrix $\alpha I + A - BB^* \left(W^{\alpha}\right)^{-1}$ has eigenvalues with strictly negative real parts; therefore, the real parts of the eigenvalues of $A - BB^* \left(W^{\alpha}\right)^{-1}$ are strictly smaller than $-\alpha$.

By taking $F = -B^* \left(W^{\alpha}\right)^{-1}$, the proof is completed.

We remark that the proof also provides a <u>*procedure*</u>. We choose $\alpha > |A|$ and we solve the linear system

$$(\alpha I + A) X + X (\alpha I + A^*) = BB^*,$$

for a symmetric matrix $X$ (we have to solve a system of $\frac{n(n+1)}{2}$ equations with the same number of unknowns).

We denote the solution obtained by $W^{\alpha}$; it is invertible and $F = -B^* \left(W^{\alpha}\right)^{-1}$ solves the problem, ensuring stabilization with eigenvalues having real parts smaller than $-\alpha$.

We shall use further in this text the following definitions:

**Definition 6.2** *If* rank$\left(B\, AB \ldots A^{n-1}B\right) = n$, *then the system defined by* $(A, B)$ *is called* <u>*controllable*</u> (let us remark that the matrix $\left(B\, AB \ldots A^{n-1}B\right)$ has $n$ rows and $n \times m$ columns, so that the rank condition is not very restrictive: among the $n \times m$ columns, $n$ independent ones should exist).

**Definition 6.3** *If there exists $F$ such that the eigenvalues of the matrix $A+BF$ have strictly negative real parts, then the system defined by $(A, B)$ is called* <u>*stabilizable*</u>.

Taking into account the above definitions, Theorem 6.1 states that controllability implies stabilizability.

We consider now the special case when $m = 1$ (we have only one control parameter).

In this case, the condition $\mathbf{rank}\left(B\, AB \ldots A^{n-1}B\right) = n$ means that the matrix $\left(B\, AB \ldots A^{n-1}B\right)$, having $n$ rows and $n$ columns, is invertible. We denote by $v^*$ the last row of the inverse; $v^*$ is obtained from the system

$$v^* b = 0, \quad v^* A b = 0, \ldots, \quad v^* A^{n-1} b = 1$$

(we wrote $b$ instead of $B$, to emphasize the fact that the matrix $B$ reduces to a vector $b$).

Given a polynomial $\pi(s) = s^n + \sum_{k=1}^{n} \beta_k s^{k-1}$, we calculate

$$\pi(A) = A^n + \sum_{k=1}^{n} \beta_k A^{k-1}.$$

We choose $q^* = v^* \pi(A)$. We shall show that $\pi$ is the characteristic polynomial of the matrix $A - bq^*$, hence $F = -q^*$ solves the stabilization problem, if the roots of $\pi$ have strictly negative real parts.

The proof is based on the following facts.

If $S = \begin{pmatrix} v^* \\ v^*A \\ \vdots \\ v^*A^{n-1} \end{pmatrix}$, then $S\,b = \begin{pmatrix} 0 \\ 0 \\ \vdots \\ 1 \end{pmatrix}$, $S\,A = \begin{pmatrix} v^*A \\ v^*A^2 \\ \vdots \\ v^*A^n \end{pmatrix}$;

the Hamilton–Cayley theorem asserts that if $p$ is the characteristic polynomial of matrix $A$, i.e. $p(s) = \det(sI - A)$, then $p(A) = 0$.

If we write

$$p(s) = s^n + \sum_{k=1}^{n} \alpha_k s^{k-1},$$

then we have

$$A^n = -\sum_{k=1}^{n} \alpha_k A^{k-1}$$

and we can write

$$S A = \begin{pmatrix} v^* A \\ v^* A^2 \\ \vdots \\ -v^* \sum_{k=1}^{n} \alpha_k A^{k-1} \end{pmatrix} = \begin{pmatrix} v^* A \\ v^* A^2 \\ \vdots \\ -\sum_{k=1}^{n} \alpha_k v^* A^{k-1} \end{pmatrix} = \tilde{A} S,$$

where $\tilde{A}$ is the Companion matrix associated to $A$:

$$\tilde{A} = \begin{pmatrix} 0 & 1 & 0 & \dots & 0 \\ 0 & 0 & 1 & \dots & 0 \\ \dots & \dots & \dots & \dots & \dots \\ 0 & 0 & 0 & \dots & 1 \\ -\alpha_1 & -\alpha_2 & -\alpha_3 & \dots & -\alpha_n \end{pmatrix}.$$

We deduce

$$S(A + BF)S^{-1} = S\left(A - b q^*\right)S^{-1} = S A S^{-1} - S b q^* S^{-1} =$$

$$= \tilde{A} - \begin{pmatrix} 0 \\ 0 \\ \vdots \\ 1 \end{pmatrix} q^* S^{-1}.$$

On the other hand, we have

$$q^* = v^*\pi(A) = v^*A^n + \sum_{k=1}^{n} v^*\beta_k A^{k-1}.$$

From the Hamilton–Cayley theorem it follows

$$A^n = -\sum_{k=1}^{n} \alpha_k A^{k-1},$$

$$v^*A^n = -\sum_{k=1}^{n} \alpha_k v^* A^{k-1},$$

hence

$$q^* = \sum_{k=1}^{n} (\beta_k - \alpha_k) v^* A^{k-1} = r^* S,$$

where

$$r^* = (\beta_1 - \alpha_1, \dots, \beta_n - \alpha_n).$$

We deduce

$$S(A + BF)S^{-1} = \tilde{A} - \begin{pmatrix} 0 \\ 0 \\ \vdots \\ 1 \end{pmatrix} r^* S S^{-1} = \tilde{A} - \begin{pmatrix} 0 \\ 0 \\ \vdots \\ 1 \end{pmatrix} r^* =$$

$$= \tilde{A} - \begin{pmatrix} 0 & 0 & \dots & 0 \\ 0 & 0 & \dots & 0 \\ \dots & \dots & \dots & \dots \\ \beta_1 - \alpha_1 & \beta_2 - \alpha_2 & \dots & \beta_n - \alpha_n \end{pmatrix} =$$

$$= \begin{pmatrix} 0 & 1 & 0 & \cdots & 0 \\ 0 & 0 & 1 & \cdots & 0 \\ \dots & \dots & \dots & \dots & \dots \\ -\beta_1 & -\beta_2 & \cdots & \cdots & -\beta_n \end{pmatrix} = \hat{A}.$$

The characteristic polynomial of matrix $\hat{A}$ is just the polynomial $\pi$ given initially, hence $\pi$ is the characteristic polynomial of $S(A + BF)S^{-1}$. Since

$$A + BF \text{ and } S(A+BF)S^{-1}$$

have the same characteristic polynomial, it follows that the characteristic polynomial of $A + BF$ is just $\pi$.

Let us make the following remark. Consider $A$ of type $n \times n$, $B$ of type $n \times 1$ and assume $\det\left(B\ AB \ldots A^{n-1}B\right) \neq 0$. Consider the matrix $S$ defined above. The proof was based essentially on the fact that $S$ is *invertible*. That follows from the fact that

$$SB = e_n,\ SAB = \widetilde{A}SB = \widetilde{A}e_n,$$

$$SA^2B = SAAB = \widetilde{A}SAB = \widetilde{A}^2SB = \widetilde{A}^2e_n$$

and, in the same way, $SA^kB = \widetilde{A}^k e_n$. We deduce that

$$\mathrm{S}\left(B\ AB \ldots A^{n-1}B\right) = \left(e_n\ \ \widetilde{A}e_n \ldots \widetilde{A}^{n-1}e_n\right);$$

it can be checked directly that

$$\det\left(e_n\ \ \widetilde{A}e_n \ldots \widetilde{A}^{n-1}e_n\right) \neq 0,$$

wherefrom it follows that $\det S \neq 0$, hence $S$ is invertible and the proof ends.

If in the linear control system $x' = Ax + Bu$ we perform the change of variables $\widetilde{x} = Sx$, then

$$\widetilde{x}' = Sx' = SAx + SBu = SAS^{-1}\widetilde{x} + SBu,$$

hence $\widetilde{x}' = \widetilde{A}\widetilde{x} + e_n u$ and it can be seen that the system obtained in this way corresponds to an affine differential equation of n-th order, with the control $u$ in the right-hand side.

## 6.2 Stabilization of Linear Systems by Using a Controller

We shall study now the stabilization problem for the case when the state is not available for measurement and we have to use in the control only a measured output of the form $y = C\,x$.

We shall assume that the system defined by $(A, B)$ is controllable and, moreover,

$$rank\begin{pmatrix} C \\ CA \\ \vdots \\ CA^{n-1} \end{pmatrix} = n.$$

This last property is called *observability* of the pair $(C, A)$.

The controllability property ensures the existence of a matrix $F$ such that the eigenvalues of $A + BF$ have strictly negative real parts.

The observability property is equivalent to

$$rank\left( C^*, \quad A^*C^*, \quad \ldots \quad , (A^*)^{n-1}C^* \right) = n,$$

hence there exists $H^*$ such that $A^* + C^*H^*$ has eigenvalues with strictly negative real parts; we deduce that $A + HC$ has eigenvalues with strictly negative real parts.

By means of the matrices $F$ and $H$ we define the Kalman–Luenberger controller, described by the system

$$x_c' = A_c\,x_c + B_c\,u_c\,, \quad y_c = C_c\,x_c\,,$$

where

$$A_c = A + B\,F + H\,C\,, \quad B_c = -H\,, \quad C_c = F$$

and we couple this controller with the system, by taking

$$u_c = y = C\,x\,,\quad u = y_c = F\,x_c\,.$$

The system obtained in this way can be written as

$$\begin{cases} x' = A\,x + B\,F\,x_c \\ x_c' = (A + B\,F + H\,C)\,x_c - H\,C\,x\,. \end{cases}$$

Denoting $z = x - x_c$, we have

$$\begin{cases} z' = (A + H\,C)\,z \\ x' = (A + B\,F)\,x - B\,F\,z\,. \end{cases}$$

The matrix of this last system is

$$\begin{pmatrix} A + H\,C & 0 \\ -B\,F & A + B\,F \end{pmatrix};$$

its eigenvalues are those of the matrices $A + H\,C$ and $A + B\,F$, hence they have strictly negative real parts. It follows that $x$ and $z$ decrease exponentially to zero, and so does $x_c$.

Let us remark that we have stabilized the system by using the control $u = F\,x_c$; the state $x_c$ of the controller is determined by the measured output $y = C\,x.$, so that we may say that the system has been stabilized by a *dynamic output feed-back*.

## 6.3 Stabilization in an Economic Growth Model

Consider an economic growth model governed by the law

$$Y' = \lambda\,(Z - Y),$$

which expresses the assumption that the evolution velocity of the product is proportional to the difference between demand and the actual level of product. In its turn, demand is of the form

$$Z = cY + I + A + G$$

where, besides the former components (see **2.4** ) a new one, $G$, is present, expressing the intervention of the state (government).

For investment $I$ we adopt the law $I' = -k(I - vY')$.

We have

$$Y' = \lambda(cY + I + A + G - Y) = -\lambda sY + \lambda I + \lambda A + \lambda G,$$

$$I' = -kI - kv\lambda sY + kv\lambda I + kv\lambda A + kv\lambda G.$$

We assume $A = \hat{A}$ constant. At equilibrium we have

$$I' = 0,\ Y' = 0,\ \text{hence } I = 0,\ Z = Y,\ \hat{Y} = \frac{A}{s}$$

if the term $G$ is absent; if $G$ is constant, equal to $\hat{G}$, then the equilibrium becomes $\hat{Y} = \dfrac{\hat{A} + \hat{G}}{s}$.

If we denote by $y$ the deviation of $Y$ from equilibrium: $y = Y - \dfrac{\hat{A} + \hat{G}}{s}$ and by $u$ the correction of government action $u = G - \hat{G}$, then we obtain

$$y' = -\lambda s\left(y + \frac{\hat{A} + \hat{G}}{s}\right) + \lambda I + \lambda\left(u + \hat{G}\right) + \lambda\hat{A} =$$

$$= -\lambda sy + \lambda I + \lambda u,$$

$$I' = -kv\lambda s\left(y + \frac{\hat{A} + \hat{G}}{s}\right) + k(v\lambda - 1)I + kv\lambda\left(\hat{A} + \hat{G} + u\right).$$

Thus, we obtain the system

$$\begin{cases} y' = -\lambda sy + \lambda I + \lambda u \\ I' = -kv\lambda sy + k(v\lambda - 1)I + kv\lambda u. \end{cases}$$

The matrix of the system is

$$M = \begin{pmatrix} -\lambda s & \lambda \\ -k v \lambda s & k(v\lambda - 1) \end{pmatrix}$$

and the vector multiplying the control is given by $b = \begin{pmatrix} \lambda \\ k v \lambda \end{pmatrix}$.

We check the controllability of the system by computing the determinant of the matrix

$$(b \;\; M b) = \begin{pmatrix} \lambda & -\lambda^2 s + k v \lambda^2 \\ k v \lambda & -k v \lambda^2 s + k^2 v \lambda (v\lambda - 1) \end{pmatrix};$$

we obtain

$$\det(b \;\; M b) = -k v \lambda^3 s + k^2 v \lambda^2 (v \lambda - 1) +$$
$$+ k v \lambda^3 s - k^2 v^2 \lambda^3 = -k^2 v \lambda^2 \neq 0,$$

hence we can choose $u$ as a linear function of $y$ and $I$ such that, using this type of government action, we could obtain any desired characteristic polynomial, hence any stability performances associated with the characteristic polynomial.

If we wish to use a feed-back by $y$ only (assuming that the decision-maker has information at any time only on $y$) then we have to define a dynamic controller.

In this case, $C = (1, 0)$; in order to check observability, we have to calculate the determinant

$$\begin{vmatrix} C \\ C M \end{vmatrix} = \begin{vmatrix} 1 & 0 \\ \lambda & \lambda^2 (k v - s) \end{vmatrix} = \lambda^2 (k v - s).$$

This determinant is non-zero if $k v \neq s$, what we admit in the sequel.

We choose now $H = \begin{pmatrix} h_1 \\ h_2 \end{pmatrix}$ such that the matrix

$$\begin{pmatrix} -\lambda s & \lambda \\ -k v \lambda s & k(v\lambda - 1) \end{pmatrix} + \begin{pmatrix} h_1 & 0 \\ h_2 & 0 \end{pmatrix}$$

had eigenvalues with strictly negative real parts and we choose $F = (f_1 \;\; f_2)$ such that the matrix

$$\begin{pmatrix} -\lambda s & \lambda \\ -k v \lambda s & k(v\lambda - 1) \end{pmatrix} + \begin{pmatrix} \lambda f_1 & \lambda f_2 \\ k v \lambda f_1 & k v \lambda f_2 \end{pmatrix}$$

had also eigenvalues with strictly negative real parts, and we define the Kalman–Luenberger controller.

## 6.4 A Monetary Policy Model

### A) Variables of the model

$P(t)$ – average price

$\pi(t) = \dfrac{P'(t)}{P(t)}$ – inflation rate

$M(t)$ – nominal money supply

$\mu(t) = \dfrac{M'(t)}{M(t)}$ – growth rate of nominal money balances

$E(t)$ – excess supply

$y(t)$ – real income

$y^*$ – full-employment level of output

$m(t) = \dfrac{M(t)}{P(t)}$ – real money supply

$h(t)$ – real cash balances

$z(t)$ – real demand deposits

$I(t)$ – reference interest rate

$i(t)$ – interest rate on demand deposits

## B) Economic assumptions

$$E(t) = y(t) - y^* \tag{6.1}$$

$$\pi'(t) = -b\,E(t);\; b > 0 \tag{6.2}$$

Relation **(6.2)** expresses a proportionality between the variation of the inflation rate and excess supply, with a negative proportionality coefficient.

$$m(t) = h(t) + z(t) \tag{6.3}$$

$$h(t) = \alpha_1\, y(t) - \alpha_2\, I(t);\;\; \alpha_1, \alpha_2 > 0 \tag{6.4}$$

$$z(t) = \alpha_3\, y(t) - \alpha_4\left[I(t) - i(t)\right];\;\; \alpha_3, \alpha_4 > 0 \tag{6.5}$$

Relations **(6.4)** and **(6.5)** express the fact that the demands for $h$ and $z$ depend positively on income and negatively on the opportunity cost of holding each asset.

$$y(t) = \alpha_7 - \alpha_8\left[I(t) - \pi(t)\right];\;\; \alpha_7, \alpha_8 > 0 \tag{6.6}$$

Relation **(6.6)** expresses a negative dependence of real income with respect to the real interest rate.

$$m'(t) = \left[\mu(t) - \pi(t)\right] m(t) \tag{6.7}$$

Relation **(6.7)** expresses the equilibrium of the money market. This relation represents one of the two differential equations that describe, eventually, the evolution of the system. To derive the other equation, relations **(6.3)–(6.6)** are used, in the following way:

Substituting **(6.4)** and **(6.5)** into **(6.3)**, we get

$$m(t) = \alpha_5\, y(t) - \alpha_6\, I(t) + \alpha_4\, i(t),$$

where $\alpha_5 = \alpha_1 + \alpha_3$, $\alpha_6 = \alpha_2 + \alpha_4$. Herefrom, we obtain

$$I(t) = \frac{1}{\alpha_6}\left[- m(t) + \alpha_5\, y(t) + \alpha_4\, i(t)\right] \qquad \textbf{(6.3')}$$

Substituting **(6.3')** into **(6.6)**, we have

$$y(t) = \phi_0 + \phi_1\, \pi(t) + \phi_2\, m(t) - \phi_3\, i(t) \qquad \textbf{(6.6')}$$

where $\phi_0 = \dfrac{\alpha_6\,\alpha_7}{\alpha_6 + \alpha_5\,\alpha_8}$, $\phi_1 = \dfrac{\alpha_6\,\alpha_8}{\alpha_6 + \alpha_5\,\alpha_8}$,

$$\phi_2 = \frac{\alpha_8}{\alpha_6 + \alpha_5\,\alpha_8}, \quad \phi_3 = \frac{\alpha_4\,\alpha_8}{\alpha_6 + \alpha_5\,\alpha_8}.$$

From relations **(6.2)**, **(6.1)** and **(6.6')**, it follows

$$\pi'(t) = - b\, E(t) = - b\left[y(t) - y^*\right] =$$
$$= b\left[- \phi_0 + y^* - \phi_1\, \pi(t) - \phi_2\, m(t) + \phi_3\, i(t)\right].$$

**C)** Eventually, we obtained the following system of differential equations:

$$\begin{cases} \pi'(t) = b\left[- \phi_0 + y^* - \phi_1\, \pi(t) - \phi_2\, m(t) + \phi_3\, i(t)\right] \\ m'(t) = \left[\mu(t) - \pi(t)\right] m(t). \end{cases}$$

This can be viewed as a dynamic control system, with <u>*state variables*</u> $\pi(t)$, $m(t)$, <u>*decision variables*</u> $i(t)$, $\mu(t)$ and <u>*output variables*</u> $y$, $I$, $h$, $z$, whose evolution is of interest and can be expressed by means of the state variables and decision variables.

We can define two special strategies, one called *money supply policy*, corresponding to the situation when the evolution is influenced only by means of the decision variable $\mu$, while $i$ is kept constant and the other, called *interest rate policy*, corresponding to the situation when $\mu$ is kept constant.

In order to analyze the model, we have to define equilibria associated to *constant* values of the decision variables.

Let $\hat{\mu}$ and $\hat{\imath}$ be *given* constant values of the decision variables. If we exclude, for money supply, the equilibrium at zero, as meaningless, we deduce, for the inflation rate, $\hat{\pi} = \hat{\mu}$ and the equilibrium value of real money supply will be

$$\hat{m} = \frac{1}{\phi_2}\left[y^* - \phi_0 - \phi_1 \hat{\mu} + \phi_3 \hat{\imath}\right].$$

In this way, we can see how the equilibrium values of state variables depend on the values adopted for the decision variables. However, these equilibrium values have real existence *only* if they correspond to a *stable* equilibrium.

In order to analyze stability, we use the theorem on stability by the first (linear) approximation. We have

$$f(\pi, m, i, \mu) = \begin{pmatrix} f_1(\pi, m, i, \mu) \\ f_2(\pi, m, i, \mu) \end{pmatrix},$$

$$f_1(\pi, m, i, \mu) = b\left[y^* - \phi_0 - \phi_1 \pi - \phi_2 m + \phi_3 i\right],$$

$$f_2(\pi, m, i, \mu) = (\mu - \pi)\, m\,.$$

The matrix of the linear part is

$$A = \begin{pmatrix} \dfrac{\partial f_1}{\partial \pi}(\hat{\pi}, \hat{m}, \hat{\imath}, \hat{\mu}) & \dfrac{\partial f_1}{\partial m}(\hat{\pi}, \hat{m}, \hat{\imath}, \hat{\mu}) \\ \dfrac{\partial f_2}{\partial \pi}(\hat{\pi}, \hat{m}, \hat{\imath}, \hat{\mu}) & \dfrac{\partial f_2}{\partial m}(\hat{\pi}, \hat{m}, \hat{\imath}, \hat{\mu}) \end{pmatrix} = \begin{pmatrix} -b\,\phi_1 & -b\,\phi_2 \\ -\hat{m} & 0 \end{pmatrix}.$$

Its characteristic polynomial is

$$\begin{vmatrix} r + b\phi_1 & b\phi_2 \\ \hat{m} & r \end{vmatrix} = r^2 + b\phi_1 r - b\phi_2 \hat{m} .$$

Since $b > 0$, $\phi_2 > 0$, $\hat{m} > 0$, it follows that the roots are real and of opposite signs, hence the equilibrium ***is not stable***; it is a saddle point.

The problem of ***stabilization*** occurs, by using ***decisions*** based on informations concerning the deviations of the state variables from equilibrium.

The matrix $B$, which occurs in the theoretical analysis, is given by

$$B = \begin{pmatrix} \frac{\partial f_1}{\partial i}(\hat{\pi}, \hat{m}, \hat{\imath}, \hat{\mu}) & \frac{\partial f_1}{\partial \mu}(\hat{\pi}, \hat{m}, \hat{\imath}, \hat{\mu}) \\ \frac{\partial f_2}{\partial i}(\hat{\pi}, \hat{m}, \hat{\imath}, \hat{\mu}) & \frac{\partial f_2}{\partial \mu}(\hat{\pi}, \hat{m}, \hat{\imath}, \hat{\mu}) \end{pmatrix} = \begin{pmatrix} b\phi_3 & 0 \\ 0 & \hat{m} \end{pmatrix}.$$

Since we are interested to analyze *separately* the money supply policy and the interest rate policy, we shall discuss stabilization by using ***separately*** the columns of matrix $B$.

For the money supply policy, the condition that ensures the possibility of stabilization (**rank** $(b_2 \quad A\, b_2) = 2$, where $b_2$ is the second column of matrix $B$) is written as

$$\begin{vmatrix} 0 & -b\phi_2 \hat{m} \\ \hat{m} & 0 \end{vmatrix} = b\phi_2 \hat{m}^2 \neq 0$$

and is satisfied.

According to the general theory, we can perform stabilization by taking $\mu - \hat{\mu} = k_1(\pi - \hat{\pi}) + k_2(m - \hat{m})$.

For such a choice of the control, the equilibrium is not modified and the matrix of the linear part becomes

$$A + b_2 (k_1 \;\; k_2) = \begin{pmatrix} -b\,\phi_1 & -b\,\phi_2 \\ -\hat{m} & 0 \end{pmatrix} + \begin{pmatrix} 0 \\ \hat{m} \end{pmatrix} (k_1 \;\; k_2) =$$

$$= \begin{pmatrix} -b\,\phi_1 & -b\,\phi_2 \\ \hat{m}\,(k_1 - 1) & \hat{m}\,k_2 \end{pmatrix};$$

the characteristic polynomial becomes

$$\begin{vmatrix} r + b\,\phi_1 & b\,\phi_2 \\ \hat{m}\,(1 - k_1) & r - \hat{m}\,k_2 \end{vmatrix} =$$

$$= r^2 + (b\,\phi_1 - \hat{m}\,k_2)\,r - b\,\hat{m}\left[k_2\,\phi_1 + \phi_2\,(1 - k_1)\right].$$

To have stability, it is necessary and sufficient that

$$b\,\phi_1 - \hat{m}\,k_2 > 0\,, \quad k_2\,\phi_1 + \phi_2 - k_1\,\phi_2 < 0\,.$$

The first condition requires that $k_2 < \dfrac{b\,\phi_1}{\hat{m}}$, and the second one is equivalent to $k_1 > 1 + k_2\,\dfrac{\phi_1}{\phi_2}$.

In particular, we can choose $k_2 = 0$, $k_1 > 1$ and the stabilizing *money supply policy* will be of the form

$$\mu - \hat{\mu} = k_1\,(\pi - \hat{\pi})$$

and, since $\hat{\pi} = \hat{\mu}$, we get $\mu = k_1\,\pi + (1 - k_1)\,\hat{\mu}$. In order to stabilize, it is required that the flow of money supply follow the inflation rate according to the above simple rule.

For the *interest rate policy*, the controllability condition is rank $(b_1 \;\; A\,b_1) = 2$, where $b_1$ is the first column of matrix $B$.

This condition is written as

$$\begin{vmatrix} b\,\phi_3 & -b\,\phi_1\,\phi_3 \\ 0 & -\hat{m}\,b\,\phi_3 \end{vmatrix} = -b^2\,\phi_3^2\,\hat{m} \neq 0\,.$$

This condition is satisfied, hence we may take

$$i - \hat{\imath} = k_1(\pi - \hat{\pi}) + k_2(m - \hat{m}).$$

For such control, the matrix of the linear part becomes

$$A + b_1 (k_1 \;\; k_2) = \begin{pmatrix} -b\,\phi_1 & -b\,\phi_2 \\ -\hat{m} & 0 \end{pmatrix} + \begin{pmatrix} b\,\phi_3 \\ 0 \end{pmatrix} (k_1 \;\; k_2) =$$

$$= \begin{pmatrix} -b\,\phi_1 + b\,\phi_3\,k_1 & b\,(\phi_3\,k_2 - \phi_2) \\ -\hat{m} & 0 \end{pmatrix};$$

the characteristic equation is now

$$\begin{vmatrix} r - b\,(\phi_3\,k_1 - \phi_1) & -b\,(\phi_3\,k_2 - \phi_2) \\ \hat{m} & r \end{vmatrix} =$$

$$= r^2 - b\,(\phi_3\,k_1 - \phi_1)\,r + \hat{m}\,b\,(\phi_3\,k_2 - \phi_2) = 0$$

and stability is ensured if we put $k_1 < \dfrac{\phi_1}{\phi_3}$, $k_2 > \dfrac{\phi_2}{\phi_3}$.

In particular, we may choose $k_1 = 0$, hence

$$i - \hat{\imath} = k_2(m - \hat{m}),$$

i.e. interest rate follows the deviations of money supply from equilibrium.

## 6.5 Stabilization of Discrete – Time Systems

Recall that a discrete-time linear system $x_{t+1} = A\,x_t$ is exponentially stable if the eigenvalues of matrix $A$ are located inside the unit disk, i.e. have modulus strictly smaller than 1.

If we have a linear discrete-time system with a scalar control, of the form

$$x_{t+1} = A\,x_t + b\,u_t$$

and we take $u_t = F x_t$ ($F$ is, in this case, a row-vector), then we obtain the dynamics

$$x_{t+1} = (A + b F) x_t ;$$

the evolution will be exponentially stable if the eigenvalues of the matrix $A + b F$ will be located inside the unit disk. In this way, the stabilization problem of a linear discrete-time system by a linear state feed-back reads: given $A$ and $b$, find $F$ such that the eigenvalues of $A + b F$ be located inside the unit disk.

But we have seen in (6.1) that if the system defined by $(A, b)$ is controllable, then, for any given polynomial $\pi$, we may construct a vector $F$, such that $\pi$ is the characteristic polynomial of $A + b F$. In particular, if the roots of the polynomial $\pi$ are located inside the unit disk, then the control $u_t = F x_t$ will stabilize the discrete-time system defined by $(A, b)$.

We have thus the result that controllability of $(A, B)$ implies stabilizability also in the discrete-time case.

In the same way, the general result in the section (6.2.) implies that, if $(c, A)$ is observable, with $c$ – a row-vector, then there exists a column-vector $h$ such that the eigenvalues of the matrix $A + h c$ are located inside the unit disk.

The stabilization procedure by using a dynamic output feed-back in the discrete-time case is described by:

$$x_{t+1} = A x_t + b F \hat{x}_t$$

$$\hat{x}_{t+1} = (A + b F + h c) \hat{x}_t - H c x_t .$$

Remark that, if we denote $z_t = x_t - \hat{x}_t$, then we shall have

$$z_{t+1} = (A + H\,c)\,z_t$$

$$x_{t+1} = (A + B\,F)\,x_t - B\,F\,z_t .$$

This form shows that the eigenvalues of the matrix defining the evolution of $(x_t, z_t)$ are those of the matrices $A + B\,F$ and

$A + h\,c$,

hence they are located inside the unit disk; the procedure ensures indeed stabilization.

Consider now the nonlinear system

$x_{t+1} = f(x_t, u_t)$ with a scalar control;

take a constant control $\hat{u}$ and a corresponding equilibrium $\hat{x}$, i.e. $\hat{x} = f(\hat{x}, \hat{u})$.

Define $z_t = x_t - \hat{x}$, $v_t = u_t - \hat{u}$. We have

$$z_{t+1} = x_{t+1} - \hat{x} = f(x_t, u_t) - f(\hat{x}, \hat{u}) = f(\hat{x} + z_t, \hat{u} + v_t) - f(\hat{x}, \hat{u}).$$

If we take now $v_t = F\,z_t$, we get the evolution defined by

$$z_{t+1} = f(\hat{x} + z_t, \hat{u} + F\,z_t) - f(\hat{x}, \hat{u}) = \tilde{f}(z_t).$$

We have $\tilde{f}(0) = 0$, hence $0$ is an equilibrium for the system describing the evolution of $z_t$. The linear approximation corresponding to this equilibrium has the matrix

$$\frac{\partial \tilde{f}}{\partial z}(0) = \frac{\partial f}{\partial x}(\hat{x}, \hat{u}) + \frac{\partial f}{\partial u}(\hat{x}, \hat{u})\,F .$$

If we denote $\frac{\partial f}{\partial x}(\hat{x}, \hat{u}) = A$ and $\frac{\partial f}{\partial u}(\hat{x}, \hat{u}) = B$, then we are able to stabilize the equilibrium at $(\hat{x}, \hat{u})$ by the linear feed-back control $u_t = \hat{u} + F(x_t - \hat{x})$ if $(A, B)$ is controllable.

## 6.6 A Discrete-Time Monetary Policy Model

We shall analyze the discrete-time analogue of the model studied in **6.4**, i.e. the discrete-time dynamic control system described by the equations:

$$\begin{cases} \pi_{t+1} = (1 - b\,\phi_1)\,\pi_t + b\left[y^* - \phi_0 - \phi_2\, m_t + \phi_3\, i_t\right] \\ m_{t+1} = (1 + \mu_t - \pi_t)\, m_t\,. \end{cases}$$

We remark that, for fixed values $\hat{\imath}, \hat{\mu}$ of the decision variables, the equilibrium state (defined by the conditions $\pi_{t+1} = \pi_t$, $m_{t+1} = m_t$) coincides with that obtained in the continuous time case:

$$\hat{\pi} = \hat{\mu}\,;\quad \hat{m} = \frac{1}{\phi_2}\left[y^* - \phi_0 - \phi_1\,\hat{\mu} + \phi_3\,\hat{\imath}\right].$$

We show that this equilibrium is unstable. To this end, we use the theorem on stability by the first approximation. We have

$$f(\pi, m, i, \mu) = \begin{pmatrix} f_1(\pi, m, i, \mu) \\ f_2(\pi, m, i, \mu) \end{pmatrix},$$

$$f_1(\pi, m, i, \mu) = (1 - b\,\phi_1)\,\pi + b\left[y^* - \phi_0 - \phi_2\, m + \phi_3\, i\right],$$

$$f_2(\pi, m, i, \mu) = (1 + \mu - \pi)\, m\,.$$

The matrix of the linear part is

$$A = \begin{pmatrix} 1 - b\,\phi_1 & -\,b\,\phi_2 \\ -\,\hat{m} & 1 \end{pmatrix}.$$

Its characteristic polynomial, defined by $\mathscr{P}(r) = \det(r\,I - A)$, is

$$\mathscr{P}(r) = r^2 - (2 - b\,\phi_1)\,r + 1 - b\,\phi_1 - b\,\phi_2\,\hat{m}\,.$$

The discriminant of the equation $\mathscr{P}(r) = 0$ is

$$\Delta = (2 - b\phi_1)^2 - 4(1 - b\phi_1 - b\phi_2\hat{m}) = b^2\phi_1^2 + 4b\phi_2\hat{m} > 0,$$

hence the roots $r_1$, $r_2$, of the characteristic equation are real.

We remark that $\mathscr{P}(1) = -b\phi_2\hat{m} < 0$, hence we have the situation $r_1 < 1 < r_2$; thus, the equilibrium is not stable.

In what follows, we shall describe stabilization by linear state feed-back, using the *money supply policy*, i.e. considering $i_t = \hat{\imath}$ constant and using $\mu$ as a scalar control parameter; the corresponding matrix $B$ reduces to the vector $b_2 = \begin{pmatrix} 0 \\ \hat{m} \end{pmatrix}$.

As we saw in **6.5**, controllability of the pair $(A, b_2)$ is a sufficient condition for stabilizability; thus, we check controllability. We have

$$\begin{pmatrix} b_2 & Ab_2 \end{pmatrix} = \begin{pmatrix} 0 & -\hat{m}b\phi_2 \\ \hat{m} & \hat{m} \end{pmatrix}.$$

The determinant of this matrix is $\hat{m}^2 b\phi_2 \neq 0$, hence the controllability condition is satisfied.

According to the general theory, we can perform stabilization by taking

$$\mu - \hat{\mu} = k_1(\pi - \hat{\pi}) + k_2(m - \hat{m}).$$

For such a choice of the control, the matrix of the linear part becomes

$$A + b\begin{pmatrix} k_1 & k_2 \end{pmatrix} = \begin{pmatrix} 1 - b\phi_1 & -b\phi_2 \\ -\hat{m} & 1 \end{pmatrix} + \begin{pmatrix} 0 \\ \hat{m} \end{pmatrix}\begin{pmatrix} k_1 & k_2 \end{pmatrix} =$$

$$= \begin{pmatrix} 1 - b\phi_1 & -b\phi_2 \\ \hat{m}(k_1 - 1) & 1 + \hat{m}k_2 \end{pmatrix}$$

and the corresponding characteristic equation is

$$\mathscr{P}(r, k_1, k_2) = \begin{vmatrix} r - 1 + b\,\phi_1 & b\,\phi_2 \\ \hat{m}\,(1 - k_1) & r - (\hat{m}\,k_2 + 1) \end{vmatrix} = 0$$

or

$$\mathscr{P}(r, k_1, k_2) = r^2 - r\,(2 - b\,\phi_1 + \hat{m}\,k_2) +$$
$$+ (1 - b\,\phi_1)(\hat{m}\,k_2 + 1) + \hat{m}\,b\,\phi_2\,(k_1 - 1) = 0 .$$

Obviously, the values of $k_1$, $k_2$ can be chosen in several ways, according to given criteria. For example, we can wish that $\mu_t$ be expressed as a function of one state variable only, say $\pi_t$. Another way of determining $k_1, k_2$ is fixing a priori the eigenvalues of the matrix $A + b\,(k_1 \;\; k_2)$ (obviously, smaller than 1) and identifying the coefficients of $\mathscr{P}(r, k_1, k_2)$ with those of the second degree polynomial having as roots the given fixed values.

**1.** According to **4.5,** the necessary and sufficient conditions that the roots of $\mathscr{P}(r, k_1, k_2)$ be smaller than 1 in modulus are:

$$\left| (1 - b\,\phi_1)(\hat{m}\,k_2 + 1) + \hat{m}\,b\,\phi_2\,(k_1 - 1) \right| < 1 \qquad \textbf{(a)}$$

and

$$\left| 2 - b\,\phi_1 + \hat{m}\,k_2 \right| < 1 + (1 - b\,\phi_1)(\hat{m}\,k_2 + 1) +$$
$$+ \hat{m}\,b\,\phi_2\,(k_1 - 1). \qquad \textbf{(b)}$$

Assume $b\,\phi_1 < 1$. Choose $k_2 = 0$ and $k_1 > 1$. Condition **(a)** becomes $\hat{m}\,b\,\phi_2\,(k_1 - 1) < b\,\phi_1$, hence we have to choose

$$1 < k_1 < 1 + \frac{\phi_1}{\hat{m}\,\phi_2}.$$

We remark that condition **(b)** is satisfied. Indeed, **(b)** reduces to the inequality

$$2 - b\,\phi_1 < 1 + 1 - b\,\phi_1 + \hat{m}\,b\,\phi_2\,(k_1 - 1)$$

which is equivalent to

$\hat{m}\, b\, \phi_2 (k_1 - 1) > 0$, hence to $k_1 > 1$.

In conclusion, choosing

$$\mu_t = k_1\, \pi_t + \hat{\mu}\,(1 - k_1), \text{ with } k_1 \in \left(1,\, 1 + \frac{\phi_1}{\hat{m}\, \phi_2}\right),$$

we can stabilize the system.

**2.** Let us choose $r_1 = \dfrac{1}{2}$ and $r_2 = \dfrac{1}{4}$ as desired values for the roots of $\mathscr{P}(r, k_1, k_2)$ and identify the coefficients of $\mathscr{P}(r, k_1, k_2)$ with those of the polynomial $r^2 - \dfrac{3}{4} r + \dfrac{1}{8}$ (the general result in **6.1** and controllability of the pair $(A,\, b_2)$ ensure the existence of a row-vector $(k_1 \;\; k_2)$ such that the polynomial $r^2 - \dfrac{3}{4} r + \dfrac{1}{8}$, having the roots $r_1 = \dfrac{1}{2}$ and $r_2 = \dfrac{1}{4}$, be the characteristic polynomial of $A + b\,(k_1 \;\; k_2)$).

We obtain the system:

$$\begin{cases} 2 - b\,\phi_1 + \hat{m}\, k_2 = \dfrac{3}{4} \\ (1 - b\,\phi_1)(\hat{m}\, k_2 + 1) + \hat{m}\, b\, \phi_2 (k_1 - 1) = \dfrac{1}{8}. \end{cases}$$

Solving this system with respect to the unknowns $k_1$, $k_2$, we get the stabilizing state feed-back

$$\mu_t - \hat{\mu} = k_1(\pi_t - \hat{\mu}) + k_2(m_t - \hat{m}),$$

or

$$\mu_t = k_1\, \pi_t + k_2\, m_t + \hat{\mu}\,(1 - k_1) - k_2\, \hat{m}.$$

## 6.7 Notes and References

The concept of controllability has been introduced by R. Kalman and thoroughly studied in a series of papers, among which we mention [14].

A detailed discussion may be found in a special appendix of the book of V. M. Popov [25]; this last source influenced essentially our development.

For the stabilization problem, an important reference is also the paper of V. M. Wonham [30]. See also the monograph [29] by Wonham.

A discussion of stabilization procedures can be found in A. Varga [28].

For stabilization by using a controller, an important reference is D. G. Luenberger [21].

The stabilization in the model of economic growth is inspired from a discussion in R. G. D. Allen [1].

The monetary policy model has been inspired by [5]. A discussion of the model in a discrete-time setting has been performed in [8].

CHAPTER 7

# Optimal Stabilization

## 7.1 Linear-Quadratic Optimization on Infinite Horizon. Continuous Time

We have seen in the previous chapter that, given a linear control system, there is much freedom in the choice of a feed-back gain that stabilizes it.

This situation suggests to try to make an optimal choice. Such optimal choice assumes a specific performance allowing to compare various controls.

To define such performance index, we have to recall that, in the linear model, the state represents usually the deviation of the actual evolution from the equilibrium. The cost of such deviation is measured in the simplest way by a quadratic form with respect to the state; in a similar way, a simple measure of the control correction will be a quadratic form with respect to the control. This simple reasoning allows to state the problem of optimal stabilization as a linear-quadratic optimization problem on infinite horizon, which will be precisely stated below.

We consider the linear dynamics

$$x' = A x + B u$$

and we look for a control $u$ such that the evolution of the system be exponentially stable and, moreover, the quadratic cost

$$\int_0^\infty [\mu(u,u) + (C x, C x)]\, dt$$

be minimized.

**Theorem 7.1** *We consider the system*

$$A^* R + R A - \frac{1}{\mu} R B B^* R + C^* C = 0$$

(Algebraic Matrix Riccati Equation).

*We assume that this system admits a solution $R \geq 0$ such that the eigenvalues of the matrix*

$$A - \frac{1}{\mu} B B^* R$$

*have strictly negative real parts. Then the linear feed-back control*

$$u = -\frac{1}{\mu} B^* R x$$

*is stabilizing and achieves the minimum cost.*

***Proof*** We substitute into the expression of the cost the term $C^* C$, in accordance with the algebraic Riccati equation and get

$$J(u) = \int_0^\infty \Big[\mu(u,u) + \frac{1}{\mu}\left(R B B^* R x, x\right) - \left(A^* R x, x\right) -$$

$$- (R A x, x)\Big]\, dt = \int_0^\infty \Big[\mu(u,u) + \frac{1}{\mu}\left(B^* R x, B^* R x\right) -$$

$$- (R\,x, x' - B\,u) - \big(R(x' - B\,u), x\big)\Big]\,dt =$$

$$= \int_a^\infty \Big[\mu(u,u) + \frac{1}{\mu}\big(B^* R\,x, B^* R\,x\big) +$$

$$+ \big(B^* R\,x, u\big) + \big(u, B^* R\,x\big)\Big]\,dt - \int_0^\infty \frac{d}{dt}(R\,x, x)\,dt =$$

$$= (R\,x_0, x_0) + \int_a^\infty \left(\sqrt{\mu}\,u + \frac{1}{\sqrt{\mu}} B^* R\,x, \sqrt{\mu}\,u + \frac{1}{\sqrt{\mu}} B^* R\,x\right) dt.$$

We deduce that for any admissible $u$ (for which the integral converges and $\lim_{t\to\infty} x(t) = 0$) we have $\mathcal{J}(u) \geq (R\,x_0, x_0)$ and equality is obtained if $u = -\frac{1}{\mu} B^* R\,x$.

Since, for such choice of $u$, the evolution is given by

$$x' = \left(A - \frac{1}{\mu} B\,B^* R\right) x$$

and this evolution admits the evaluation $|x(t)| \leq \beta e^{-\alpha t}|x_0|$, we deduce that $u$ is an admissible control and hence it achieves optimal stabilization; we have used here the assumption that the eigenvalues of matrix $A - \frac{1}{\mu} B\,B^* R$ have strictly negative real parts.

**Remark 7.2** We did not use in the proof the fact that $R \geq 0$; in fact, this property follows from the condition imposed on the matrix $A - \frac{1}{\mu} B\,B^* R$. Indeed, the algebraic Riccati equation can be rewritten as

$$R\left(A - \frac{1}{\mu} B\,B^* R\right) + \left(A^* - \frac{1}{\mu} R\,B\,B^*\right) R + \frac{1}{\mu} R\,B\,B^* R + C^* C = 0.$$

Herefrom we deduce

$$e^{\left(A^*-\frac{1}{\mu}RBB^*\right)t}R\left(A-\frac{1}{\mu}BB^*R\right)e^{\left(A-\frac{1}{\mu}BB^*R\right)t}+$$

$$+e^{\left(A^*-\frac{1}{\mu}RBB^*\right)t}\left(A^*-\frac{1}{\mu}RBB^*\right)Re^{\left(A-\frac{1}{\mu}BB^*R\right)t}+$$

$$+e^{\left(A^*-\frac{1}{\mu}RBB^*\right)t}\left[\frac{1}{\mu}RBB^*R+C^*C\right]e^{\left(A-\frac{1}{\mu}BB^*R\right)t}=0$$

what can be written as

$$\frac{d}{dt}\left[e^{\left(A^*-\frac{1}{\mu}RBB^*\right)t}Re^{\left(A-\frac{1}{\mu}BB^*R\right)t}\right]+$$

$$+e^{\left(A^*-\frac{1}{\mu}RBB^*\right)t}\left[\frac{1}{\mu}RBB^*R+C^*C\right]e^{\left(A-\frac{1}{\mu}BB^*R\right)t}=0.$$

We integrate from 0 to $\infty$, what is allowed, due to the evaluation

$$\left|e^{\left(A-\frac{1}{\mu}RBB^*\right)t}\right|\le\beta e^{-\alpha t}$$

and deduce

$$R=\int_0^\infty e^{\left(A^*-\frac{1}{\mu}RBB^*\right)t}\left[\frac{1}{\mu}RBB^*R+C^*C\right]e^{\left(A-\frac{1}{\mu}BB^*R\right)t}dt,$$

wherefrom it follows $R\ge 0$.

We shall prove now a stability theorem.

**Theorem 7.3** *We consider the Liapunov equation*

$$AP+PA^*+BB^*=-Q,\quad Q\ge 0.$$

*We assume that there exists $F$ such that the eigenvalues of $A+BF$ have strictly negative real parts and that the equation admits a solution $P\ge 0$.*

*Then the eigenvalues of matrix A have strictly negative real parts.*

***Proof*** Let $x$ be an arbitrary solution of the system $x' = A^* x$. We have

$$\frac{d}{dt}(P\,x(t), x(t)) = \left(P\,A^* x(t), x(t)\right) + (P\,x(t), A\,x(t)) =$$

$$= \left(\left(A\,P + P\,A^*\right) x(t), x(t)\right) \le -\left(B\,B^* x(t), x(t)\right) = -\left| B^* x(t) \right|^2.$$

Integrating from 0 to $t$, we obtain

$$(P\,x(t), x(t)) - (P\,x(0), x(0)) + \int_0^t \left| B^* x(\tau) \right|^2 d\tau \le 0,$$

hence

$$\int_0^t \left| B^* x(\tau) \right|^2 d\tau \le (P\,x(0), x(0)) \le \tilde{\mu} \left| x(0) \right|^2$$

and, for $t \to \infty$, we deduce

$$\int_0^\infty \left| B^* x(\tau) \right|^2 d\tau \le \tilde{\mu} \left| x(0) \right|^2.$$

By assumption, there exists $F$ such that the eigenvalues of $A + B\,F$ have strictly negative real parts; these eigenvalues coincide with those of the matrix $A^* + F^* B^*$.

We can write

$$x'(t) = A^* x(t) = \left(A^* + F^* B^*\right) x(t) - F^* B^*\, x(t),$$

hence

$$x(t) = e^{\left(A^* + F^* B^*\right) t}\, x(0) - \int_0^t e^{\left(A^* + F^* B^*\right)(t-\tau)} F^* B^*\, x(\tau)\, d\tau.$$

Since $\left| e^{\left(A^* + F^* B^*\right) t} \right| \le \beta e^{-\alpha t}$, we deduce

$$|x(t)| \le \beta e^{-\alpha t}|x(0)| + \int_0^t \beta e^{-\alpha(t-\tau)}|F|\left|B^* x(\tau)\right| d\tau .$$

Herefrom it follows

$$|x(t)|^2 \le 2\left[\beta^2 e^{-2\alpha t}|x(0)|^2 + \hat{\beta}^2\left(\int_0^t e^{-\alpha(t-\tau)}\left|B^* x(\tau)\right| d\tau\right)^2\right].$$

Using the Schwarz-Cauchy inequality, we deduce

$$|x(t)|^2 \le 2\left[\ \beta^2 e^{-2\alpha t}|x(0)|^2 + \right.$$

$$\left. + \hat{\beta}^2 \int_0^t e^{-\alpha(t-\tau)} d\tau \int_0^t e^{-\alpha(t-\tau)}\left|B^* x(\tau)\right|^2 d\tau\right] =$$

$$= 2\beta^2 e^{-2\alpha t}|x(0)|^2 + \frac{2\hat{\beta}^2}{\alpha}\int_0^t e^{-\alpha(t-\tau)}\left|B^* x(\tau)\right|^2 d\tau .$$

Herefrom, integrating from 0 to ∞, we obtain

$$\int_0^\infty |x(t)|^2 dt \le \frac{\beta^2}{\alpha}|x(0)|^2 +$$

$$+ \frac{2\hat{\beta}^2}{\alpha}\int_0^\infty \left[\int_0^t e^{-\alpha(t-\tau)}\left|B^* x(\tau)\right|^2 d\tau\right] dt =$$

$$= \frac{\beta^2}{\alpha}|x(0)|^2 + \frac{2\hat{\beta}^2}{\alpha^2}\int_0^\infty \left(\int_0^\infty e^{-\alpha t} dt\right)\left|B^* x(\tau)\right|^2 d\tau =$$

$$= \frac{\beta^2}{\alpha}|x(0)|^2 + \frac{2\hat{\beta}^2}{\alpha^2}\int_0^\infty \left|B^* x(\tau)\right|^2 d\tau .$$

Using the estimate $\int_0^\infty \left|B^* x(\tau)\right|^2 d\tau \le \tilde{\mu}|x(0)|^2$, we deduce

$$\int_0^\infty |x(t)|^2 dt \le \left(\frac{\beta^2}{\alpha} + \frac{2\hat{\beta}^2}{\alpha}\tilde{\mu}\right)|x(0)|^2 = \hat{\mu}|x(0)|^2 .$$

Since $x(t) = e^{A^* t}x(0)$, we can write

$$\int_0^\infty \left(e^{A^* t}x(0), e^{A^* t}x(0)\right) dt \le \hat{\mu}|x(0)|^2 .$$

It is important to remark that this inequality holds for any $x(0)$.

Denote $V(x) = \int_0^\infty \left(e^{A^*\tau} x, e^{A^*\tau} x\right) d\tau$.

For an arbitrary solution $x$ of the system, we denote $\tilde{V}(t) = V\big(x(t)\big)$. We have

$$\tilde{V}(t) = \int_0^\infty \left(e^{A^*\tau} e^{A^* t} x(0), e^{A^*\tau} e^{A^* t} x(0)\right) d\tau =$$

$$= \int_0^\infty \left(e^{A^*(t+\tau)} x(0), e^{A^*(t+\tau)} x(0)\right) d\tau =$$

$$= \int_t^\infty \left(e^{A^* s} x(0), e^{A^* s} x(0)\right) ds = \int_t^\infty |x(s)|^2 ds .$$

It follows $\tilde{V}'(t) = -|x(t)|^2$.

Since $V\big(x(t)\big) \le \hat{\mu} |x(t)|^2$, we deduce $-|x(t)|^2 \le -\frac{1}{\hat{\mu}} \tilde{V}(t)$,

hence

$$\tilde{V}'(t) \le -\frac{1}{\hat{\mu}} \tilde{V}(t), \text{ or } \tilde{V}'(t) + \frac{1}{\hat{\mu}} \tilde{V}(t) \le 0,$$

hence

$$e^{\frac{1}{\hat{\mu}} t} \tilde{V}'(t) + \frac{1}{\hat{\mu}} e^{\frac{1}{\hat{\mu}} t} \tilde{V}(t) \le 0, \quad \frac{d}{dt}\left[e^{\frac{1}{\hat{\mu}} t} \tilde{V}(t)\right] \le 0,$$

$$\tilde{V}(t) \le e^{-\frac{1}{\hat{\mu}} t} \tilde{V}(0) = e^{-\frac{1}{\hat{\mu}} t} V\big(x(0)\big) \le \hat{\mu} e^{-\frac{1}{\hat{\mu}} t} |x(0)|^2 .$$

On the other hand, it can be checked that

$$V(x) = \int_0^\infty \left|e^{A^*\tau} x\right|^2 d\tau \ge \int_0^\delta \left|e^{A^*\tau} x\right|^2 d\tau \ge \nu |x|^2;$$

the last inequality uses the fact that $\delta$ can be chosen small enough. It follows that

$$|x(t)|^2 \le \frac{\hat{\mu}}{\hat{\nu}} e^{-\frac{1}{\hat{\mu}} t} |x(0)|^2 .$$

This behaviour of the solutions implies that the eigenvalues of the matrix $A^*$ have strictly negative real parts; obviously, the same conclusion holds for the eigenvalues of matrix $A$.

We shall use the above theorem in order to define an iterative procedure for obtaining the solution of the algebraic Riccati equation.

**Theorem 7.4** *If there exist $F$ and $H$ such that the eigenvalues of $A + BF$ and of $A + HC$ have strictly negative real parts, then the equation*

$$A^*R + RA - \frac{1}{\mu} RBB^*R + C^*C = 0$$

*admits a solution $R$ such that the eigenvalues of $A - \frac{1}{\mu} BB^*R$ had strictly negative real parts.*

***Proof*** We shall define the solution by an iterative process. By assumption, there exists $F$ such that the eigenvalues of $A + BF$ have strictly negative real parts.

We denote $W_1 = -F$ and write the Liapunov equation

$$(A - BW_1)^* R_1 + R_1 (A - BW_1) + C^*C + \mu W_1^* W_1 = 0.$$

Since the eigenvalues of $A - BW_1$ have strictly negative real parts, using the reasoning of Remark 7.2, we deduce that $R_1 \geq 0$.

We choose now $W_2 = \frac{1}{\mu} B^*R_1$ and check the fact that the eigenvalues of $A - BW_2$ have strictly negative real parts.

We have $A - BW_2 = A - \frac{1}{\mu} BB^*R_1$, hence

$$R_1\left(A-\frac{1}{\mu}BB^*R_1\right)+\left(A-\frac{1}{\mu}BB^*R_1\right)^*R_1=$$

$$=R_1A+A^*R_1-\frac{2}{\mu}R_1BB^*R_1=$$

$$=R_1(A-BW_1)+(A-BW_1)^*R_1+$$

$$+R_1BW_1+W_1^*B^*R_1-\frac{2}{\mu}R_1BB^*R_1=$$

$$=-C^*C-W_1^*W_1+R_1BW_1+W_1^*B^*R_1-\frac{2}{\mu}R_1BB^*R_1=$$

$$=-C^*C-\mu\left(W_1^*-\frac{1}{\mu}R_1B\right)\left(W_1-\frac{1}{\mu}B^*R_1\right)-\frac{1}{\mu}R_1BB^*R_1.$$

We deduce

$$R_1(A-BW_2)+(A-BW_2)^*R_1+C^*C+$$

$$+\mu\left(W_1^*-W_2^*\right)(W_1-W_2)+\mu W_2^*W_2=0.$$

Since the eigenvalues of

$$A-BW_2-B(W_1-W_2)=A-BW_1$$

have strictly negative real parts, we can apply Theorem 7.3 to conclude that the eigenvalues of $A-BW_2$ have strictly negative real parts. We can now continue the procedure, by choosing $R_2$ as a solution of the Liapunov equation

$$R_2(A-BW_2)+(A-BW_2)^*R_2+C^*C+\mu W_2^*W_2=0.$$

As above, we have $R_2\geq 0$ and taking $W_3=\frac{1}{\mu}B^*R_2$, we get $A-BW_3$ having eigenvalues with strictly negative real parts.

Taking into account the relation

$$R_1(A-BW_2)+(A-BW_2)^*R_1+C^*C+\mu W_2^*W_2+$$

$$+\mu\left(W_1^*-W_2^*\right)(W_1-W_2)=0$$

it follows

$$(R_1 - R_2)(A - BW_2) + (A - BW_2)^*(R_1 - R_2) +$$
$$+ \mu(W_1^* - W_2^*)(W_1 - W_2) = 0 .$$

Since $A - BW_2$ has eigenvalues with strictly negative real parts, we deduce $R_1 \geq R_2$ (using the representation formula given in Remark 7.2).

The procedure continues and we obtain a sequence $R_k \geq 0$ and a sequence $W_k$ such that

$$W_{k+1} = \frac{1}{\mu} B^* R_k \text{ and } 0 \leq R_{k+1} \leq R_k .$$

This inequality means that, for any $x$, we have

$$0 \leq (R_{k+1}\, x, x) \leq (R_k\, x, x);$$

therefore, the sequence $(R_k\, x, x)$ is decreasing and bounded from below, hence it has a limit. Herefrom it follows that, for any $x$ and $y$, the sequence $(R_k\, x, y)$ has a limit; indeed, we have

$$(R_k\,(x + y), x + y) = (R_k\, x, x) + 2\,(R_k\, x, y) + (R_k\, y, y)$$

and we deduce

$$\lim_{k\to\infty} (R_k\, x, y) = \frac{1}{2}\Big[\lim_{k\to\infty} (R_k\,(x + y), x + y) -$$
$$- \lim_{k\to\infty} (R_k\, x, x) - \lim_{k\to\infty} (R_k\, y, y)\Big].$$

By a convenient choice of the vectors $x$ and $y$, we obtain the convergence of the entries of the matrices $R_k$ to the entries of a matrix $R \geq 0$.

Herefrom we deduce that the sequence $W_k$ has also a limit $W = \frac{1}{\mu} B^* R$. Taking the limit in the equation satisfied by $R_k$, we get

$$R\left(A - \frac{1}{\mu} B B^* R\right) + \left(A - \frac{1}{\mu} B B^* R\right)^* R + C^* C + \frac{1}{\mu} R B B^* R = 0,$$

what shows that $R$ is a solution of the Riccati equation. We still have to verify that the eigenvalues of $A - \frac{1}{\mu} B B^* R$ have strictly negative real parts.

We proceed as in the proof of Theorem 7.3 and we deduce that if $x' = \left(A - B W^*\right) x$, then

$$\int_0^\infty | C x(t) |^2 dt \le \tilde{\mu} | x(0) |^2, \int_0^\infty \left| W^* x(t) \right|^2 dt \le \tilde{\mu} | x(0) |^2.$$

Further, we can write

$$x'(t) = (A + H C) x(t) - B W^* x(t) - H C x(t)$$

and we repeat the reasoning of Theorem 7.3 to conclude that the eigenvalues of

$$A - B W^* = A - \frac{1}{\mu} B B^* R$$

have strictly negative real parts.

This ends the proof of Theorem 7.4.

Theorem 7.1 and Theorem 7.4 solve completely the linear-quadratic optimization problem on infinite horizon.

**Remark 7.5** The solution $R$ obtained by the above iterative procedure is the maximal solution of the Riccati equation.

Indeed, if $\tilde{R}$ is a solution of the equation, then we can write

$$\tilde{R}\left(A - \frac{1}{\mu} B B^* R\right) + \left(A - \frac{1}{\mu} B B^* R\right)^* \tilde{R} + \frac{1}{\mu} \tilde{R} B B^* R +$$

$$+ \frac{1}{\mu} R B B^* \tilde{R} - \frac{1}{\mu} \tilde{R} B B^* \tilde{R} + C^* C = 0,$$

hence

$$\widetilde{R}\left(A - \frac{1}{\mu} B B^* R\right) + \left(A - \frac{1}{\mu} B B^* R\right)^* \widetilde{R} -$$

$$- \frac{1}{\mu}\left(\widetilde{R} - R\right) B B^* \left(\widetilde{R} - R\right) + C^* C + \frac{1}{\mu} R B B^* R = 0 .$$

Herefrom

$$\left(R - \widetilde{R}\right)\left(A - \frac{1}{\mu} B B^* R\right) + \left(A - \frac{1}{\mu} B B^* R\right)^* \left(R - \widetilde{R}\right) +$$

$$+ \frac{1}{\mu}\left(\widetilde{R} - R\right) B B^* \left(\widetilde{R} - R\right) = 0.$$

Since the eigenvalues of $A - \frac{1}{\mu} B B^* R$ have strictly negative real parts, it follows that $R - \widetilde{R} \geq 0$ and therefore $R \geq \widetilde{R}$ and $R$ is maximal. From this maximality we deduce the uniqueness of the stabilizing solution of the Algebraic Matrix Riccati Equation.

**Remark 7.6** The Liapunov equations, which have to be solved at each step, are linear equations, that can be solved by usual procedures. To this end, there exist efficient computer codes based on reducing matrices to triangular form by orthogonal transformations.

## 7.2 Application to a Price Model

We can analyze the model of price evolution, considered in **1.4**, i.e.

$$p' = \lambda\left[\,(a - \alpha)\,p + b - \beta\,\right],$$

by assuming that the term $b - \beta$ can be controlled by a decision-maker, by complementary demand or complementary supply, respectively.

Therefore, we consider the scalar control system, described by

$$p' = \lambda[(a-\alpha)p + u].$$

For a given value $\hat{u}$ of the control, we have the equilibrium

$$\hat{p} = -\frac{\hat{u}}{a-\alpha}, \quad a-\alpha < 0.$$

By introducing new variables, representing deviations from equilibrium: $x = p - \hat{p}$, $v = u - \hat{u}$, we obtain

$$x' = \lambda[(a-\alpha)x + v].$$

By associating to this control system a cost of the form

$$J(v) = \int_0^\infty \left(q^2 x^2 + \mu v^2\right) dt,$$

we obtain an optimal stabilization problem.

The corresponding algebraic Riccati equation is

$$\frac{\lambda^2}{\mu} r^2 - 2\lambda(a-\alpha)r - q^2 = 0.$$

When $q = 0$, hence when there is no penalty for the deviation from equilibrium, then the optimal control is obviously zero and corresponds to the root $r = 0$.

If $q > 0$, then the Riccati equation has two real roots, $r_1 > 0$ and $r_2 < 0$ (since $r_1 r_2 = -\dfrac{q^2\mu}{\lambda^2} < 0$).

The optimal control corresponds to the positive root: $\tilde{v} = -\dfrac{\lambda}{\mu} r_1 x$,

where

$$r_1 = \frac{\sqrt{\mu^2(a-\alpha)^2 + q^2\mu} - \mu(a-\alpha)}{\lambda}.$$

The optimal cost is $r_1 x_0^2$.

## 7.3 Optimal Stabilization in Discrete Time

We consider a linear problem with quadratic cost function, in discrete time, on infinite horizon.

The evolution of the process is described by

$$x_{t+1} = A\,x_t + B\,u_t$$

and the cost is

$$J(u) = \sum_{t=0}^{\infty}\left[(Q\,x_t, x_t) + (R\,u_t, u_t)\right], \text{ with } R > 0,\ Q \geq 0.$$

A control $u_t$ is admissible if the corresponding solution $x_t$ converges to zero and the series is convergent; among all these admissible controls we look for one for which the cost is minimized.

**Proposition 7.7** *Let us assume that the nonlinear system*

$$R + B^* P\,B = V^2$$

$$A^* P\,B = W^* V$$

$$P = A^* P\,A + Q - W^* W$$

*admits a solution with* $P \geq 0$, *such that the eigenvalues of the matrix* $A + B\,F$, *with* $F = -V^{-1}\,W$, *are located inside the unit circle (with modulus strictly smaller than* 1). *Then the control* $u_t = F\,x_t$ *is admissible and provides the minimum cost.*

***Proof*** The fact that such a control is admissible follows directly: if $u_t = F\,x_t$, then the evolution is $u_t = (A + B\,F)\,x_t$, hence $x_t = (A + B\,F)^t\,x_0$ and if the eigenvalues of matrix $A + B\,F$

have absolute values smaller than 1 then there exist $0 < q < 1$, $\beta > 1$ such that $|x_t| \le \beta q^t |x_0|$ for all $t \ge 0$, hence $|u_t| \le \hat{\beta} q^t |x_0|$ and

$$\sum_{t=0}^{\infty} [(Q x_t, x_t) + (R u_t, u_t)] \le \sum_{t=0}^{\infty} \mu q^{2t} |x_0|^2 = \frac{\mu |x_0|^2}{1 - q^2}.$$

Let us consider now an arbitrary admissible control and transform the expression of the cost, using the solution $(P, V, W)$ of the system. We have

$$\sum_{t=0}^{\infty} [(Q x_t, x_t) + (R u_t, u_t)] =$$

$$= \sum_{t=0}^{\infty} \left[ \left( \left(P + W^* W - A^* P A\right) x_t, x_t \right) + \left( \left(V^2 - B^* P B\right) u_t, u_t \right) \right] =$$

$$= \sum_{t=0}^{\infty} [(P x_t, x_t) + (W x_t, W x_t) - (P A x_t, A x_t) +$$

$$+ (V u_t, V u_t) - (P B u_t, B u_t)] =$$

$$= \sum_{t=0}^{\infty} [(P x_t, x_t) + (W x_t, W x_t) - (P (x_{t+1} - B u_t), x_{t+1} - B u_t) +$$

$$+ (V u_t, V u_t) - (P B u_t, B u_t)].$$

We deduce

$$J(u) = \sum_{t=0}^{\infty} [(W x_t, W x_t) + (V u_t, V u_t) + (P x_t, x_t) -$$

$$- (P x_{t+1}, x_{t+1}) - 2 (P B u_t, B u_t) + (P x_{t+1}, B u_t) + (P B u_t, x_{t+1})] =$$

$$= (P x_0, x_0) + \sum_{t=0}^{\infty} [(W x_t, W x_t) + (V u_t, V u_t) - 2 (P B u_t, B u_t) +$$

$$+ \left(P\left(A x_t + B u_t\right), B u_t\right) + \left(P B u_t, A x_t + B u_t\right)\Big] =$$

$$= \left(P x_0, x_0\right) + \sum_{t=0}^{\infty} \Big[\left(W x_t, W x_t\right) + \left(V u_t, V u_t\right) +$$

$$+ \left(B^* P A x_t, u_t\right) + \left(A^* P B u_t, x_t\right)\Big] =$$

$$= \left(P x_0, x_0\right) + \sum_{t=0}^{\infty} \Big[\left(W x_t, W x_t\right) + \left(V u_t, V u_t\right) +$$

$$+ \left(W^* V u_t, x_t\right) + \left(V W x_t, u_t\right)\Big] =$$

$$= \left(P x_0, x_0\right) + \sum_{t=0}^{\infty} \left(W x_t + V u_t, W x_t + V u_t\right).$$

It follows that for any admissible control we have $J(u) \geq \left(P x_0, x_0\right)$, equality being achieved for

$$u_t = -V^{-1} W x_t = F x_t .$$

The question occurs now in a natural way, in what circumstances does a solution of the nonlinear system exist and how can we compute it?

We shall show that, if there exists matrices $F_0$, $H_0$ such that $A + B F_0$, $A + H_0 C$ have eigenvalues with modulus strictly smaller than 1, then there exists a solution $(P, V, W)$ of the system

$$R + B^* P B = V^2$$

$$A^* P B = W^* V$$

$$P = A^* P A + C^* C - W^* W$$

such that $P \geq 0$, $A + B F$ has eigenvalues strictly smaller than 1 in modulus, where $F = -V^{-1} W$.

The solution is obtained as the limit of a sequence defined recurrently by the relations:

$$A_k = A + B F_k,$$

$$P_{k+1} = A_k^* P_{k+1} A_k + C^* C + F_k^* R F_k,$$

$$V_{k+1}^2 = R + B^* P_{k+1} B,$$

$$W_{k+1}^* V_{k+1} = A^* P_{k+1} B,$$

$$F_{k+1} = -V_{k+1}^{-1} W_{k+1},$$

the eigenvalues of the matrix $A + B F_k$ being strictly smaller than 1 in modulus.

We start with $F_0$ such that $A_0 = A + B F_0$ has eigenvalues strictly smaller than 1 in modulus, then we define $P_1$, $V_1$, $W_1$, $F_1$ and the process continues. We have to check that we can define $P_{k+1} \geq 0$ if $A_k$ has eigenvalues with modulus strictly smaller than 1 and that, if we defined $P_{k+1}$, hence also $F_{k+1}$, then the eigenvalues of $A_{k+1}$ have modulus strictly smaller than 1. Finally, we have to prove that the sequences $P_k, V_k, W_k$ are convergent and that their limits provide the solution of the system.

We begin by proving some propositions interesting by themselves.

**Proposition 7.8** *Let $A$ be a matrix such that its eigenvalues are located inside the unit circle. Then the equation $P = A^* P A + Q$ has a solution for any $Q$ and, if $Q \geq 0$, then the solution $P$ satisfies $P \geq 0$.*

***Proof*** Since the eigenvalues of matrix $A$ are strictly smaller than 1 in modulus, the series $\sum_{k=0}^{\infty}\left(A^*\right)^k Q A^k$ is convergent.

We show that this series gives the solution of the equation.

Denoting $P = \sum_{k=0}^{\infty}\left(A^*\right)^{k+1} Q A^k$, we have indeed

$$A^* P A = \sum_{k=0}^{\infty}\left(A^*\right)^{k+1} Q A^{k+1} = \sum_{k=1}^{\infty}\left(A^*\right)^k Q A^k = P - Q$$

what shows that the series provides a solution of the equation.

Let us assume now that $P$ is a solution of the equation; we have

$$\left(A^*\right)^k P A^k = \left(A^*\right)^{k+1} P A^{k+1} + \left(A^*\right)^k Q A^k .$$

Summing up from $k = 0$ to $\infty$ (the convergence being assured by the assumption concerning the matrix $A$), it follows

$$\sum_{k=0}^{\infty}\left(A^*\right)^k P A^k = \sum_{k=0}^{\infty}\left(A^*\right)^{k+1} P A^{k+1} + \sum_{k=0}^{\infty}\left(A^*\right)^k Q A^k ,$$

hence

$$P + \sum_{k=1}^{\infty}\left(A^*\right)^k P A^k = \sum_{k=1}^{\infty}\left(A^*\right)^k P A^k + \sum_{k=0}^{\infty}\left(A^*\right)^k Q A^k$$

what shows that $P = \sum_{k=0}^{\infty}\left(A^*\right)^k Q A^k$.

The conclusion of the proposition follows automatically; moreover, it can be seen that the solution of the equation is unique.

**Proposition 7.9** *Let us consider the equation*

$$P = A^* P A + C^* C.$$

*If the equation has a solution* $P \geq 0$ *and if there exists* $H_0$ *such that the eigenvalues of* $A + H_0 C$ *have modulus strictly smaller than* 1, *then the eigenvalues of* $A$ *have modulus strictly smaller than* 1.

***Proof*** Let $x_0$ be an arbitrary vector; we define the sequence

$$x_{t+1} = A\, x_t, \text{ hence } x_t = A^t\, x_0.$$

We define $V_t = (P\, x_t, x_t)$; we have

$$V_{t+1} = (P\, x_{t+1}, x_{t+1}) = (P\, A\, x_t, A\, x_t) = (A^* P\, A\, x_t, x_t) =$$

$$= (P\, x_t, x_t) - (C^* C\, x_t, x_t) = V_t - |C\, x_t|^2.$$

Summing up from 0 to $T - 1$, we deduce

$$V_T - V_0 = -\sum_{t=0}^{T-1} |C\, x_t|^2;$$

$$\sum_{t=0}^{T-1} |C\, x_t|^2 = V_0 - V_T \leq V_0,$$

hence $\sum_{t=0}^{\infty} |C\, x_t|^2$ converges and, moreover, $\sum_{t=0}^{\infty} |C\, x_t|^2 \leq \mu |x_0|^2$.

Further, we can write

$$x_{t+1} = A\, x_t = (A + H_0\, C)\, x_t - H_0\, C\, x_t.$$

We deduce

$$x_1 = (A + H_0\, C)\, x_0 - H_0\, C\, x_0,$$

$$x_2 = (A + H_0\, C)\, x_1 - H_0\, C\, x_1 =$$

$$= (A + H_0\, C)^2\, x_0 - (A + H_0\, C)\, H_0\, C\, x_0 - H_0\, C\, x_1.$$

We check by induction that

$$x_t = (A + H_0 C)^t x_0 - \sum_{j=0}^{t-1} (A + H_0 C)^{t-j-1} H_0 C x_j .$$

The formula is valid for $t = 1$, $t = 2$; then we have

$$\begin{aligned} x_{t+1} &= (A + H_0 C) x_t - H_0 C x_t = \\ &= (A + H_0 C) \Big[ (A + H_0 C)^t x_0 - \\ &\quad - \sum_{j=0}^{t-1} (A + H_0 C)^{t-j-1} H_0 C x_j \Big] - H_0 C x_t = \\ &= (A + H_0 C)^{t+1} x_0 - \sum_{j=0}^{t-1} (A + H_0 C)^{t-j} H_0 C x_j - H_0 C x_t = \\ &= (A + H_0 C)^{t+1} x_0 - \sum_{j=0}^{t} (A + H_0 C)^{t-j} H_0 C x_j \end{aligned}$$

what shows that, if the formula is valid for $x_t$, then it remains valid for $x_{t+1}$.

Taking into account that the eigenvalues of $A + H_0 C$ are strictly inside the unit circle, we deduce that

$$\left| (A + H_0 C)^t \right| \le \beta q^t , \ 0 < q < 1.$$

Herefrom it follows that

$$| x_t | \le \beta q^t | x_0 | + \hat{\beta} \sum_{j=0}^{t-1} q^{t-j-1} | C x_j |,$$

$$\begin{aligned} | x_t |^2 &\le 2 \beta^2 q^{2t} | x_0 |^2 + 2 \hat{\beta}^2 \left( \sum_{j=0}^{t-1} q^{t-j-1} | C x_j | \right)^2 \le \\ &\le 2 \beta^2 q^{2t} | x_0 |^2 + 2 \hat{\beta}^2 \left[ \sum_{j=0}^{t-1} q^{t-j-1} \sum_{j=0}^{t-1} q^{t-j-1} | C x_j |^2 \right] \le \end{aligned}$$

$$\leq 2\beta^2 q^{2t} |x_0|^2 + \frac{2\hat{\beta}^2}{1-q} \sum_{j=0}^{t-1} q^{t-j-1} |C x_j|^2 .$$

Further, we obtain

$$\sum_{t=0}^{\infty} |x_t|^2 \leq 2\beta^2 |x_0|^2 \sum_{t=0}^{\infty} q^{2t} + \frac{2\hat{\beta}^2}{1-q} \sum_{t=0}^{\infty} \sum_{j=0}^{t-1} q^{t-j-1} |C x_j|^2 =$$

$$= \frac{2\beta^2 |x_0|^2}{1-q^2} + \frac{2\hat{\beta}^2}{1-q} \sum_{j=0}^{\infty} \left( \sum_{t=j+1}^{\infty} q^{t-j-1} \right) |C x_j|^2 =$$

$$= \frac{2\beta^2 |x_0|^2}{1-q^2} + \frac{2\hat{\beta}^2}{(1-q)^2} \sum_{j=0}^{\infty} |C x_j|^2 \leq \hat{\mu} |x_0|^2 .$$

The convergence of the series implies that

$$\lim_{t\to\infty} |x_t|^2 = 0, \ \lim_{t\to\infty} A^t = 0,$$

what is possible only if the eigenvalues of matrix $A$ have modulus strictly smaller than 1.

We remark that the same proof works if

$$P = A^* P A + C^* C + Q, \text{ with } Q \geq 0.$$

From Proposition 7.8 it follows that, given $A_k$ with eigenvalues strictly smaller than 1 in modulus and $F_k$, we can find $P_{k+1}$ in a unique way and $P_{k+1} \geq 0$.

Let us assume now that we defined $P_{k+1}, V_{k+1}, W_{k+1}, F_{k+1}$; we shall prove that the eigenvalues of matrix $A_{k+1} = A + B F_{k+1}$ are strictly inside the unit circle. To this end, we write the relation for $P_{k+1}$, using $A_{k+1}$. We have

$$P_{k+1} = \left(A^* + F_k^* B^*\right) P_{k+1} \left(A + B F_k\right) + C^* C + F_k^* R F_k =$$

$$= \left[A^* + F_{k+1}^* B^* + \left(F_k^* - F_{k+1}^*\right) B^*\right] P_{k+1} \left[A + B F_{k+1} + \right.$$

$$
\begin{aligned}
&+ B\left(F_k - F_{k+1}\right)\big] + C^*C + F_k^*\, R\, F_k = \\
&= \left(A^* + F_{k+1}^*\, B^*\right) P_{k+1} \left(A + B\, F_{k+1}\right) + \\
&+ \left(A^* + F_{k+1}^*\, B^*\right) P_{k+1}\, B \left(F_k - F_{k+1}\right) + \\
&+ \left(F_k^* - F_{k+1}^*\right) B^*\, P_{k+1} \left(A + B\, F_{k+1}\right) + \\
&+ \left(F_k^* - F_{k+1}^*\right) B^*\, P_{k+1}\, B \left(F_k - F_{k+1}\right) + C^*C + F_k^*\, R\, F_k = \\
&= A_{k+1}^* P_{k+1}\, A_{k+1} + W_{k+1}^*\, V_{k+1} \left(F_k - F_{k+1}\right) + \\
&+ F_{k+1}^* \left(V_{k+1}^2 - R\right)\left(F_k - F_{k+1}\right) + \left(F_k^* - F_{k+1}^*\right) V_k\, W_{k+1} + \\
&+ \left(F_k^* - F_{k+1}^*\right)\left(V_{k+1}^2 - R\right) F_{k+1} + C^*\, C + F_k^*\, R\, F_k + \\
&+ \left(F_k^* - F_{k+1}^*\right)\left(V_{k+1}^2 - R\right)\left(F_k - F_{k+1}\right) = \\
&= A_{k+1}^* P_{k+1}\, A_{k+1} - F_{k+1}^*\, V_{k+1}^2 \left(F_k - F_{k+1}\right) + F_{k+1}^* \left(V_{k+1}^2 - R\right) F_k - \\
&- F_{k+1}^* \left(V_{k+1}^2 - R\right) F_{k+1} - \left(F_k^* - F_{k+1}^*\right) V_{k+1}^2\, F_{k+1} + \\
&+ F_k^* \left(V_{k+1}^2 - R\right) F_{k+1} - F_{k+1}^* \left(V_{k+1}^2 - R\right) F_{k+1} + \\
&+ F_k^* \left(V_{k+1}^2 - R\right) F_k - F_{k+1}^* \left(V_{k+1}^2 - R\right) F_k - \\
&- F_k^* \left(V_{k+1}^2 - R\right) F_{k+1} + F_{k+1}^* \left(V_{k+1}^2 - R\right) F_{k+1} + C^*C + F_k^* R\, F_k = \\
&= A_{k+1}^* P_{k+1}\, A_{k+1} - F_{k+1}^*\, V_{k+1}^2\, F_k + F_{k+1}^*\, V_{k+1}^2\, F_{k+1} - \\
&- F_k^*\, V_{k+1}^2\, F_{k+1} + F_{k+1}^*\, V_{k+1}^2\, F_{k+1} - F_{k+1}^*\, V_{k+1}^2\, F_{k+1} + \\
&+ F_{k+1}^*\, R\, F_{k+1} + F_k^*\, V_{k+1}^2\, F_k - F_k^*\, R\, F_k + C^*C + F_k^*\, R\, F_k = \\
&= A_{k+1}^*\, P_{k+1}\, A_{k+1} + \left(F_k^* - F_{k+1}^*\right) V_{k+1}^2 \left(F_k - F_{k+1}\right) + \\
&+ C^*C + F_{k+1}^* R\, F_{k+1}.
\end{aligned}
$$

Thus we got

$$
\begin{aligned}
P_{k+1} = {}& A_{k+1}^* P_{k+1} A_{k+1} + \left(F_k^* - F_{k+1}^*\right) V_{k+1}^2 \left(F_k - F_{k+1}\right) + \\
& + C^*C + F_{k+1}^* R\, F_{k+1}.
\end{aligned}
$$

We also have

$$A_{k+1} = A + B\,F_{k+1} = A + B\,V_{k+1}^{-1}\,V_{k+1}\left(F_{k+1} - F_k\right) + B\,F_k,$$

hence

$$A_{k+1} - B\,V_{k+1}^{-1}\,V_{k+1}\left(F_{k+1} - F_k\right) = A_k.$$

Since the eigenvalues of matrix $A_k$ have modulus strictly smaller than 1, we deduce that the pair $\left(A_{k+1}, V_{k+1}\left(F_{k+1} - F_k\right)\right)$ satisfies the condition stated in Proposition 7.9, hence the eigenvalues of matrix $A_{k+1}$ have modulus strictly smaller than 1.

In this way, we proved the possibility of performing the iterative construction for $P_k, V_k, W_k, F_k$, starting from $F_0$.

We shall prove that the sequence $P_k$ converges. We have:

$$P_{k+1} = A_k^*\,P_{k+1}\,A_k + C^*C + F_k^*R\,F_k.$$

On the other hand,

$$P_k = A_k^*\,P_k\,A_k + \left(F_{k-1} - F_k\right)^*\,V_k^2\left(F_{k-1} - F_k\right) + C^*C + F_k^*R\,F_k.$$

We deduce that

$$P_k - P_{k+1} = A_k^*\left(P_k - P_{k+1}\right) + \left(F_{k-1} - F_k\right)^*\,V_k^2\left(F_{k-1} - F_k\right).$$

Since the eigenvalues of $A_k$ have modulus strictly smaller than 1, we deduce, on the basis of Proposition 7.8 that $P_k - P_{k+1} \geq 0$, hence the sequence $P_k$ is decreasing and bounded from below ($P_k \geq 0$), what ensures its convergence.

Recall that

$$V_k^2 = R + B^*\,P_k\,B, \quad W_k^*\,V_k = A^*\,P_k\,B, \quad F_k = -V_k^{-1}\,W_k.$$

Since $P_k \geq 0$, we have $V_k^2 \geq R > 0$. We deduce that $V_k$ with $V_k > 0$ exists and is invertible. It follows that $W_k = V_k^{-1}\,B^*\,P_k\,A$.

From here we see that

$$F_k = -V_k^{-2} B^* P_k A = -\left(R + B^* P_k B\right)^{-1} B^* P_k A.$$

From $\lim_{k\to\infty} P_k = P$ it follows that

$$\lim_{k\to\infty} \left(R + B^* P_k B\right)^{-1} = \left(R + B^* P B\right)^{-1}$$

(the sequence $\left|\left(R + B^* P B\right)^{-1}\right|$ is bounded by $\left|R^{-1}\right|$), hence

$$\lim_{k\to\infty} F_k = -\left(R + B^* P B\right)^{-1} B^* P A = F.$$

Since $A_k = A + B F_k$ it results that $\lim_{k\to\infty} A_k = A + B F$ and it follows that

$$P = (A + B F)^* P (A + B F) + C^* C + F^* R F =$$

$$= \left(A^* + F^* B^*\right)(P A + P B F) + C^* C + F^* R F =$$

$$= A^* P A + A^* P B F + F^* B^* P A + F^* B^* P B F +$$

$$+ C^* C + F^* R F =$$

$$= A^* P A + A^* P B F + F^* B^* P A + C^* C + F^*\left(B^* P B + R\right) F$$

Replacing the formula for $F$, we obtain:

$$P = A^* P A + C^* C - 2 A^* P B\left(R + B^* P B\right)^{-1} B^* P A +$$

$$+ A^* P B\left(R + B^* P B\right)^{-1}\left(R + B^* P B\right)\left(R + B^* P B\right)^{-1} B^* P A.$$

We obtained in this way for $P$ the equation:

$$A^* P A - P - A^* P B\left(R + B^* P B\right)^{-1} B^* P A + C^* C = 0$$

which is termed as the *algebraic matrix Riccati equation*.

Let us denote $V^2 = R + B^* P B$ and $W = V^{-1} B^* P A$.

We see now that the equation for $P$ may be written as:

$$A^* P A - P - W^* W + C^* C = 0.$$

We finally obtained a solution $(V,P,W)$ to the system

$$R + B^* P B = V^2$$

$$A^* P B = W^* V$$

$$P = A^* P A + Q - W^* W,$$

as required.

We have still to check that the eigenvalues of $A + BF$ have modulus strictly smaller than 1. We have

$$P = (A + BF)^* P (A + BF) + C^* C + F^* R F.$$

The conclusion will follow as in the proof of Proposition 7.9.

We consider $x_{t+1} = (A + BF) x_t$, hence $x_t = (A + BF)^t x_0$, $v_t = (P x_t, x_t)$: it follows

$$v_{t+1} = (P(A + BF) x_t, (A + BF) x_t) =$$

$$= ((A + BF)^* P (A + BF) x_t, x_t) =$$

$$= (P x_t, x_t) - |C x_t|^2 - |R^{1/2} F x_t|^2,$$

$$v_{t+1} - v_t = -|C x_t|^2 - |R^{1/2} F x_t|^2.$$

By summation from 0 to $T - 1$, we obtain

$$\sum_{t=0}^{T-1} |C x_t|^2 + \sum_{t=0}^{T-1} |R^{1/2} F x_t|^2 = v_0 - v_T \leq v_0 \leq \mu |x_0|^2.$$

Assuming the existence of a matrix $H_0$ such that the eigenvalues of $A + H_0 C$ were strictly inside the unit circle, we can write

$$x_{t+1} = (A + BF) x_t =$$

$$= (A + H_0 C) x_t - H_0 C x_t + B R^{-1/2} R^{1/2} F x_t.$$

We take into account that $\left|(A + H_0 C)^t\right| \le \beta q^t$ and we repeat the reasoning of the proof of Proposition 7.9.

Let us summarize the conclusion we reached:

**Theorem 7.10** *Let us consider an evolution given by*

$$x_{t+1} = A x_t + B u_t$$

*and the cost-function*

$$J(u) = \sum_{t=0}^{\infty}\left[\left|C x_t\right|^2 + (R u_t, u_t)\right], \text{ with } R > 0.$$

*We assume the existence of matrices $F_0$, $H_0$ such that $A + B F_0$, $A + H_0 C$ have eigenvalues with modulus strictly smaller than 1.*

*Then the sequences defined by*

$$P_{k+1} = (A + B F_k)^* P_{k+1} (A + B F_k) + C^* C + F_k^* R F_k ,$$

$$V_{k+1}^2 = R + B^* P_{k+1} B ,$$

$$W_{k+1}^* V_{k+1} = A^* P_{k+1} B ,$$

$$F_{k+1} = - V_{k+1}^{-1} W_{k+1} ,$$

*starting with $F_0$, are convergent and the control $u_t = F x_t$ stabilizes the system (in the sense that $\left|x_t\right| \le \beta q^t \left|x_0\right|$, $0 < q < 1$) and achieves the minimum cost. The minimum value of the cost is $(P x_0, x_0)$; here $F = \lim_{k\to\infty} F_k$, $P = \lim_{k\to\infty} P_k$.*

Finally, let us make the following important remark. If, during the iterative process, we stop at a step $k$ and choose the feed-back control $u_t = F_k x_t$, then the evolution will be given by

$$x_{t+1} = A x_t + B u_t = (A + B F_k) x_t = A_k x_t$$

and, hence, it will satisfy

$$|x_t| \le \beta q^t |x_0|, \; 0 < q < 1.$$

Moreover, the cost associated to this evolution is

$$\sum_{t=0}^{\infty} \left[ \left( C^* C x_t, x_t \right) + \left( R u_t, u_t \right) \right] =$$

$$= \sum_{t=0}^{\infty} \left[ \left( C^* C x_t, x_t \right) + \left( R F_k x_t, F_k x_t \right) \right] =$$

$$= \sum_{t=0}^{\infty} \left( \left( C^* C + F_k^* R F_k \right) x_t, x_t \right) =$$

$$= \sum_{t=0}^{\infty} \left[ \left( P_{k+1} x_t, x_t \right) - \left( A_k^* P_{k+1} A_k x_t, x_t \right) \right] =$$

$$= \sum_{t=0}^{\infty} \left[ \left( P_{k+1} x_t, x_t \right) - \left( P_{k+1} x_{t+1}, x_{t+1} \right) \right] = \left( P_{k+1} x_0, x_0 \right).$$

Since we have $P_{k+1} \le P_k$, it can be seen that, at each iteration, we get a stabilizing feed-back which improves the cost.

## 7.4 Optimal Stabilization in a Discrete-Time Model of Price Evolution

We consider the discrete-time model of price evolution described in **1.5**. This time, we shall denote the price at time $t$ by $y_t$. We have then the following free evolution:

$$y_{t+1} = \lambda (\alpha - \beta) + \left[ 1 - \lambda (b - a) \right] y_t \qquad \textbf{(7.1)}$$

We can transform this evolution in a controlled one assuming, for instance, that $\alpha - \beta$ is no longer constant, but can be controlled by a decision-maker, by influencing correspondingly the supply or demand. Thus we have the evolution:

$$y_{t+1} = \left[1 - \lambda\left(b - a\right)\right] y_t + v_t \tag{7.2}$$

where $v_t$ is the control.

For a desired equilibrium value $\hat{y}$ of the price, we obtain a corresponding equilibrium value for the control; the equilibrium condition

$$\hat{y} = \hat{y}\left[1 - \lambda\left(b - a\right)\right] + \lambda\,\hat{v} \text{ implies } \hat{v} = \left(b - a\right)\hat{y}.$$

Denoting $x_t = y_t - \hat{y}$, $u_t = v_t - \hat{v}$, we obtain the evolution:

$$x_{t+1} = \left[1 - \lambda\left(b - a\right)\right] x_t + \lambda\, u_t \tag{7.3}$$

We associate to this evolution the cost:

$$J(u) = \sum_{t=0}^{\infty}\left(c^2 x_t^2 + r^2 u_t^2\right) \tag{7.4}$$

and state the problem of finding the control $u$ that minimizes this cost. Here $x_t$ represents the deviation of the actual price from the desired one and $u_t$ represents the effort of control; the coefficients $c$ and $r$ represent weights associated to the above quantities, reflecting preferences of the decision-maker.

**Remark 7.11** We remark that the evolution given by relation **(7.3)** is of the same type as that given by **(7.2)**. Nevertheless, we preferred to work with the variables $x$ and $u$ instead of $y$ and $v$ respectively, in order to have the usual form of a stabilization problem.

**Remark 7.12** The evolutions given by **(7.2)** and **(7.3)** are stable if $\lambda(b-a) < 2$, since then the modulus of the coefficient $1-\lambda(b-a)$ is strictly smaller than 1 (see **1.5**).

In order to find the optimal control, we associate to our problem (which consists of relations **(7.3)** and **(7.4)**) the nonlinear system:

$$r^2 + \lambda^2 p = V^2$$

$$\lambda\left[1-\lambda(b-a)\right]p = WV$$

$$p = \left[1-\lambda(b-a)\right]^2 p + c^2 - W^2 .$$

We have

$$V = \sqrt{r^2 + \lambda^2 p} ;$$

$$W = \frac{\lambda\left[1-\lambda(b-a)\right]p}{\sqrt{r^2 + \lambda^2 p}} .$$

Denoting $\lambda(b-a) = d$ and substituting $V$ and $W$ into the third equation, we get

$$p\left[1-(1-d)^2\right] - c^2 + \frac{\lambda^2(1-d)^2 p^2}{r^2 + \lambda^2 p} = 0$$

or

$$\lambda^2 p^2 - \left(d^2 r^2 - 2dr^2 + c^2\lambda^2\right)p - c^2 r^2 = 0 ,$$

which is an equation of second degree with respect to $p$. This equation has real roots $p_1$, $p_2$ and $p_1 p_2 = -\dfrac{c^2 r^2}{\lambda^2} < 0$, therefore the equation has a unique positive root, say $p_1$.

We get then the optimal feed-back control in the form $\tilde{u}_t = F x_t$, where

$$F = -\frac{W}{V} = -\frac{\lambda(1-d)p_1}{r^2 + \lambda^2 p_1} .$$

## 7.5 Notes and References

A source for the linear–quadratic optimization problem can be considered to be the paper of R. E. Kalman [15].

For further developments and the discussion of the discrete–time case, see the monograph of Popov [25].

The iterative procedure to solve the algebraic Riccati equation has been proposed by Kleiman in [16].

The Kalman-Szegö equations, which appear in the solution of the discrete–time optimal stabilization problem, were obtained in connection with the existence of a Liapunov function in the absolute stability problem. The connection with the linear-quadratic optimization problem was mentioned first by V. M. Popov in 1966, in his book on hyperstability [25].

The Riccati equation approach to solve the optimal stabilization problem may be found, for example, in the book by Kwakernaak and Sivan [18].

CHAPTER 8

# Linear-Quadratic Optimization on Finite Horizon

## 8.1 Continuous Time

An optimal control problem, which can be viewed as a variant for stabilization, is the following.

We consider a linear controlled evolution:

$$x' = A\,x + B\,u \tag{8.1}$$

Here $x$ can be interpreted as a deviation from equilibrium, and $u$ - as a control correction. The control aims to get $x$ close to zero, therefore a penalty is imposed for the deviation of $x$ (or for an output $y = C\,x$) from zero, measured by a quadratic form $(Q\,x, x)$, which is positive semidefinite, for instance $|C\,x|^2$.

At the same time, the effort of control meant to get $x$ close to zero is also measured, e.g. by $|u|^2$.

A compromise between the penalty for the deviation of $x$ and the cost of control can be of the form $(Q\,x, x) + \mu\,(u, u)$; since the optimization is performed on a time horizon $[\,0,T\,]$, we adopt as a total cost criterion the integral

$$\int_0^T \left[(Q\,x(t),x(t)) + \mu\,(u(t),u(t))\right] dt\,.$$

On the other hand, since the time horizon is finite, it is natural to impose a penalty for the final value of the deviation, what leads to the quadratic functional

$$J(u) = (G\,x(T),x(T)) + \\ + \int_0^T \left[(Q\,x(t),x(t)) + \mu\,(u(t),u(t))\right] dt \tag{8.2}$$

The problem is to find a control $u$ which minimizes the cost, starting from an initial state $x_0$.

To solve this problem, more precisely, to find necessary optimality conditions, we consider an optimal solution $(\tilde{u},\tilde{x})$ and its variations of the form $u = \tilde{u} + \varepsilon\,v$, $x = \tilde{x} + \varepsilon\,z$. We have

$$x' = \tilde{x}' + \varepsilon\,z' = A\,(\tilde{x} + \varepsilon\,z) + B\,(\tilde{u} + \varepsilon\,v).$$

Since $\tilde{x}' = A\,\tilde{x} + B\,\tilde{u}$, it follows $z' = A\,z + B\,v$.

Since $x(0) = x_0 = \tilde{x}(0)$, we deduce $z(0) = 0$.

We denote

$$\tilde{J}(\varepsilon) = J(\tilde{u} + \varepsilon\,v) = \left(G(\tilde{x}(T) + \varepsilon\,z(T)), \tilde{x}(T) + \varepsilon\,z(T)\right) +$$

$$+ \int_0^T \left[\left(Q(\tilde{x}(t) + \varepsilon\,z(t)), \tilde{x}(t) + \varepsilon\,z(t)\right) +\right.$$

$$\left. + \mu\,(\tilde{u}(t) + \varepsilon\,v(t), \tilde{u}(t) + \varepsilon\,v(t))\right] dt =$$

$$= J(\tilde{u}) + \varepsilon\,\left\{(G\,z(t), \tilde{x}(T)) + (G\,\tilde{x}(T), z(t)) +\right.$$

$$+\int_0^T \left[(Q\,\tilde{x}(t), z(t)) + (Q\,z(t), \tilde{x}(t)) + \right.$$

$$\left. + \mu\,(v(t), \tilde{u}(t)) + \mu\,(\tilde{u}(t), v(t))\right] dt\Big\} + \varepsilon^2 J(v).$$

Since $J(\tilde{u} + \varepsilon\, v) \geq J(\tilde{u})$ for any $\varepsilon$, it follows that $\tilde{J}(\varepsilon) \geq \tilde{J}(0)$ for any $\varepsilon$; on the basis of the Fermat theorem, we deduce $\tilde{J}'(0) = 0$, hence

$$(G\,z(T), \tilde{x}(T)) + (G\,\tilde{x}(T), z(T)) +$$

$$+\int_0^T \left[(Q\,\tilde{x}(t), z(t)) + (Q\,z(t), \tilde{x}(t)) + \right.$$

$$\left. + \mu\,(v(t), \tilde{u}(t)) + \mu\,(\tilde{u}(t), v(t))\right] dt = 0. \tag{8.3}$$

The fact that this equality holds for any $v$ will give us informations about the optimal solution $(\tilde{u}, \tilde{x})$.

We define the function $p$ as a solution of the Cauchy problem

$$p' = -A^* p + Q\,\tilde{x}(t), \quad p(T) = -G\,\tilde{x}(T).$$

We use this function to rewrite the equality obtained above. We have

$$\int_0^T (Q\,\tilde{x}(t), z(t))\,dt = \int_0^T \left(p'(t) + A^* p(t), z(t)\right) dt =$$

$$= \int_0^T \left[(p'(t), z(t)) + (p(t), A\,z(t))\right] dt =$$

$$= \int_0^T \left[(p'(t), z(t)) + (p(t), z'(t) - B\,v(t))\right] dt =$$

$$= \int_0^T \frac{d}{dt}(p(t), z(t))\,dt - \int_0^T \left(B^* p(t), v(t)\right) dt =$$

$$= (p(T), z(T)) - (p(0), z(0)) - \int_0^T \left(B^* p(t), v(t)\right) dt.$$

But $z(0) = 0$ and $p(T) = -G\,\tilde{x}(T)$. We deduce

$$(G\,\tilde{x}(T), z(T)) + \int_0^T (Q\,\tilde{x}(t), z(t))\,dt = -\int_0^T \left(B^* p(t), v(t)\right) dt\,.$$

Therefore, condition **(8.3)** becomes

$$\int_0^T \left(\mu\,\tilde{u}(t) - B^* p(t), v(t)\right) dt = 0 \text{ for any } v.$$

Choosing $v = \mu\,\tilde{u} - B^* p$, we obtain

$$\int_0^T \left|\mu\,\tilde{u}(t) - B^* p(t)\right|^2 dt = 0\,,$$

hence $\tilde{u}(t) = \dfrac{1}{\mu} B^* p(t)$ for any $t$.

The result obtained leads us to state

**Theorem 8.1** *Let* $\tilde{u}$ *be the optimal control and* $\tilde{x}$ *- the corresponding solution of the problem defined by* **(8.1)** *and* **(8.2)**. *Let* $p$ *be the solution of the problem*

$$p' = -A^* p + Q\,\tilde{x}(t)\,,\ p(T) = -G\,\tilde{x}(T)\,.$$

*Then the optimal control is given by*

$$\tilde{u}(t) = \frac{1}{\mu} B^* p(t)\,.$$

**Corollary 8.2** *The optimal solution is defined by the solution of the two-point boundary value problem*

$$x' = A\,x + \frac{1}{\mu} B\,B^* p$$

$$p' = -A^* p + Q\,x$$

$$x(0) = x_0\,,\ p(T) = -G\,x(T)\,.$$

Thus, we have been led to the study of such problems. We shall approach such a problem by trying to *reduce* it to *Cauchy problems*.

To this end, we look for a matrix-valued function $R$, such that $p(t) = -R(t)\,x(t)$, where $(x,p)$ is a solution of the boundary value problem. We have

$$p'(t) = -R'(t)\,x(t) - R(t)\,x'(t) =$$

$$= -R'(t)\,x(t) - R(t)\left[A\,x(t) - \frac{1}{\mu} B\,B^{*}R(t)\,x(t)\right].$$

On the other hand,

$$p'(t) = -A^{*}p(t) + Q\,x(t) = A^{*}R(t)\,x(t) + Q\,x(t).$$

We deduce that, if $(x,p)$ is a solution of the boundary value problem, then $R$ satisfies

$$\begin{cases} R'(t) + R(t)\,A + A^{*}R(t) - \dfrac{1}{\mu} R(t)\,B\,B^{*}R(t) + Q = 0 \\ R(t) = G \end{cases} \tag{8.4}$$

(Matrix Riccati Differential Equation)

Conversely, if $R$ is a solution of the Cauchy problem (8.4), if $x$ is a solution of the problem

$$x' = \left[A - \frac{1}{\mu} B\,B^{*}R(t)\right]x\,,\ x(0) = x_0$$

and $p(t) = -R(t)\,x(t)$, then the pair $(x,p)$ is a solution of the boundary value problem.

In this way, the solution of the boundary value problem is reduced to the solution of two Cauchy problems: one for $R$ and another for $x$. We remark that, for $R$, the final value at $T$ is given and we look for the solution for $t < T$, while for $x$ the value at 0 (initial value) is given and we look for the solution for $t > 0$.

Let us make the following important remark: equation (8.4)

is a nonlinear equation (in fact, a system of $\frac{n(n+1)}{2}$ equations, since $R(t)$ is a symmetric matrix). For such a system, the fundamental existence and uniqueness theorem has a *local* character. We shall obtain the solution of the boundary value problem only if we shall be able to prove that the solution $R$ is defined on the whole interval $0 \leq t \leq T$. We shall deduce this conclusion from the specific structure of the system, by using a general fact from the theory of differential equations: if the solution is not defined globally, then it cannot be bounded. We shall prove that $R$ is bounded, and this will imply global existence for $t < T$.

From (8.4) we deduce, by multiplication

$$e^{A^*t} R'(t)\, e^{At} + e^{A^*t} R(t)\, A\, e^{At} + e^{A^*t} A^*\, R(t)\, e^{At} -$$

$$- \frac{1}{\mu} e^{A^*t} R(t)\, B\, B^* R(t)\, e^{At} + e^{A^*t} Q\, e^{At} = 0,$$

what can be written as

$$\frac{d}{dt}\left[e^{A^*t} R(t)\, e^{At}\right] - \frac{1}{\mu} e^{A^*t} R(t)\, B\, B^* R(t)\, e^{At} + e^{A^*t} Q\, e^{At} = 0;$$

integrating from $t$ to $T$, we obtain

$$e^{A^*T} R(T)\, e^{AT} - e^{A^*t} R(t)\, e^{At} + \int_t^T e^{A^*\tau} Q\, e^{A\tau}\, d\tau -$$

$$- \frac{1}{\mu} \int_t^T e^{A^*\tau} R(\tau)\, B\, B^* R(\tau)\, e^{A\tau}\, d\tau = 0.$$

Premultiplying by $e^{-A^*t}$ and postmultiplying by $e^{-At}$, we get

$$R(t) + \frac{1}{\mu} \int_t^T e^{A^*(\tau - t)} R(\tau)\, B\, B^* R(\tau)\, e^{A(\tau - t)}\, d\tau =$$

$$= e^{A^*(T-t)} R(T)\, e^{A(T-t)} + \int_t^T e^{A^*(\tau - t)} Q\, e^{A(\tau - t)}\, d\tau.$$

Since $R(T) \geq 0$, $Q \geq 0$ and the right-hand side remains bounded for $0 \leq t \leq T$, we deduce that $R(t)$ is *bounded above* for $0 \leq t \leq T$.

We shall also show that $R(t) \geq 0$ for $0 \leq t \leq T$. To do so, we rewrite equation **(8.4)** (the matrix Riccati equation) in the form

$$R' + R\left(A - \frac{1}{\mu} B B^* R\right) + \left(A^* - \frac{1}{\mu} R B B^*\right) R +$$

$$+ \frac{1}{\mu} R B B^* R + Q = 0 .$$

We consider the matrix-valued function $X(t,T)$ satisfying

$$X(T,T) = I, \; \frac{d}{dt}[X(t,T)] = \left[A - \frac{1}{\mu} B B^* R(t)\right] X(t,T).$$

Using this function, we deduce

$$X^*(t,T) R'(t) X(t,T) + X^*(t,T)\left[A - \frac{1}{\mu} B B^* R(t)\right] X(t,T) +$$

$$+ X^*(t,T)\left[A^* - \frac{1}{\mu} R(t) B B^*\right] R(t) X(t,T) +$$

$$+ \frac{1}{\mu} X^*(t,T) R(t) B B^* R(t) X(t,T) + X^*(t,T) Q X(t,T) = 0,$$

what can be rewritten as

$$\frac{d}{dt}\left[X^*(t,T) R(t) X(t,T)\right] +$$

$$+ X^*(t,T)\left[\frac{1}{\mu} R(t) B B^* R(t) + Q\right] X(t,T) = 0 .$$

We integrate again from $t$ to $T$ and obtain

$$R(T) - X^*(t,T) R(t) X(t,T) +$$

$$+ \int_t^T X^*(\tau,T)\left[\frac{1}{\mu} R(\tau) B B^* R(\tau) + Q\right] X(\tau,T) d\tau = 0 .$$

Due to a general property of linear systems, the matrix $X(t,T)$ is invertible and $X^{-1}(t,T) = X(T,t)$. We deduce

$$R(t) = \left[X^*(t,T)\right]^{-1} R(T)\left[X(t,T)\right]^{-1} +$$

$$+ \left[X^*(t,T)\right]^{-1}\left[\int_t^T X^*(\tau,T)\left[\frac{1}{\mu} R(\tau)\,B\,B^* R(\tau) +\right.\right.$$

$$\left.\left. + Q\right] X(\tau,T)\,d\tau\right]\left[X(t,T)\right]^{-1}.$$

Since $R(T) = G \geq 0$ and $Q(t) \geq 0$, it follows herefrom that $R(t) \geq 0$.

From the inequalities obtained it follows that the solution $R$ of **(8.4)** is defined on $[\,0,T\,]$ and, taking $p(t) = -\,R(t)\,x(t)$, we obtain a solution of the boundary value problem.

Let us remark that we can also obtain the solution $R$ starting from the system

$$X' - A\,X + \frac{1}{\mu} B\,B^* P$$

$$P' = -A^*\,P + Q\,X.$$

For this system, we consider the solution of the Cauchy problem with $X(T) = I$, $P(T) = -G$. Then the function $t \mapsto X(t)$ will be *invertible* for all $0 \leq t \leq T$ and if we put $R(t) = -\,P(t)\,X^{-1}(t)$, then we see that $R(T) = G$ and that $R$ is a solution of the Riccati equation **(8.4)**.

At the same time, the optimal control $\tilde{u}$ can be written, by means of the solution $R$ of the Riccati equation, in the form

$$\tilde{u}(t) = -\frac{1}{\mu} B^* R(t)\,\tilde{x}(t).$$

We can now check that, if we choose the control in the above form, then we obtain indeed the optimal value of the cost. To this end, we shall compute the cost associated to an arbitrary control, using the solution $R$ of the Riccati equation. Thus, we have

$$J(u) = \big(R(T)\,x(T), x(T)\big) + \int_0^T \Big[\mu\,(u(t), u(t)) -$$

$$- \left(\left(R'(t) + R(t)\,A + A^*R(t) - \frac{1}{\mu}\,R(t)\,B\,B^*R(t)\right) x(t), x(t)\right)\Big]\,dt =$$

$$= \big(R(T)\,x(T), x(T)\big) + \int_0^T \mu\,(u(t), u(t))\,dt -$$

$$- \int_0^T \big[(R'(t)\,x(t), x(t)) + (A\,x(t), R(t)\,x(t)) + (R(t)\,x(t), A\,x(t)) -$$

$$- \frac{1}{\mu}\Big(B^*R(t)\,x(t), B^*R(t)\,x(t)\Big)\Big]\,dt =$$

$$= \big(R(T)\,x(T), x(T)\big) + \int_0^T \Big[\mu\,(u(t), u(t)) +$$

$$+ \frac{1}{\mu}\Big(B^*R(t)\,x(t), B^*R(t)\,x(t)\Big)\Big]\,dt -$$

$$- \int_0^T \big[(R'(t)\,x(t), x(t)) + (x'(t) - B\,u(t), R(t)\,x(t)) +$$

$$+ (R(t)\,x(t), x'(t) - B\,u(t))\big]\,dt =$$

$$= \big(R(T)\,x(T), x(T)\big) + \int_0^T \Big[\mu\,(u(t), u(t)) +$$

$$+ \frac{1}{\mu}\Big(B^*R(t)\,x(t), B^*R(t)\,x(t)\Big) + \Big(B^*R(t)\,x(t), u(t)\Big) +$$

$$+ \Big(u(t), B^*R(t)\,x(t)\Big)\Big]\,dt - \int_0^T \frac{d}{dt}\,(R(t)\,x(t), x(t))\,dt.$$

Finally, we obtain

$$J(u) = \left(R(0)\,x_0, x_0\right) + \int_0^T \left(\sqrt{\mu}\,u(t) + \frac{1}{\sqrt{\mu}}\,B^* R(t)\,x(t), \sqrt{\mu}\,u(t) + \right.$$

$$\left. + \frac{1}{\sqrt{\mu}}\,B^* R(t)\,x(t)\right) dt\,.$$

We deduce that $J(u) \geq \left(R(0)\,x_0, x_0\right)$, equality being reached for

$$\tilde{u}(t) = -\frac{1}{\mu}\,B^* R(t)\,x(t)\,.$$

Let us summarize the conclusions in:

**Theorem 8.3** *The Cauchy problem*

$$R' + R\,A + A^* R - \frac{1}{\mu}\,R\,B\,B^* R + Q = 0, \;\; R(T) = G$$

*admits a solution $R$ defined for $t \in \left[\,0, T\,\right]$.*

*If we consider the solution $\tilde{x}$ of the Cauchy problem*

$$x' = \left[A - \frac{1}{\mu}\,B\,B^* R(t)\right] x\,, \;\; x(0) = x_0$$

*and define the control*

$$\tilde{u}(t) = -\frac{1}{\mu}\,B^* R(t)\,\tilde{x}(t)\,,$$

*then $\tilde{u}$ minimizes the cost-function $J(u)$.*

## 8.2 Applications

**8.2.1** As in **7.2**, we consider the price evolution model

$$p' = \lambda\left[\,(b - a)\,p + \beta - \alpha\,\right], \tag{8.5}$$

assuming that the term $\beta - \alpha$ can be controlled by a decision maker; finally, we get the evolution

$$x' = \lambda\left[(b - a)\,x + v\right],$$

where $x$ and $v$ are deviations from equilibrium (see **7.2**).

We wish to choose $v$ so that to minimize a cost of the form

$$J(v) = g\, x^2(T) + \int_0^T \left[q\, x^2(t) + \mu\, v^2(t)\right] dt\,,\ g \geq 0,\ q \geq 0. \tag{8.6}$$

According to Theorem 8.3, the optimal control correction will be given by

$$\tilde{v}(t) = -\frac{\lambda}{\mu}\, r(t)\, x(t)\,,$$

where $r$ is a solution of the problem

$$r' + 2\lambda\,(b - a)\, r - \frac{\lambda^2}{\mu} r^2 + q = 0,\ r(T) = g. \tag{8.7}$$

This problem can be solved so that we obtain a simple formula for the solution. To this end, we shall describe the general method of solving scalar Riccati equations.

We consider a differential equation of the form

$$x' = a(t)\, x^2 + b(t)\, x + c(t)\,.$$

We assume that a particular solution $\tilde{x}$ of this equation has been found and that $\tilde{x}$ is defined in an interval $(t_0, t_1)$.

By means of this solution, we shall be able to represent any other solution of the equation, using the following procedure. We denote by $x$ an arbitrary solution, different from $\tilde{x}$, and define $y = x - \tilde{x}$. We have

$$y' = x' - \tilde{x}' = a\, x^2 + b\, x + c - a\, \tilde{x}^2 - b\, \tilde{x} - c =$$

$$= a\,(x - \tilde{x})(x + \tilde{x}) + b\,(x - \tilde{x}) = a\, y\,(y + 2\,\tilde{x}) + b\, y\,,$$

hence $y' = (b + 2\,a\,\tilde{x})\, y + a\, y^2$.

We see that $y$ satisfies an equation of the same type, but without the term of degree zero in the polynomial defining the equation. Such an equation is called *Bernoulli type equation*.

Since, by assumption, $y \neq 0$, we divide by $-y^2$ and obtain

$$-\frac{y'}{y^2} = -(b + 2a\tilde{x})\frac{1}{y} - a.$$

We denote now $z = \frac{1}{y}$ and deduce $z' = -(b + 2a\tilde{x})z - a$, hence $z$ satisfies an affine differential equation. Therefore, we can write

$$z(t) = e^{-\int_{t_0}^{t}[b(s)+2a(s)\tilde{x}(s)]\,ds} - \int_{t_0}^{t} e^{-\int_{\tau}^{t}[b(s)+2a(s)\tilde{x}(s)]\,ds} a(\tau)\,d\tau$$

and, further,

$$x(t) = \tilde{x}(t) + \frac{1}{z(t)},$$

what gives a representation of the solution $x$ by means of the given solution $\tilde{x}$.

In the special case of the Riccati equation associated to the optimal price evolution problem, we can easily find even two particular solutions, namely the <u>*constant*</u> solutions, given by

$$\frac{\lambda^2}{\mu}r^2 - 2\lambda(b-a)r - q = 0. \tag{8.8}$$

If the final value $r(T)$ equals one of the roots (the positive one), then, by uniqueness, the corresponding solution will coincide with the constant one. In the following, we shall assume that this is not the case.

We denote by $r_1, r_2$ the roots of equation **(8.8)**. These roots are real and of opposite signs, since their product $-\frac{q\mu}{\lambda^2}$ is negative. Suppose $r_1 < 0$, $r_2 > 0$.

We define $y(t) = r(t) - r_1$. Then $y' = r'$; substituting into **(8.7)**, we obtain

$$y' + 2\lambda(b-a)(y+r_1) - \frac{\lambda^2}{\mu}(y+r_1)^2 + q =$$

$$= y' + \left[2\lambda(b-a) - \frac{2\lambda^2}{\mu} r_1\right] y - \frac{\lambda^2}{\mu} y^2 +$$

$$+ \left(-\frac{\lambda^2}{\mu} r_1^2 + 2\lambda(b-a) r_1 + q\right) =$$

$$= y' + \left[2\lambda(b-a) - \frac{2\lambda^2}{\mu} r_1\right] y - \frac{\lambda^2}{\mu} y^2 = 0 .$$

Dividing by $y^2$, we obtain

$$-\frac{y'}{y^2} = \left[2\lambda(b-a) - \frac{2\lambda^2}{\mu} r_1\right] y - \frac{\lambda^2}{\mu}. \tag{8.9}$$

From **(8.8)**, we deduce $r_1 + r_2 = \dfrac{2\lambda(b-a)\mu}{\lambda^2}$, hence, denoting

$2\lambda(b-a) - \dfrac{2\lambda^2}{\mu} = \rho$, we have

$$\rho = \frac{\lambda^2}{\mu}(r_1 + r_2 - 2r_1) = \frac{\lambda^2}{\mu}(r_2 - r_1);$$

finally, denoting $r_2 - r_1 = n$, we have $\rho = \dfrac{\lambda^2}{\mu} n$.

Denoting $z = \dfrac{1}{y}$, we obtain from **(8.9)**

$$z' = \rho z - \frac{\lambda^2}{\mu} = \frac{\lambda^2}{\mu}[n z - 1]. \tag{8.10}$$

Equation **(8.10)** is an affine equation, with constant coefficients, having the particular solution $\hat{z} = \dfrac{1}{n}$. The associated linear equation $z' = \rho z$ has the general solution $\tilde{z}(t) = K e^{\rho t}$, hence the general solution of **(8.10)** can be written as

$$z(t) = K e^{\rho t} + \frac{1}{n}. \tag{8.11}$$

The final condition $r(T) = g$ implies $z(T) = \dfrac{1}{g - r_1}$.

Herefrom, we obtain

$$K = e^{-\rho T}\left[\frac{1}{g-r_1} - \frac{1}{n}\right] = e^{-\rho T}\,\frac{r_2 - g}{n\left(g - r_1\right)}. \tag{8.12}$$

Returning to the variable $r(t)$, we obtain

$$r(t) = \frac{1}{K\,e^{\rho t} + \dfrac{1}{n}} + r_1. \tag{8.13}$$

According to the general theory, the optimal solution of the problem defined by **(8.5)** and **(8.6)** is given by the solution of the Cauchy problem

$$\tilde{x}' = \left[\lambda\,(b-a) - \frac{\lambda^2}{\mu}\,r(t)\right]x\,;\ x(0) = x_0, \tag{8.14}$$

while $\tilde{v}(t) = -\dfrac{\lambda}{\mu}\,r(t)\,\tilde{x}(t)$.

To write the explicit form for $\tilde{x}(t)$, we distinguish two cases: **a)** $r_2 = g$, **b)** $r_2 \neq g$.

In case **a)**, we have $r(t) \equiv r_2 = g$ and equation **(8.14)** becomes

$$x' = \left[\lambda\,(b-a) - \frac{\lambda^2}{\mu}\,r_2\right]x, \tag{8.15}$$

which is a linear differential equation with constant coefficients. We remark that

$$\lambda\,(b-a) - \frac{\lambda^2}{\mu}\,r_2 = \frac{\lambda^2}{2\,\mu}\left(r_1 + r_2 - 2\,r_2\right) = -\frac{\lambda^2}{2\,\mu}\,n = -\frac{\rho}{2}.$$

Therefore, the solution of **(8.15)** will be

$$\tilde{x}(t) = x_0\,e^{-\frac{\rho}{2}t} = x_0\,e^{-\frac{\lambda^2 n}{2\mu}t}. \tag{8.16}$$

In case **b)**, equation **(8.14)** becomes

$$x' = \left[\lambda (b-a) - \frac{\lambda^2}{\mu} r_1 - \frac{\lambda^2}{\mu} \frac{1}{K e^{\rho t} + \frac{1}{n}}\right] x =$$

$$= \left[\frac{\rho}{2} - \frac{\lambda^2}{\mu} \frac{1}{K e^{\rho t} + \frac{1}{n}}\right] x . \tag{8.17}$$

This is a linear equation, whose solution is

$$x(t) = x_0 \, e^{\int_0^t \left[\frac{\rho}{2} - \frac{\lambda^2}{\mu} \frac{1}{K e^{\rho \tau} + \frac{1}{n}}\right] d\tau} =$$

$$= x_0 \, e^{\frac{\rho}{2} t} \, e^{\int_0^t \left[-\frac{\lambda^2}{\mu} \frac{1}{K e^{\rho \tau} + \frac{1}{n}}\right] d\tau} . \tag{8.18}$$

To compute the integral

$$I = \int \frac{-\frac{\lambda^2}{\mu} d\tau}{K e^{\rho \tau} + \frac{1}{n}} = \int \frac{-\frac{\lambda^2}{\mu} e^{-\rho \tau} d\tau}{K + \frac{e^{-\rho \tau}}{n}},$$

we perform the change of variables $u = \frac{e^{-\rho \tau}}{n}$ and obtain

$$I = \int \frac{du}{K + u} = \ln |K + u| .$$

Returning to the variable $\tau$, we have

$$I = \ln \left| K + \frac{e^{-\rho \tau}}{n} \right|, \text{ and}$$

$$\int_0^t \frac{-\frac{\lambda^2}{\mu} d\tau}{K e^{\rho \tau} + \frac{1}{n}} = \ln \left| K + \frac{e^{-\rho \tau}}{n} \right| \Bigg|_0^t =$$

$$= \ln \left| \frac{K + \frac{e^{-\rho t}}{n}}{K + \frac{1}{n}} \right| = \ln \left| \frac{n K + e^{-\rho t}}{n K + 1} \right| .$$

Keeping in mind that $nK = e^{-\rho T}\dfrac{r_2 - g}{g - r_1}$, we can conclude that, whatever $T \geq t \geq 0$,

$$nK + e^{-\rho t} = e^{-\rho T}\frac{r_2 - g}{g - r_1} + e^{-\rho t} \geq$$

$$\geq e^{-\rho T}\left[\frac{r_2 - g}{g - r_1} + 1\right] = e^{-\rho T}\frac{n}{g - r_1} > 0,$$

hence

$$\left|\frac{nK + e^{-\rho t}}{nK + 1}\right| = \frac{nK + e^{-\rho t}}{nK + 1} = \frac{e^{-\rho T}(r_2 - g) + e^{-\rho t}(g - r_1)}{e^{-\rho T}(r_2 - g) + g - r_1}.$$

Returning to relation **(8.18)**, we obtain

$$\widetilde{x}(t) = x_0\, e^{\frac{\rho}{2}t} \cdot \frac{e^{-\rho T}(r_2 - g) + e^{-\rho t}(g - r_1)}{e^{-\rho T}(r_2 - g) + g - r_1} =$$

$$= x_0 \cdot \frac{e^{-\rho\left(T - \frac{t}{2}\right)}(r_2 - g) + e^{-\frac{\rho}{2}t}(g - r_1)}{e^{-\rho T}(r_2 - g) + g - r_1}. \qquad \textbf{(8.19)}$$

**8.2.2** We consider the Phillips model with multiplier, discussed in **1.6**. The evolution of product (income) $Y$ is given by

$$Y' = -\lambda s Y + \lambda A, \text{ where } \lambda > 0,\ s = 1 - c > 0.$$

If $A$ is constant, equal to $\hat{A}$, then the equilibrium is given by $\hat{Y} = \dfrac{\hat{A}}{s}$ and it is <u>*stable*</u>, since $\lambda s > 0$. The choice of $A$ at the level $\hat{A}$ determines a value $\hat{Y}$ of the product. Although this equilibrium value is stable, we can try to improve the growth performance by using corrections on $A$.

Denoting $Y - \hat{Y} = x$, $A - \hat{A} = u$, it follows

$$x' = -\lambda s x + \lambda u.$$

We associate a cost of the form:

$$J(u) = g\,x^2(T) + \int_0^T \left[q\,x^2(t) + \mu\,u^2(t)\right] dt\,.$$

According to the general theory, we have an optimal feed-back control of the form

$$\tilde{u}(t) = -\frac{1}{\mu}\lambda\,r(t)\,\tilde{x}(t),$$

where $r$ is the solution of the problem

$$r' - 2\lambda s r - \frac{\lambda^2}{\mu} r^2 + q = 0\,,\ r(T) = g$$

and $\tilde{x}$ - the solution of the problem

$$x' = -\left(\lambda s + \frac{\lambda^2}{\mu} r(t)\right) x\,,\ x(0) = x_0\,.$$

**8.2.3** We consider now the Phillips model with accelerator, discussed in **2.4**. The model is described by the system

$$Y' = \lambda\,(I + A - sY)$$

$$I' = -\,k\left[I - vY'\right].$$

We introduce a new variable $i = I - k v Y$. We have

$$i' = -\,k I = -\,k\,(i + k v Y),$$

$$Y' = \lambda\,(k v - s)\,Y + \lambda\,i + \lambda\,A\,.$$

If $A = \hat{A}$, then the equilibrium is given by $\hat{I} = 0$, $\hat{Y} = \dfrac{\hat{A}}{s}$, of the same type as that in the above model. As above, we denote

$$u = A - \hat{A}\,,\quad x_1 = i + \frac{k v}{s}\hat{A}\,,\quad x_2 = Y - \frac{\hat{A}}{s}$$

and we obtain the system

$$\begin{cases} x_1' = -\,k\,x_1 - k^2\,v\,x_2 \\ x_2' = \lambda\,x_1 + \lambda\,(k v - s)\,x_2 + \lambda\,u\,, \end{cases}$$

having the matrix $\begin{pmatrix} -k & -k^2 v \\ \lambda & \lambda(kv-s) \end{pmatrix}$.

We associate a quadratic cost in which we impose a penalty only on the deviation of the product from equilibrium and on the control effort, by choosing

$$G = \begin{pmatrix} g & 0 \\ 0 & 0 \end{pmatrix}, \quad Q = \begin{pmatrix} q & 0 \\ 0 & 0 \end{pmatrix}.$$

This time, the matrix Riccati equation will represent a system of three nonlinear differential equations; for solving it, we can apply a numerical procedure.

We can also associate the fourth order system

$$\begin{pmatrix} x_1 \\ x_2 \end{pmatrix}' = \begin{pmatrix} -k & -k^2 v \\ \lambda & \lambda(kv-s) \end{pmatrix}\begin{pmatrix} x_1 \\ x_2 \end{pmatrix} + \frac{1}{\mu}\begin{pmatrix} 0 \\ \lambda \end{pmatrix}(0 \quad \lambda)\begin{pmatrix} p_1 \\ p_2 \end{pmatrix}$$

$$\begin{pmatrix} p_1 \\ p_2 \end{pmatrix}' = \begin{pmatrix} k & -\lambda \\ k^2 v & -\lambda(kv-s) \end{pmatrix}\begin{pmatrix} p_1 \\ p_2 \end{pmatrix} + \begin{pmatrix} q & 0 \\ 0 & 0 \end{pmatrix}\begin{pmatrix} x_1 \\ x_2 \end{pmatrix}.$$

This is a linear system with constant coefficients; we shall define, according to the general scheme, the solutions satisfying the conditions

$$\begin{pmatrix} \tilde{x}_1(T) \\ \tilde{x}_2(T) \\ \tilde{p}_1(T) \\ \tilde{p}_2(T) \end{pmatrix} = \begin{pmatrix} 1 \\ 0 \\ -g \\ 0 \end{pmatrix}; \quad \begin{pmatrix} \hat{x}_1(T) \\ \hat{x}_2(T) \\ \hat{p}_1(T) \\ \hat{p}_2(T) \end{pmatrix} = \begin{pmatrix} 0 \\ 1 \\ 0 \\ 0 \end{pmatrix}$$

and we shall have

$$R(t) = -\begin{pmatrix} \tilde{p}_1(t) & \hat{p}_1(t) \\ \tilde{p}_2(t) & \hat{p}_2(t) \end{pmatrix}\begin{pmatrix} \tilde{x}_1(t) & \hat{x}_1(t) \\ \tilde{x}_2(t) & \hat{x}_2(t) \end{pmatrix}^{-1}.$$

The optimal control will be $\tilde{u}(t) = -\frac{1}{\mu}(0 \quad \lambda)\,R(t)\,x(t)$.

To conclude, let us remark that, for the system with variables $x_1$, $x_2$, $p_1$, $p_2$, the characteristic equation has a special form. This property is generally valid, since the matrix of the system has the structure

$$\begin{pmatrix} A & \frac{1}{\mu}BB^* \\ Q & -A^* \end{pmatrix}.$$

Indeed, we have

$$\pi(r) = det\begin{pmatrix} rI - A & -\frac{1}{\mu}BB^* \\ -Q & rI + A^* \end{pmatrix},$$

$$\pi(-r) = det\begin{pmatrix} -rI - A & -\frac{1}{\mu}BB^* \\ -Q & -rI + A^* \end{pmatrix}.$$

Taking into account the properties of determinants, we deduce

$$\pi(-r) = det\begin{pmatrix} rI + A & \frac{1}{\mu}BB^* \\ -Q & -rI + A^* \end{pmatrix}(-1)^n =$$

$$= det\begin{pmatrix} rI + A & -\frac{1}{\mu}BB^* \\ -Q & rI - A^* \end{pmatrix} =$$

$$= det\begin{pmatrix} rI + A^* & -Q \\ -\frac{1}{\mu}BB^* & rI - A \end{pmatrix};$$

further, we deduce that $\pi(-r) = 0$ is equivalent to

$$det\begin{pmatrix} -Q & rI + A^* \\ rI - A & -\frac{1}{\mu}BB^* \end{pmatrix} = 0,$$

what is equivalent to

$$det\begin{pmatrix} rI - A & -\frac{1}{\mu}BB^* \\ -Q & rI + A^* \end{pmatrix} = 0,$$

hence to $\pi(r) = 0$.

Therefore, if $r$ is an eigenvalue, then $-r$ is also an eigenvalue, hence the characteristic equation has the special form in which the terms of odd degree are missing (biquadratic equation).

## 8.3 Discrete Time

We consider now the discrete-time analogue of the problem studied in **8.1**.

We have the dynamics

$$x_{t+1} = A_t x_t + B_t u_t \tag{8.20}$$

and the cost

$$J(u) = (G x_T, x_T) + \sum_{t=t_0}^{T-1} (Q_t x_t, x_t) + (R_t u_t, u_t), \tag{8.21}$$

with the natural assumptions $G \geq 0$, $Q_t \geq 0$, $R_t \geq \rho I$, $\rho > 0$.

We associate to this problem the nonlinear system with unknowns $V_t$, $W_t$, $P_t$

$$R_t + B_t^* P_{t+1} B_t = V_t^2$$

$$A_t^* P_{t+1} B_t = W_t^* V_t$$

$$P_t = A_t^* P_{t+1} A_t + Q_t - W_t^* W_t$$

$$P_T = G \geq 0.$$

We show, by backward induction, that the solution of this

system is defined for all $t$ $(t = 0,\dots,T-1)$.

We assume that $P_{t+1} \geq 0$. Then $R_t + B_t^* P_{t+1} B_t \geq \rho I$ and we can find a symmetric, <u>invertible</u> $V_t$ such that

$$V_t^2 = R_t + B_t^* P_{t+1} B_t .$$

Further, we put $W_t^* = A_t^* P_{t+1} B_t V_t^{-1}$ and then $P_t$ is determined from the last equation. To be able to continue the reasoning, we have to prove that the matrix $P_t$ determined in this way satisfies $P_t \geq 0$.

Let us define $F_t = -V_t^{-1} W_t$. We calculate:

$$\left(A_t^* + F_t^* B_t^*\right) P_{t+1} \left(A_t + B_t F_t\right) =$$

$$= A_t^* P_{t+1} A_t + F_t^* B_t^* P_{t+1} A_t + A_t^* P_{t+1} B_t F_t + F_t^* B_t^* P_{t+1} B_t F_t =$$

$$= A_t^* P_{t+1} A_t + F_t^* V_t W_t + W_t^* V_t F_t + F_t^* \left(V_t^2 - R_t\right) F_t .$$

But $V_t F_t = -W_t$, $F_t^* V_t = -W_t^*$ and we can write

$$\left(A_t^* + F_t^* B_t^*\right) P_{t+1} \left(A_t + B_t F_t\right) =$$

$$= A_t^* P_{t+1} A_t - W_t^* W_t - W_t^* W_t + W_t^* W_t - F_t^* R_t F_t =$$

$$= P_t - Q_t - F_t^* R_t F_t ,$$

hence

$$P_t = \left(A_t + B_t F_t\right)^* P_{t+1} \left(A_t + B_t F_t\right) + Q_t + F_t^* R_t F_t ,$$

what shows that $P_t \geq 0$.

Therefore, the inductive process may be continued and we conclude that the system admits a solution with $P_t \geq 0$ and symmetric, invertible $V_t$.

We can use this solution in order to write the cost in a convenient way. Thus, we have

$$J(u) = \left(P_T\, x_T, x_T\right) + \sum_{t=t_0}^{T} \Big[\Big(\big(P_t - A_t^* P_{t+1} A_t + W_t^* W_t\big)\, x_t, x_t\Big) +$$

$$+ \Big(\big(V_t^2 - B_t^* P_{t+1} B_t\big)\, u_t, u_t\Big)\Big] =$$

$$= \left(P_T\, x_T, x_T\right) + \sum_{t=t_0}^{T-1} \big[(P_t\, x_t, x_t) - (P_{t+1}\, A_t\, x_t, A_t\, x_t) +$$

$$+ (W_t\, x_t, W_t\, x_t) + (V_t\, u_t, V_t\, u_t) - (P_{t+1}\, B_t\, u_t, B_t\, u_t)\big] =$$

$$= \left(P_T\, x_T, x_T\right) + \sum_{t=t_0}^{T-1} \big[(P_t\, x_t, x_t) -$$

$$- \big(P_{t+1}\,(x_{t+1} - B_t\, u_t), (x_{t+1} - B_t\, u_t)\big) +$$

$$+ (W_t\, x_t, W_t\, x_t) + (V_t\, u_t, V_t\, u_t) - (P_{t+1}\, B_t\, u_t, B_t\, u_t)\big] =$$

$$= \left(P_{t_0}\, x_{t_0}, x_{t_0}\right) + \sum_{t=t_0}^{T-1} \big[(P_{t+1}\, B_t\, u_t, x_{t+1}) + (P_{t+1}\, x_{t+1}, B_t\, u_t) -$$

$$- 2\,(P_{t+1}\, B_t\, u_t, B_t\, u_t) + (W_t\, x_t, W_t\, x_t) + (V_t\, u_t, V_t\, u_t)\big] =$$

$$= \left(P_{t_0}\, x_{t_0}, x_{t_0}\right) + \sum_{t=t_0}^{T-1} \big[(P_{t+1}\, B_t\, u_t, A_t\, x_t + B_t\, u_t) +$$

$$+ (P_{t+1}\, A_t\, x_t + P_{t+1}\, B_t\, u_t, B_t\, u_t) -$$

$$- 2\,(P_{t+1}\, B_t\, u_t, B_t\, u_t) + (W_t\, x_t, W_t\, x_t) + (V_t\, u_t, V_t\, u_t)\big] =$$

$$= \left(P_{t_0}\, x_{t_0}, x_{t_0}\right) + \sum_{t=t_0}^{T-1} \Big[\big(u_t, B_t^* P_{t+1}\, A_t\, x_t\big) + \big(B_t^* P_{t+1}\, A_t\, x_t, u_t\big) +$$

$$+ (W_t\, x_t, W_t\, x_t) + (V_t\, u_t, V_t\, u_t)\Big] =$$

$$= \left(P_{t_0}\, x_{t_0}, x_{t_0}\right) + \sum_{t=t_0}^{T-1} \big[(u_t, V_t\, W_t\, x_t) + (V_t\, W_t\, x_t, u_t) +$$

$$+ (W_t\, x_t, W_t\, x_t) + (V_t\, u_t, V_t\, u_t)\big] =$$

$$= \left(P_{t_0}\, x_{t_0}, x_{t_0}\right) + \sum_{t=t_0}^{T-1} \left(V_t\, u_t + W_t\, x_t, V_t\, u_t + W_t\, x_t\right).$$

The last expression for the cost shows that $J(u) \geq \left(P_{t_0} x_{t_0}, x_{t_0}\right)$ and that equality is obtained if we choose

$$u_t = -V_t^{-1} W_t x_t = F_t x_t . \tag{8.22}$$

In this way, the linear feed-back control **(8.22)** minimizes the cost.

## 8.4 Applications in Discrete Time

We shall discuss now the discrete time analogues of the models studied in **8.2**.

**8.4.1** The simple price evolution model is the same as that discussed in **7.4**:

$$p_{t+1} = \left[1 - \lambda(b-a)\right] p_t + \lambda(\alpha - \beta).$$

Replacing the term $(\alpha - \beta)$ by a control variable and then performing a change of variables (see **7.4**), we obtain the dynamics

$$x_{t+1} = \left[1 - \lambda(b-a)\right] x_t + \lambda u_t ,$$

where $x_t$ represents the deviation of $p_t$ from the equilibrium and $u_t$-the deviation of the control variable from the value corresponding to the equilibrium.

We associate a cost of the form

$$J(u) = g\, x_T^2 + \sum_{t=0}^{T-1} \left(q^2 x_t^2 + r^2 u_t^2\right).$$

The corresponding nonlinear system is

$$r^2 + \lambda^2 P_{t+1} = V_t^2$$

$$\lambda\left[1 - \lambda(b-a)\right] P_{t+1} = W_t V_t$$

$$P_t = [1 - \lambda(b-a)]^2 P_{t+1} + q^2 - W_t^2 .$$

It follows

$$V_t = \sqrt{r^2 + \lambda^2 P_{t+1}}, \quad W_t = \frac{\lambda[1 - \lambda(b-a)]}{\sqrt{r^2 + \lambda^2 P_{t+1}}} P_{t+1},$$

hence

$$P_t = [1 - \lambda(b-a)]^2 P_{t+1} + q^2 - \frac{\lambda^2 [1 - \lambda(b-a)]^2}{r^2 + \lambda^2 P_{t+1}} P_{t+1}^2 .$$

Given $P_T = g$, we have a recurrence relation for determining the sequence $\{P_t\}$, then the sequences $\{V_t\}$ and $\{W_t\}$.

The optimal control is $u_t = F_t x_t$, where

$$F_t = -\frac{W_t}{V_t} = -\frac{\lambda[1 - \lambda(b-a)] P_{t+1}}{r^2 + \lambda^2 P_{t+1}}$$

and, in this way, we have a procedure for determining the linear feed-back which gives optimal control correction.

We remark that the determination of the optimal solution $\tilde{x}_t$ $(t = 1,\dots,T)$ and $\tilde{u}_t$ $(t = 0,\dots,T-1)$ comprises two phases.

In the first phase, we determine, by backward recurrence, the values $F_{T-1}, F_{T-2}, \dots, F_0$, using the procedure described above.

In the second phase, we determine, starting from $t = 0$, the optimal control $\tilde{u}_0 = F_0 x_0$, then

$$\tilde{x}_1 = [1 - \lambda(b-a)] x_0 + \lambda \tilde{u}_0 ,$$

and, in general, for $t < T$

$$\tilde{u}_t = F_t \tilde{x}_t , \quad \tilde{x}_{t+1} = [1 - \lambda(b-a)] \tilde{x}_t + \lambda \tilde{u}_t .$$

**8.4.2** We treat in the same way the discrete analogue of the Phillips model:

$$Y_{t+1} - Y_t = \lambda (Z_t - Y_t) \tag{8.23}$$

$$Z_t = C_t + A_t \tag{8.24}$$

$$C_t = c Y_t . \tag{8.25}$$

If $A_t = \hat{A}$ constant, then we have the equilibrium $\hat{Y} = \frac{\hat{A}}{s}$, where $s = 1 - c$.

If we denote $x_t = Y_t - \hat{Y}$, $u_t = A_t - \hat{A}$, then we have

$x_{t+1} - x_t = \lambda (c x_t - x_t + u_t)$ i.e.

$x_{t+1} = (1 - \lambda s) x_t + \lambda u_t$.

We associate again a quadratic cost of the form

$$J(u) = g x_T^2 + \sum_{t=0}^{T-1} \left(q^2 x_t^2 + r^2 u_t^2\right)$$

and the optimal solution can be found as in **8.4.1**.

**8.4.3** We consider the discrete analogue of the Phillips model with accelerator. By taking into account investments, we have, besides relations **(8.23)**, **(8.25)**, the equations

$Z_t = C_t + I_t + A_t$,

$I_{t+1} - I_t = -k\left[I_t - v(Y_{t+1} - Y_t)\right]$.

For $A_t = \hat{A}$, we have again the equilibrium $\hat{I} = 0$, $\hat{Y} = \frac{\hat{A}}{s}$.

Performing the change of variables $y_t = Y_t - \hat{Y}$, $u_t = A_t - \hat{A}$, we get the dynamics

$y_{t+1} = (1 - \lambda s) y_t + \lambda I_t + \lambda u_t$

$$I_{t+1} = -k\,v\,\lambda\,s\,y_t + (1 + k\,v\,\lambda - k)\,I_t + \lambda\,k\,v\,u_t.$$

We associate a quadratic cost, in which we impose a penalty only on $y_t$ and $u_t$; we take

$$G = \begin{pmatrix} g & 0 \\ 0 & 0 \end{pmatrix};\; Q = \begin{pmatrix} q & 0 \\ 0 & 0 \end{pmatrix};\; R = r.$$

In this case, $P_t$ will be a $2 \times 2$ matrix.

The first equation of the associated system will be

$$r + (\lambda \quad \lambda\,v\,k)\,P_{t+1}\begin{pmatrix} \lambda \\ \lambda\,v\,k \end{pmatrix} = V_t^2.$$

The second equation is

$$\begin{pmatrix} 1 - \lambda\,s & -k\,v\,s\,\lambda \\ \lambda & 1 + k\,v\,\lambda - k \end{pmatrix} P_{t+1}\begin{pmatrix} \lambda \\ \lambda\,k\,v \end{pmatrix} = \begin{pmatrix} w_t^1 \\ w_t^2 \end{pmatrix} V_t$$

and the third equation is

$$P_t = \begin{pmatrix} 1 - \lambda\,s & -k\,v\,s\,\lambda \\ \lambda & 1 + k\,v\,\lambda - k \end{pmatrix} P_{t+1}\begin{pmatrix} 1 - \lambda\,s & \lambda \\ -k\,v\,s\,\lambda & 1 + k\,v\,\lambda - k \end{pmatrix} +$$

$$+ \begin{pmatrix} q & 0 \\ 0 & 0 \end{pmatrix} + \begin{pmatrix} w_t^1 \\ w_t^2 \end{pmatrix}\begin{pmatrix} w_t^1 & w_t^2 \end{pmatrix}.$$

The final condition is $P_T = \begin{pmatrix} g & 0 \\ 0 & 0 \end{pmatrix}$.

The optimal control will be given by

$$\tilde{u}_t = -\frac{1}{V_t}\left(w_t^1\,y_t + w_t^2\,I_t\right).$$

To illustrate the procedure of determining the optimal control, we shall perform the first step. We have

$$V_{T-1}^2 = r + (\lambda \quad \lambda\,v\,k)\begin{pmatrix} g & 0 \\ 0 & 0 \end{pmatrix}\begin{pmatrix} \lambda \\ \lambda\,v\,k \end{pmatrix} = r + g\,\lambda^2,$$

hence $V_{T-1} = \sqrt{r + g\,\lambda^2}$, then

$$\begin{pmatrix} w_{T-1}^1 \\ w_{T-1}^2 \end{pmatrix} = \frac{1}{\sqrt{r + g\lambda^2}} \begin{pmatrix} g\lambda(1-\lambda s) \\ g\lambda^2 \end{pmatrix}.$$

We have

$$F_{T-1} = -V_{T-1}^{-1} W_{T-1} = \frac{1}{r + g\lambda^2}\left(g\lambda(\lambda s - 1) \quad -g\lambda^2\right)$$

and

$$\widetilde{u}_{T-1} = \frac{1}{r + g\lambda^2}\left[g\lambda(\lambda s - 1)y - g\lambda^2 I\right].$$

Further, we calculate $P_{T-1}$ from the third equation and pass to the second step, and so on.

## 8.5 A Tracking Problem

We consider the dynamic system

$$x_{t+1} = A_t x_t + B_t u_t + F_t,$$

where $x$ is the $n$-dimensional column-vector of state variables, $u$ is the $m$-dimensional column-vector of control variables, $F$ is the $p$-dimensional column-vector of exogenous variables, $A$ is an $n \times n$ matrix and $B$ is an $n \times m$ matrix.

We associate the cost-function

$$J(u) = (x_T - \hat{x}_T)^* Q_T (x_T - \hat{x}_T) + \\ + \sum_{t=0}^{T-1}\left[(x_t - \hat{x}_t)^* Q_t (x_t - \hat{x}_t) + u_t^* K_t u_t\right],$$

where $\hat{x}_t$ are of the same dimension as $x_t$ and stand for <u>*targets*</u>, $Q_t$ are symmetric, positive definite $n \times n$ matrices and $K_t$ are symmetric, strictly positive definite $m \times m$ matrices.

The optimal control problem consists in finding a control $u = (u_0, u_1, \ldots, u_{T-1})$ which minimizes $J(u)$.

To solve this problem, we consider the auxiliary system

$\eta_{t+1} = A_t \eta_t + F_t;\ \eta_0 = 0.$

Using the solution of this system, we define the vectors

$y_t = x_t - \eta_t;\ \hat{y}_t = \hat{x}_t - \eta_t$

and obtain the linear control problem

$$y_{t+1} = A_t y_t + B_t u_t \tag{8.26}$$

$$\min_u J(u) = (y_T - \hat{y}_T)^* Q_T (y_T - \hat{y}_T) + \\ + \sum_{t=0}^{T-1} \left[ (y_t - \hat{y}_t)^* Q_t (y_t - \hat{y}_t) + u_t^* K_t u_t \right]. \tag{8.27}$$

We define the system of nonlinear matrix equations

$$\begin{cases} K_t + B_t^* S_{t+1} B_t = V_t^2 \\ A_t^* S_{t+1} B_t = W_t^* V_t \\ A_t^* S_{t+1} A_t - S_t + Q_t = W_t^* W_t, \end{cases} \tag{8.28}$$

with final condition $S_T = Q_T$, as well as the linear system

$$g_t = (A_t + B_t H_t)^* g_{t+1} + Q_t \hat{y}_t, \quad t = 0, 1, \ldots, T-1, \tag{8.29}$$

with final condition $g_T = Q_T \hat{y}_T$.

In **(8.29)**, we denoted $H_t = -V_t^{-1} W$, where $V_t$, $W_t$ are solutions of **(8.28)**.

It can be proved (by backward induction) that the system **(8.28)** has a solution $(V_t, W_t, S_t)$, $t = 0, 1, \ldots, T-1$, such that $V_t$ is strictly positive definite (hence invertible) and $S_t$ is symmetric and positive definite (the reasoning is the same as in 8.3).

We shall prove that

$$\tilde{u}_t = H_t y_t + \left(K_t + B_t^* S_{t+1} B_t\right)^{-1} B_t^* g_{t+1}, \; t = 0,1,\ldots,T-1$$

is optimal for problem **(8.27)**.

To do so, we evaluate the cost $J(u)$, using systems **(8.28)** and **(8.29)**. We have

$$J(u) = y_T^* Q_T y_T + \sum_{t=0}^{T-1}\left(y_t^* Q_t y_t + u_t^* K_t u_t\right) + \hat{y}_T^* Q_T \hat{y}_T -$$

$$- y_T^* Q_T \hat{y}_T - \hat{y}_T^* Q_T y_T + \sum_{t=0}^{T-1} \hat{y}_t^* Q_t \hat{y}_t - \sum_{t=0}^{T-1}\left(\hat{y}_t^* Q_t y_t + y_t^* Q_t \hat{y}_t\right).$$

We denote

$$\Sigma_1 = y_T^* Q_T y_T + \sum_{t=0}^{T-1}\left(y_t^* Q_t y_t + u_t^* K_t u_t\right)$$

and

$$\Sigma_2 = y_T^* Q_T \hat{y}_T + \hat{y}_T^* Q_T y_T + \sum_{t=0}^{T-1}\left(\hat{y}_t^* Q_t y_t + y_t^* Q_t \hat{y}_t\right).$$

Then we can write

$$J(u) = \Sigma_1 - \Sigma_2 + \sum_{t=0} \hat{y}_t^* Q_t \hat{y}_t$$

We remark that the last term is independent of $\tilde{u}$ , so that the optimal control $\tilde{u}$ is that which minimizes $\Sigma_1 - \Sigma_2$.

For $\Sigma_1$ we find the following evaluation

$$\Sigma_1 = y_T^* S_T y_T + \sum_{t=0}^{T-1}\left[y_t^*\left(W_t^* W_t + S_t - A_t^* S_{t+1} A_t\right) y_t + u_t^* K_t u_t\right] =$$

$$= y_T^* S_T y_T + \sum_{t=0}^{T-1}\left[y_t^* W_t^* W_t y_t + y_t^* S_t y_t -\right.$$

$$- y_t^* A_t^* S_{t+1} A_t y_t + u_t^* K_t u_t \Big] =$$

$$= y_T^* S_T y_T + \sum_{t=0}^{T-1} \Big[ y_t^* W_t^* W_t y_t + y_t^* S_t y_t -$$

$$- (y_{t+1} - B_t u_t)^* S_{t+1} (y_{t+1} - B_t u_t) + u_t^* K_t u_t \Big] =$$

$$= y_T^* S_T y_T + \sum_{t=0}^{T-1} \Big( y_t^* W_t^* W_t y_t + y_t^* S_t y_t - y_{t+1}^* S_{t+1} y_{t+1} -$$

$$- u_t^* B_t^* S_{t+1} B_t u_t + u_t^* B_t^* S_{t+1} y_{t+1} + y_{t+1}^* S_{t+1} B_t y_t + u_t^* K_t u_t \Big) =$$

$$= y_0^* S_0 y_0 + \sum_{t=0}^{T-1} \Big[ y_t^* W_t^* W_t y_t + u_t^* B_t^* S_{t+1} (A_t y_t + B_t u_t) +$$

$$+ \big( y_t^* A_t^* + u_t^* B_t^* \big) S_{t+1} B_t u_t - u_t^* B_t^* S_{t+1} B_t u_t + u_t^* K_t u_t \Big] =$$

$$= y_0^* S_0 y_0 + \sum_{t=0}^{T-1} \Big[ y_t^* W_t^* W_t y_t + u_t^* V_t^* W_t y_t +$$

$$+ y_t^* W_t^* V_t u_t + u_t^* \big( B_t^* S_{t+1} B_t + K_t \big) u_t \Big] =$$

$$= y_0^* S_0 y_0 + \sum_{t=0}^{T-1} \Big[ y_t^* W_t^* W_t y_t +$$

$$+ u_t^* V_t^* W_t y_t + y_t^* W_t^* V_t u_t + u_t^* V_t^* V_t u_t \Big] =$$

$$= y_0^* S_0 y_0 + \sum_{t=0}^{T-1} (V_t u_t + W_t y_t)^* (V_t u_t + W_t y_t) .$$

As concerns $\Sigma_2$, we have

$$\Sigma_2 = g_T^* y_T + y_T^* g_T + \sum_{t=0}^{T-1} \Big[ \Big( g_t^* - g_{t+1}^* (A_t + B_t H_t) \Big) y_t +$$

$$+ y_t^* \Big( g_t - \big( A_t^* + H_t^* B_t^* \big) g_{t+1} \Big) \Big] =$$

$$= g_T^* y_T + y_T^* g_T + \sum_{t=0}^{T-1} \Big( g_t^* y_t - g_{t+1}^* A_t y_t - g_{t+1}^* B_t H_t y_t +$$

$$+ y_t^* g_t - y_t^* A_t^* g_{t+1} - y_t^* H_t^* B_t^* g_{t+1}\Big) =$$

$$= g_T^* y_T + y_T^* g_T + \sum_{t=0}^{T-1}\left(g_t^* y_t + y_t^* g_t\right) -$$

$$- \sum_{t=0}^{T-1}\left[g_{t+1}^* \left(y_{t+1} - B_t u_t\right) + \left(y_{t+1}^* - u_t^* B_t^*\right) g_{t+1}\right] -$$

$$- \sum_{t=0}^{T-1}\left(g_{t+1}^* B_t H_t y_t + y_t^* H_t^* B_t^* g_{t+1}\right) =$$

$$= g_T^* y_T + y_T^* g_T + \sum_{t=0}^{T-1}\left(g_t^* y_t + y_t^* g_t\right) - \sum_{t=0}^{T-1}\left(g_{t+1}^* y_{t+1} + y_{t+1}^* g_{t+1}\right) +$$

$$+ \sum_{t=0}^{T-1}\left(g_{t+1}^* B_t u_t - u_t^* B_t^* g_{t+1}\right) -$$

$$- \sum_{t=0}^{T-1}\left(g_{t+1}^* B_t H_t y_t + y_t^* H_t^* B_t^* g_{t+1}\right) =$$

$$= g_0^* y_0 + y_0^* g_0 + \sum_{t=0}^{T-1}\left(g_{t+1}^* B_t u_t + u_t^* B_t^* g_{t+1}\right) -$$

$$- \sum_{t=0}^{T-1}\left(g_{t+1}^* B_t H_t y_t + y_t^* H_t^* B_t^* g_{t+1}\right).$$

Then

$$\Sigma_1 - \Sigma_2 = y_0^* S_0 y_0 + \sum_{t=0}^{T-1}\left[\left(V_t u_t + W_t y_t\right)^* \left(V_t u_t + W_t y_t\right)\right] -$$

$$- g_0^* y_0 - y_0^* g_0 - \sum_{t=0}^{T-1}\left(g_{t+1}^* B_t u_t + u_t^* B_t^* g_{t+1}\right) +$$

$$+ \sum_{t=0}^{T-1}\left(g_{t+1}^* B_t H_t y_t + y_t^* H_t^* B_t^* g_{t+1}\right) =$$

$$= y_0^* S_0 y_0 - g_0^* y_0 - y_0^* g_0 +$$

$$+\sum_{t=0}^{T-1}\left[V_t u_t + W_t y_t - \left(V_t^*\right)^{-1} B_t^* g_{t+1}\right]^* \left[V_t u_t + W_t y_t - \right.$$

$$\left. - \left(V_t^*\right)^{-1} B_t^* g_{t+1}\right] - \sum_{t=0}^{T-1}\left[g_{t+1}^* B_t V_t^{-1}\left(V_t^*\right)^{-1} B_t^* g_{t+1}\right].$$

The only term which depends on $u$ is

$$\sum_{t=0}^{T-1}\left[V_t u_t + W_t y_t - \left(V_t^*\right)^{-1} B_t^* g_{t+1}\right]^* \left[V_t u_t + W_t y_t - \left(V_t^*\right)^{-1} B_t^* g_{t+1}\right].$$

We remark that it is non-negative, so that its minimum value is obtained if

$$V_t u_t + W_t y_t - \left(V_t^*\right)^{-1} B_t^* g_{t+1} = 0, \quad t = 0, 1, \ldots, T-1.$$

It follows that

$$\tilde{u}_t = -V_t^{-1} W_t y_t - V_t^{-1}\left(V_t^*\right)^{-1} B_t^* g_{t+1} =$$

$$= H_t y_t + \left(K_t + B_t^* S_{t+1} B_t\right)^{-1} B_t^* g_{t+1}.$$

**Remark 8.4** We used in this proof the fact that $V_t$ is invertible.

**Remark 8.5** In order to solve system **(8.29)**, we need to know all the values for $F_t$, $t = 1, \ldots, T$.

In what follows, we shall apply the above theoretical results to a macro model for the monetary policy of a central bank.

The model contains four state variables, all of them expressed as logarithmic differences of cumulative indices; these states correspond to *indices of financial transactions, broad money, current interest rate* and *prices*. The only control variable is the "*call rate*", as the main instrument of the central bank to influence the financial market.

There are also two exogenous inputs, corresponding to the indices of world financial markets activity and of the prices of raw materials, including oil.

Considering logarithmic differences allows to use regressions to build a linear discrete time model of the form

$$x_{t+1} = A\,x_t + B\,u_t + C\,f_t,$$

where $f_t$ are the exogenous inputs.

We denote by $Z_t$ the total amount of financial transactions in the time interval indexed by $t$; the other variables are: broad money, denoted by $M$, the price index $P$ and the short-term interest rate $R$.

As we already mentioned, we shall work with logarithmic differences. Thus, we use the following notations:

$$z_t = \ln\left(\frac{Z_t}{Z_{t-1}}\right);\ r_t = \ln\left(\frac{R_t}{R_{t-1}}\right);\ m_t = \ln\left(\frac{M_t}{M_{t-1}}\right)$$

$$p_t = \ln\left(\frac{P_t}{P_{t-1}}\right);\ z_t^* = \ln\left(\frac{Z_t^*}{Z_{t-1}^*}\right);\ p_t^* = \ln\left(\frac{P_t^*}{P_{t-1}^*}\right).$$

The last two variables are the exogenous inputs, referring to external financial activity and external price, respectively.

The regressions expressing the dependence between the above variables are the following:

$$z_t = \gamma_1\, z_{t-1} + \gamma_2\, r_{t-1} + \gamma_3\,(m - p)_{t-1} + \gamma_4\, z_t^*, \tag{8.30}$$

$$m_t = \mu_1\, r_{t-1} + \mu_2\, i_t, \tag{8.31}$$

where $i_t$ stands for the control variable (the "call rate").

$$(m-p)_t = \lambda_1 r_t + \lambda_2 z_t + \lambda_3 (m-p)_{t-1}, \tag{8.32}$$

$$p_t = \Gamma_1 z_t + \Gamma_2 p_{t-1} + \Gamma_3 p_t^* . \tag{8.33}$$

To obtain a standard discrete time dynamic control system of the form

$$x_{t+1} = A x_t + B u_t + C f_t,$$

we shall transform equations **(8.30)** – **(8.33)** as follows: substituting **(8.30)** into **(8.33)**, we obtain

$$p_t = a_{41} z_{t-1} + a_{42} m_{t-1} + a_{43} r_{t-1} +$$

$$+ a_{44} p_{t-1} + c_{41} z_t^* + c_{42} p_t^* \tag{8.34}$$

where

$$a_{41} = \Gamma_1 \gamma_1, \quad a_{42} = \Gamma_1 \gamma_3, \quad a_{43} = \Gamma_1 \gamma_2,$$
$$a_{44} = \Gamma_2 - \Gamma_1 \gamma_3, \quad c_{41} = \Gamma_1 \gamma_4, \quad c_{42} = \Gamma_3 .$$

We use then **(8.32)** to express $r_t$ as

$$r_t = \frac{1}{\lambda_1}\left[m_t - p_t - \lambda_2 z_t - \lambda_3 m_{t-1} - \lambda_3 p_{t-1}\right]. \tag{8.32'}$$

Substituting in the right-hand side of **(8.32')** the expressions for $z_t$, $m_t$ and $p_t$, given by **(8.31)**, **(8.32)** and **(8.33)**, respectively, we get

$$r_t = a_{31} z_{t-1} + a_{32} m_{t-1} + a_{33} r_{t-1} +$$

$$+ a_{34} p_{t-1} + b_3 i_t + c_{31} z_t^* , \tag{8.35}$$

where

$$a_{31} = -\frac{\lambda_2 \gamma_1}{\lambda_1}, \quad a_{32} = -\frac{\lambda_3 + \lambda_2 \gamma_3}{\lambda_1}, \quad a_{33} = -\frac{\mu_1 - \lambda_2 \gamma_1}{\lambda_1},$$
$$a_{34} = \frac{\lambda_2 \gamma_3 - \lambda_3}{\lambda_1}, \quad b_3 = \frac{\mu_2}{\lambda_1}, \quad c_{31} = -\frac{\lambda_2 \gamma_4}{\lambda_1}.$$

We rewrite equation **(8.30)** in the form

$$z_t = a_{11} z_{t-1} + a_{12} m_{t-1} + a_{13} r_{t-1} +$$

$$+ a_{14} p_{t-1} + c_{11} z_t^* \tag{8.36}$$

where

$$a_{11} = \gamma_1, \quad a_{12} = \gamma_3, \quad a_{13} = \gamma_2,$$
$$a_{14} = -\gamma_3, \quad c_{11} = \gamma_4 .$$

Finally, denoting $i_t$ by $u_{t-1}$ and writing $a_{23}$ for $\mu_1$ and $b_2$ for $\mu_2$, we have

$$m_t = a_{23} r_{t-1} + b_2 u_{t-1}. \tag{8.37}$$

Denoting by $x$ the column-vector consisting of the variables $z$, $m$, $r$ and $p$, and by $f$ - the column-vector formed by the exogenous variables $z^*$ and $p^*$, the system of equations **(8.34)** - **(8.37)** can be written in the following matrix form:

$$x_{t+1} = A x_t + B u_t + C f_{t+1}, \; t = 0,1,\ldots,T-1,$$

where

$$A = \begin{pmatrix} a_{11} & a_{12} & a_{13} & a_{14} \\ 0 & 0 & a_{23} & 0 \\ a_{31} & a_{32} & a_{33} & a_{34} \\ a_{41} & a_{42} & a_{43} & a_{44} \end{pmatrix}; \quad B = \begin{pmatrix} 0 \\ b_2 \\ b_3 \\ 0 \end{pmatrix}; \quad C = \begin{pmatrix} c_{11} & 0 \\ 0 & 0 \\ c_{31} & 0 \\ c_{41} & c_{42} \end{pmatrix}.$$

Denoting $C f_t = F_{t-1}$, $t = 1,\ldots,T$, we obtain the standard form

$$x_{t+1} = A x_t + B u_t + F_t .$$

We attach to this system a cost-function of the central bank

$$\sum_{t=0}^{T-1} \rho i_t^2 + \sum_{t=1}^{T} \left[ (m_t - \hat{m}_t)^2 q_2 + (r_t - \hat{r}_t)^2 q_3 + (p_t - \hat{p}_t)^2 q_4 \right].$$

We can rewrite the above cost-function in matrix form, as

$$J(u) = \sum_{t=0}^{T-1} u_t^* K u_t + \sum_{t=1}^{T} \left[ (x_t - \hat{x}_t)^* Q (x_t - \hat{x}_t) \right].$$

where

$$Q = \begin{pmatrix} 0 & 0 & 0 & 0 \\ 0 & q_2 & 0 & 0 \\ 0 & 0 & q_3 & 0 \\ 0 & 0 & 0 & q_4 \end{pmatrix} \text{ and } K = \rho.$$

The sequences $\hat{m}_t, \hat{r}_t, \hat{p}_t$ represent targets; $\hat{r}_t, \hat{p}_t$ may be negative, if we want interest rates and prices to decrease, while $\hat{m}_t > 0$, if we want broad money to increase. The weights $q_2$, $q_3$, $q_4$ are positive, and so is $\rho$.

As such, the problem corresponds to the theory at hand.

## 8.6 A Simple Differential Game

Differential games represent a natural continuation of the problems of optimal control of dynamic systems. In a game, there exist two types of control, with opposite objectives. We shall illustrate such a game, by considering a simple situation with linear dynamics and quadratic cost.

We consider two controlled evolutions

$$x' = A(t)\,x + B(t)\,u$$

$$y' = C(t)\,y + D(t)\,v.$$

The objective of the first evolution is that, at the final moment, $x(T)$ be as close as possible to $y(T)$, while the objective of the second one is that $y(T)$ be as far as possible from $x(T)$.

In order to take also into account the situation when these objectives do not refer to the whole state, but only to some of its coordinates, or, more generally, to a given output, we consider a cost of the form

$$J(u,v) = \big(K(x(T)-y(T)), K(x(T)-y(T))\big) +$$
$$+ \int_0^T \big[(L(t)\,u(t), u(t)) - (M(t)\,v(t), v(t))\big]\,dt .$$

The first term in the expression of the cost represents the square of the distance between $K\,x(T)$ and $K\,y(T)$, while the second one describes the effort of control; since the first "player" aims to minimize the output depending on the final states, and the second one aims to maximize it, while both wish to minimize the effort of control, the term $\big(M(t)\,v(t), v(t)\big)$ occurs with the sign "–".

In such problems, in which opposite interests confront, we can not speak any longer about minimum or maximum.

We consider that the game has an optimal solution $(\tilde{u},\tilde{v})$ if $J(\tilde{u},v) \le J(\tilde{u},\tilde{v}) \le J(u,\tilde{v})$, whatever the pair of controls $(u,v)$.

If $(\tilde{u},\tilde{v})$ is optimal, then $\tilde{u}$ achieves the minimum when the second player adopts the strategy $\tilde{v}$, and $\tilde{v}$ achieves the maximum when the first player adopts the strategy $\tilde{u}$.

In what follows, we shall find conditions that allow effective determination of the optimal strategies.

We begin by performing some changes of variables.

Let $X(T,t)$ be the solution of the problem

$$\frac{d}{dt}X(T,t) = -\,X(T,t)\,A(t), \quad X(T,T) = I .$$

We remark that the matrix $X^*(T,t)$ is the solution of the system $\frac{d}{dt}X^*(T,t) = -\,A^*(t)\,X^*(T,t)$ which coincides at $T$ with the unit matrix.

From the general theory of differential equations it follows that such solution is defined on the whole interval $[t_0, T]$.

Analogously, we consider $Y(T,t)$ as a solution of the problem

$$\frac{d}{dt} Y(T,t) = -Y(T,t)C(t), \quad Y(T,T) = I.$$

We denote $z(t) = K[X(T,t)x(t) - Y(T,t)y(t)]$. It follows

$$z'(t) = K[X(T,t)x'(t) - X(T,t)A(t)x(t) -$$

$$- Y(T,t)y'(t) + Y(T,t)C(t)y(t)] =$$

$$= K[X(T,t)B(t)u(t) - Y(T,t)D(t)v(t)].$$

We also have

$$z(T) = K[X(T) - Y(T)],$$

hence the cost can be written as

$$J(u,v) = (z(T), z(T)) +$$

$$+ \int_0^T [(L(t)u(t), u(t)) - (M(t)v(t), v(t))]\, dt.$$

We admit that $L(t) > 0$, $M(t) > 0$, what implies the existence of the inverses $L^{-1}(t)$, $M^{-1}(t)$.

We denote $G(t) = K X(T,t) B(t)$, $H(t) = K Y(T,t) D(t)$ and we consider the function $P$ defined as a solution of the problem

$$P' = P[G(t)L^{-1}(t)G^*(t) - H(t)M^{-1}(t)H^*(t)]P, \; P(T) = I.$$

Since the above system is nonlinear, we can not assert from the beginning that a solution $P$ exists on the whole interval $[t_0, T]$.

If we admit that $G(t)L^{-1}(t)G^*(t) \geq H(t)M^{-1}(t)H^*(t)$, then it will follow that the function

$$Q(t) = I + \int_t^T \left[G(\tau)L^{-1}(\tau)G^*(\tau) - H(\tau)M^{-1}(\tau)H^*(\tau)\right]d\tau$$

is invertible, since $Q(t) \geq I$ for $t \in [\,t_0, T\,]$.

We have the general formula

$$\frac{d}{dt}\left[Q^{-1}(t)\right] = -\,Q^{-1}(t)\,Q'(t)\,Q^{-1}(t)$$

and

$$Q'(t) = -\left[G(t)L^{-1}(t)G^*(t) - H(t)M^{-1}(t)H^*(t)\right].$$

We deduce that

$$\frac{d}{dt}Q^{-1}(t) = Q^{-1}(t)\left[G(t)L^{-1}(t)G^*(t) - H(t)M^{-1}(t)H^*(t)\right]Q^{-1}(t)$$

and, since $Q(T) = I$, we have $Q^{-1}(T) = I$.

We remark that the function $Q^{-1}$ is the solution of the same Cauchy problem as $P$, hence $P(t) = Q^{-1}(t)$, what shows that $P$ is defined on the whole interval $[\,t_0, T\,]$.

We define now $R(t) = (P(t)\,z(t), z(t))$ and calculate its derivative. We have

$$R'(t) = (P'(t)\,z(t), z(t)) + (P(t)\,z'(t), z(t)) + (P(t)\,z(t), z'(t)) =$$

$$= \left(P(t)\left[G(t)L^{-1}(t)G^*(t) - H(t)M^{-1}(t)H^*(t)\right]P(t)\,z(t), z(t)\right) +$$

$$+ \left(P(t)\left[G(t)\,u(t) - H(t)\,v(t)\right], z(t)\right) +$$

$$+ (P(t)\,z(t), G(t)\,u(t) - H(t)\,v(t)) =$$

$$= \left(L^{-1}(t)G^*(t)P(t)\,z(t), G^*(t)P(t)\,z(t)\right) -$$

$$- \left(M^{-1}(t)\, H^*(t)\, P(t)\, z(t), H^*(t)\, P(t)\, z(t)\right) +$$

$$+ 2\left(u(t), G^*(t)\, P(t)\, z(t)\right) - 2\left(v(t), H^*(t)\, P(t)\, z(t)\right) =$$

$$= \left(L^{-1}(t)\left[G^*(t)\, P(t)\, z(t) + L(t)\, u(t)\right], G^*(t)\, P(t)\, z(t) + L(t)\, u(t)\right) -$$

$$- \left(M^{-1}(t)\left[H^*(t) P(t)\, z(t) + M(t)\, v(t)\right], H^*(t) P(t)\, z(t) + M(t)\, v(t)\right) -$$

$$- (u(t), L(t)\, u(t)) + (v(t), M(t)\, v(t)).$$

Therefore

$$R'(t) + (L(t)\, u(t), u(t)) - (M(t)\, v(t), v(t)) =$$

$$= \left(L^{-1}(t)\left[G^*(t)\, P(t)\, z(t) + L(t)\, u(t)\right], G^*(t)\, P(t)\, z(t) + L(t)\, u(t)\right) -$$

$$- \left(M^{-1}(t)\left[H^*(t) P(t)\, z(t) + M(t)\, v(t)\right], H^*(t) P(t)\, z(t) + M(t)\, v(t)\right).$$

Integrating from $t_0$ to $T$, we obtain

$$R(T) - R(t_0) + \int_{t_0}^{T} \left[(L(t)\, u(t), u(t)) - (M(t)\, v(t), v(t))\right] dt =$$

$$= \int_{t_0}^{T} \left[\left(L^{-1}\left(G^* P z + L\, u\right), G^* P z + L\, u\right) -\right.$$

$$\left. - \left(M^{-1}\left(H^* P z + M\, v\right), H^* P z + M\, v\right)\right] dt.$$

But

$$R(T) = (P(T)\, z(T), z(T)) = (z(T), z(T)),$$

hence

$$J(u, v) = (P(t_0)\, z(t_0), z(t_0)) +$$

$$+ \int_{t_0}^{T} \left[\left(L^{-1}\left(G^* P\, z + L\, u\right), G^* P\, z + L\, u\right) -\right.$$

$$-\left(M^{-1}\left(H^* P z + M v\right), H^* P z + M v\right)\Big] dt.$$

From this expression of the cost we can see that if we take $\tilde{u} = -L^{-1} G^* P z$, $\tilde{v} = M^{-1} H^* P z$ then

$$J(\tilde{u}, \tilde{v}) = \left(P(t_0)\, z(t_0), z(t_0)\right),$$

$$J(\tilde{u}, v) = \left(P(t_0)\, z(t_0), z(t_0)\right) -$$

$$-\int_{t_0}^{T} \left(M^{-1}\left(H^* P z + M v\right), H^* P z + M v\right) dt,$$

hence

$$J(\tilde{u}, v) \leq \left(P(t_0)\, z(t_0), z(t_0)\right) = J(\tilde{u}, \tilde{v});$$

likewise,

$$J(u, \tilde{v}) = \left(P(t_0)\, z(t_0), z(t_0)\right) +$$

$$+\int_{t_0}^{T} \left(L^{-1}\left(G^* P z + L u\right), G^* P z + L u\right) dt,$$

hence $J(u, \tilde{v}) \geq J(\tilde{u}, \tilde{v})$.

In this way, it follows that $(\tilde{u}, \tilde{v})$ is an optimal solution.

Returning to the variables $x$ and $y$, we deduce

$$\tilde{u}(t) = -L^{-1}(t)\, G^*(t)\, P(t)\, K\left[X(T,t)\, x(t) - Y(T,t)\, y(t)\right],$$

$$\tilde{v}(t) = -M^{-1}(t)\, H^*(t) P(t)\, K\left[X(T,t)\, x(t) - Y(T,t)\, y(t)\right].$$

Thus, the solution is written explicitly, starting from the data of the problem.

We remark that, in order to act optimally, the two players have to know, at any moment $t$, the states $x(t)$ and $y(t)$.

## 8.7 Notes and References

Linear-quadratic optimization is a standard topics in control theory and there are many books where the problem is considered, with different approaches. We shall not mention here all these sources, but we have to stress the importance of the pioneering paper of R. E. Kalman [15] and the inspiring book of V. M. Popov [25].

The application in **8.5** was stimulated by [22].

The differential game studied in **8.6** is inspired from Y. C. Ho, A. E. Bryson and M.L. Baron [13].

CHAPTER 9

# Some Unconstrained Dynamic Optimization Problems

## 9.1 The Simplest Problem of the Calculus of Variations

The Calculus of Variations represents the oldest mathematical theory for optimization in the case of a controlled evolution.

We shall present the simplest problem of the Calculus of Variations as an optimal control problem.

We consider the evolution described by

$$x' = u. \tag{9.1}$$

where $u$ is a function modelling the <u>*control*</u> (the decision element).

We have the evolution on a given time interval $(t_0, t_1)$, an initial condition $x_0$, a final condition $x_1$ and a <u>*cost functional*</u> expressing the <u>*quality*</u> (performance) of the evolution, ensured by the choice of the function $u$. We consider the cost-functional being of the form

$$J(u) = \int_{t_0}^{t_1} f_0(t, x(t), u(t))\, dt \tag{9.2}$$

where $f_0$ is a function of class $\mathscr{C}^2$.

The problem consists in determining a control $u$ for which the solution starting from $x(t_0) = x_0$ satisfies $x(t_1) = x_1$ and for which the cost is minimized, i.e. $J(u)$ is minimized.

In the study of such problems, the following questions arise in a natural way.

**1)** Assuming that $\tilde{u}$ is an optimal control and $\tilde{x}$ is the corresponding evolution, what properties do have $\tilde{u}$ and $\tilde{x}$?

**2)** Among all controls $\tilde{u}$, satisfying the properties answering the first question, how do we recognize those which minimize effectively the cost?

**3)** In what conditions does exist a control minimizing the cost?

**4)** What procedures can be used to obtain approximately the control minimizing the cost?

**5)** What procedure can be used to obtain approximately the minimum value of the cost?

**6)** What procedures can be used to obtain a *suboptimal* control, which provides a value of the cost close to the minimum?

The answer to the first question is given by the theory of *necessary* optimality conditions; the answer to the second question is given by the theory of *sufficient* conditions.

The answer to the third question is given by the study of

existence problems, while the last questions are connected with the numerical study of optimization problems.

In what follows, using the classical approach of the Calculus of Variations, we shall limit ourselves to the first problem, that which leads to necessary optimality conditions, providing candidates for the rôle of optimal solution.

If such conditions allow us to determine a finite number of candidates, then the choice of optimal controls can be performed by direct checking.

The method of obtaining the necessary optimality conditions is based on the idea of considering *variations* around the optimum and reducing, in this way, the study to the simplest case, described by the Fermat theorem: *if a function admits a local minimum and if it is differentiable at this local minimum, then the derivative must vanish.*

Thus, let $\tilde{u}$ be an optimal control and $\tilde{x}$ – the corresponding solution; we chose an arbitrary $v$ and define a control of the form $u = \tilde{u} + \varepsilon v$; it follows $x(t) = \tilde{x}(t) + \varepsilon z(t)$, where $z'(t) = v(t)$ and $z(t_0) = 0$, $z(t_1) = 0$, since $x(t_0) = \tilde{x}(t_0) = x_0$ and $x(t_1) = \tilde{x}(t_1) = x_1$.

We denote $J(\tilde{u} + \varepsilon v) = I(\varepsilon)$ and we have

$$I(\varepsilon) = \int_{t_0}^{t_1} f_0(t, \tilde{x}(t) + \varepsilon z(t), \tilde{u}(t) + \varepsilon v(t))\, dt .$$

Since, by assumption, $\tilde{u}$ is a control that minimizes the cost, we have for any (sufficiently small) $\varepsilon$, $J(\tilde{u} + \varepsilon v) \geq J(\tilde{u})$, hence $I(\varepsilon) \geq I(0)$, therefore the point $\varepsilon = 0$ is a local minimum for the function $I$.

The Fermat theorem leads to the condition $I'(0) = 0$. We have

$$I'(\varepsilon) = \int_{t_0}^{t_1} \left[ \frac{\partial f_0}{\partial x}(t, \tilde{x}(t) + \varepsilon\, z(t), \tilde{u}(t) + \varepsilon\, v(t))\, z(t) + \right.$$

$$\left. + \frac{\partial f_0}{\partial u}(t, \tilde{x}(t) + \varepsilon\, z(t), \tilde{u}(t) + \varepsilon\, v(t))\, v(t) \right] dt$$

and we obtain the necessary optimality condition in the form:

$$\int_{t_0}^{t_1} \left[ \frac{\partial f_0}{\partial x}(t, \tilde{x}(t), \tilde{u}(t))\, z(t) + \frac{\partial f_0}{\partial u}(t, \tilde{x}(t), \tilde{u}(t))\, v(t) \right] dt = 0 \qquad \textbf{(9.3)}$$

for any $v$ and for any $z$ with $z' = v$, $z(t_0) = 0$, $z(t_1) = 0$.

We rewrite **(9.3)** in the form:

$$\int_{t_0}^{t_1} \frac{d}{dt} \left\{ \int_{t_0}^{t} \frac{\partial f_0}{\partial x}(s, \tilde{x}(s), \tilde{u}(s))\, ds \right\} z(t)\, dt +$$

$$+ \int_{t_0}^{t_1} \frac{\partial f_0}{\partial u}(t, \tilde{x}(t), \tilde{u}(t))\, z'(t)\, dt = 0 .$$

Integrating by parts in the first term of **(9.3)**, we obtain

$$\int_{t_0}^{t_1} \frac{d}{dt} \left\{ \int_{t_0}^{t} \frac{\partial f_0}{\partial x}(s, \tilde{x}(s), \tilde{u}(s))\, ds \right\} z(t)\, dt =$$

$$= \left\{ \int_{t_0}^{t} \frac{\partial f_0}{\partial x}(s, \tilde{x}(s), \tilde{u}(s))\, ds \right\} z(t) \Bigg|_{t_0}^{t_1} -$$

$$- \int_{t_0}^{t_1} \left[ \int_{t_0}^{t} \frac{\partial f_0}{\partial x}(s, \tilde{x}(s), \tilde{u}(s))\, ds \right] z'(t)\, dt .$$

We remark that the first term in the above sum vanishes, since $z(t_0) = z(t_1) = 0$.

Therefore **(9.3)** becomes:

$$\int_{t_0}^{t_1} \left[ \frac{\partial f_0}{\partial u}(t, \tilde{x}(t), \tilde{u}(t)) - \int_{t_0}^{t} \frac{\partial f_0}{\partial x}(s, \tilde{x}(s), \tilde{u}(s)) ds \right] z'(t) dt = 0 \quad \textbf{(9.4)}$$

for all $z \in \mathscr{C}^1$ with $z(t_0) = z(t_1) = 0$.

In order to be able to find a simpler form of the necessary optimality conditions, we shall prove the following

**Lemma 9.1** (Du Bois–Raymond)

*If* $f:\mathrm{R} \to \mathrm{R}$ *is a continuous function and if* $\int_{t_0}^{t_1} f(t)\, g'(t)\, dt = 0$ *for any* $\mathscr{C}^1$ *function* $g$, *with* $g(t_0) = 0$, $g(t_1) = 0$, *then the function* $f$ *is constant.*

***Proof*** We remark that, if $f$ is constant, then $\int_{t_0}^{t_1} f(t)\, g'(t)\, dt = 0$.

It follows that, if $\int_{t_0}^{t_1} f(t)\, g'(t)\, dt = 0$, whatever $g$ with $g(t_0) = g(t_1) = 0$, then

$$\int_{t_0}^{t_1} (f(t) - c)\, g'(t)\, dt = 0 \tag{9.5}$$

for any $c$ and any $g$ with $g(t_0) = g(t_1) = 0$.

We choose $g(t) = \int_{t_0}^{t} (f(s) - c)\, ds$. $\quad$ **(9.6)**

We have $g'(t) = f(t) - c$, $g(t_1) = \int_{t_0}^{t_1} f(s)\, ds - c\,(t_1 - t_0)$.

To have $g(t_1) = 0$, we choose $c = \dfrac{1}{t_1 - t_0} \int_{t_0}^{t_1} f(s)\, ds$.

Substituting **(9.6)** into **(9.5)**, we get

$$\int_{t_0}^{t_1} (f(s) - c)^2\, ds = 0,$$

wherefrom $f(t) = c$, whatever $t$; this ends the proof of the lemma.

Using the above lemma, we deduce from **(9.2)** that

$$\frac{\partial f_0}{\partial u}(t,\tilde{x}(t),\tilde{u}(t)) - \int_{t_0}^{t} \frac{\partial f_0}{\partial x}(s,\tilde{x}(s),\tilde{u}(s))\,ds = C. \tag{9.7}$$

We remark that the function $t \mapsto \frac{\partial f_0}{\partial u}(t,\tilde{x}(t),\tilde{u}(t))$ is differentiable.

By differentiating, we deduce from **(9.5)**

**Theorem 9.2** *If* $\tilde{u}$ *is an optimal control for the problem defined by* **(9.1)**, **(9.2)**, $x(t_0) = x_0$, $x(t_1) = x_1$, *and* $\tilde{x}$ *is the corresponding solution, then* $\tilde{x}$, $\tilde{u}$ *satisfy the equation*

$$\frac{d}{dt}\frac{\partial f_0}{\partial u}(t,\tilde{x}(t),\tilde{u}(t)) - \frac{\partial f_0}{\partial x}(t,\tilde{x}(t),\tilde{u}(t)) = 0.$$

The above equation is known as the *Euler-Lagrange equation*.

To understand the nature of this equation, let us recall that $\tilde{u} = \tilde{x}'$, hence we have

$$\frac{d}{dt}\frac{\partial f_0}{\partial u}(t,\tilde{x}(t),\tilde{x}'(t)) - \frac{\partial f_0}{\partial x}(t,\tilde{x}(t),\tilde{x}'(t)) = 0;$$

assuming that $\tilde{x}$ is $\mathscr{C}^2$ and computing the derivative with respect to $t$ by the chain rule, we get

$$\frac{\partial^2 f_0}{\partial u\,\partial t}(t,\tilde{x}(t),\tilde{x}'(t)) + \frac{\partial^2 f_0}{\partial u\,\partial x}(t,\tilde{x}(t),\tilde{x}'(t))\,\tilde{x}'(t) +$$

$$+ \frac{\partial^2 f_0}{\partial u\,\partial u}(t,\tilde{x}(t),\tilde{x}'(t))\,\tilde{x}''(t) - \frac{\partial f_0}{\partial x}(t,\tilde{x}(t),\tilde{x}'(t)) = 0.$$

In this way, we obtained a system of second order differential equations satisfied by $\tilde{x}$ or, returning to the function $\tilde{u}$, a system of $2n$ first order differential equations:

$$\begin{cases} \tilde{x}'(t) = \tilde{u}(t) \\ \dfrac{\partial^2 f_0}{\partial u\, \partial u}(t, \tilde{x}(t), \tilde{u}(t))\, \tilde{u}'(t) + \dfrac{\partial^2 f_0}{\partial u\, \partial x}(t, \tilde{x}(t), \tilde{u}(t))\, \tilde{u}(t) + \\ \qquad + \dfrac{\partial^2 f_0}{\partial u\, \partial t}(t, \tilde{x}(t), \tilde{u}(t)) - \dfrac{\partial f_0}{\partial x}(t, \tilde{x}(t), \tilde{u}(t)) = 0 \end{cases} \tag{9.8}$$

If *we admit* that the Hessian matrix $\dfrac{\partial^2 f_0}{\partial u\, \partial u}(t, \tilde{x}(t), \tilde{u}(t))$ is *invertible*, then the system **(9.8)** can be written as:

$$\begin{cases} \tilde{x}'(t) = \tilde{u}(t) \\ \tilde{u}'(t) = \left[\dfrac{\partial^2 f_0}{\partial u\, \partial u}(t, \tilde{x}(t), \tilde{u}(t))\right]^{-1} \left[\dfrac{\partial f_0}{\partial x}(t, \tilde{x}(t), \tilde{u}(t)) - \right. \\ \qquad \left. - \dfrac{\partial^2 f_0}{\partial u\, \partial t}(t, \tilde{x}(t), \tilde{u}(t)) - \dfrac{\partial^2 f_0}{\partial u\, \partial x}(t, \tilde{x}(t), \tilde{u}(t))\, \tilde{u}(t)\right] \end{cases} \tag{9.9}$$

This is a form we are familiar with, but we remark that, for this system, we have no longer *initial conditions* for $\tilde{x}$ and $\tilde{u}$, but only the initial condition $\tilde{x}(t_0) = x_0$, while instead of the initial condition for $\tilde{u}$, we have the final condition $\tilde{x}(t_1) = x_1$. The problem we obtained is a two-point boundary value problem (we met already such problem in **8.1**).

For such problems, we have no longer a general existence and uniqueness theorem.

To understand somehow the way these problems can be treated, let us take $\tilde{u}(t_0) = \tilde{u}_0$, where $\tilde{u}_0$ appears as a parameter to be determined. For each choice of $\tilde{u}_0$, the fundamental existence and uniqueness theorem provides a solution defined in a neighbourhood of $t_0$ (if we admit that $f_0$ is of class $\mathscr{C}^3$).

If the solution obtained in this way is defined on the whole interval $[t_0, t_1]$, then the condition $x(t_1) = x_1$ will provide an equation (in fact, a system of nonlinear equations) for determining $u_0$.

The difficulties arise, on the one hand, from the fact that the solutions are not always defined on the whole interval $[t_0, t_1]$ and, on the other hand, from the possibility that the system from which $\tilde{u}_0$ has to be determined admits no solution or admits several solutions.

Besides these difficulties, there are those that refer to the *computation* of the solution of the Cauchy problem or to the solution of the system of nonlinear equations.

At any event, the analysis of the properties of the optimal solution led us to the useful information that the optimal pair $(\tilde{u}, \tilde{x})$ represents the solution of a two-point boundary value problem for a system of differential equations.

Let us end this section with the description of a classical problem, namely to find, among all paths in the plane, graphs of $\mathscr{C}^1$ functions joining two given points, the one of minimum length.

This problem corresponds to $f_0(t, x, u) = \sqrt{1 + u^2}$, since the length of the path is given by $\int_\alpha^\beta \sqrt{1 + [x'(t)]^2}\, dt$.

The Euler–Lagrange equation is $\dfrac{d}{dt} \dfrac{\tilde{u}(t)}{\sqrt{1 + \tilde{u}^2(t)}} = 0$, hence $\tilde{u}(t) = c$, $\tilde{x}'(t) = c$ and we deduce $\tilde{x}(t) = c\,t + d$; we proved in this way that the shortest path is a straight line.

## 9.2 A Macroeconomic Growth Model

We denote by $K(t)$ the level of capital at time $t$; its evolution will be given by the simple law

$$K'(t) = I(t) - \mu K(t), \; K(0) = K_0, \tag{9.10}$$

where $I(t)$ represents gross investment, while $\mu$ is the coefficient of physical and moral wear and tear of capital, assumed to be constant.

The product $Y(t)$ is given by a production function with neutral technical progress

$$Y(t) = e^{\delta t} F(K(t), L(t)), \tag{9.11}$$

where $L(t)$ represents labour at time $t$. It is assumed to grow exponentially, with rate $n$: $L(t) = L_0 e^{nt}$.

We assume that the product is allotted for consumption and gross investment:

$$Y(t) = C(t) + I(t), \text{ hence } I(t) = Y(t) - C(t). \tag{9.12}$$

Substituting **(9.11)** and **(9.12)** into **(9.10)** we obtain the evolution law for capital

$$K'(t) = e^{\delta t} F(K(t), L(t)) - \mu K(t) - C(t), \; K(0) = K_0, \tag{9.13}$$

with the constraint

$$0 \le C(t) \le e^{\delta t} F(K(t), L(t)). \tag{9.14}$$

For technical reasons, it is more convenient to use per worker values; thus, we define new variables:

$$k(t) = \frac{K(t)}{L(t)}, \; c(t) = \frac{C(t)}{L(t)};$$

assuming that $F(K, L)$ is homogeneous of the first degree, i.e.

$$F(\lambda K, \lambda L) = \lambda F(K, L) \text{ for any } \lambda > 0,$$

we can write

$$y(t) = \frac{Y(t)}{L(t)} = e^{\delta t} \frac{F(K(t), L(t))}{L(t)} = e^{\delta t} F(k(t), 1).$$

We shall denote $F(k(t),1)$ by $f(k(t))$. We have

$$k'(t) = \frac{K'(t)}{L(t)} - \frac{K(t)}{L(t)} \frac{L'(t)}{L(t)} =$$

$$= e^{\delta t} \frac{F(K(t), L(t))}{L(t)} - \frac{C(t)}{L(t)} - \mu \frac{K(t)}{L(t)} - \frac{K(t)}{L(t)} \cdot n =$$

$$= e^{\delta t} f(k(t)) - c(t) - (\mu + n) k(t) \text{ and } k(0) = \frac{K_0}{L_0} = k_0.$$

Thus, we obtained the evolution

$$k'(t) = e^{\delta t} f(k(t)) - (\mu + n) k(t) - c(t), \tag{9.15}$$

with initial condition $k(0) = k_0$ and the constraint

$$0 \le c(t) \le e^{\delta t} f(k(t)). \tag{9.16}$$

We associate to the above evolution a utility function of the form $\mathscr{U}(c) = \ln c$ and we choose a performance (objective function) of the form

$$J(c) = \int_0^T \mathscr{U}(c(t)) e^{-\alpha t} \, dt. \tag{9.17}$$

The coefficient $\alpha$ is a discount rate, expressing the weight (preference) ascribed to present consumption (versus future consumption).

The performance $J(c)$ of the form **(9.17)** is usually called *macroeconomic welfare function.*

Since the evolution is examined over a finite time horizon, it is natural to impose a final level of capital: $k(T) = k_T$. **(9.18)**

The problem of optimal economic growth, which consists in maximizing $J(c)$, can be treated as a Calculus of Variations problem.

To this aim, we denote $k'(t) = u(t)$ and obtain

$c(t) = e^{\delta t} f(k(t)) - (\mu + n)\,k(t) - u(t)$.

We denote $f_0(t,k,u) = -\left[\mathcal{U}\left(e^{\delta t} f(k) - (\mu + n)\,k - u\right)\right] e^{-\alpha t}$.

In this way, we obtained the problem

$$k'(t) = u(t),\ k(0) = k_0,\ k(T) = k_T,\ \min_u \int_0^T f_0(t, k(t), u(t))\,dt.$$

(The fact that we introduced a change of sign in the definition of $f_0$ transformed the maximization problem into a minimization problem).

The Euler–Lagrange equation has the form

$$\frac{d}{dt}\frac{\partial f_0}{\partial u}(t,k(t),u(t)) - \frac{\partial f_0}{\partial k}(t,k(t),u(t)) = 0.$$

We have

$$\frac{\partial f_0}{\partial u}(t,k,u) = e^{-\alpha t}\,\mathcal{U}'\left[e^{\delta t} f(k) - (\mu + n)\,k - u\right].$$

If $\mathcal{U}(c) = \ln c$, then $\mathcal{U}'(c) = \frac{1}{c}$, hence

$$\frac{\partial f_0}{\partial u}(t,k,u) = \frac{e^{-\alpha t}}{e^{\delta t} f(k) - (\mu + n)\,k - u},$$

$$\frac{\partial f_0}{\partial u}(t,k,u) = -\,\mathcal{U}'\left[e^{\delta t} f(k) - (\mu + n)\,k - u\right] e^{-\alpha t}\left[e^{\delta t} f'(k) - \right.$$

$$\left. - (\mu + n)\right] = -\frac{e^{-\alpha t}\left[e^{\delta t} f'(k) - (\mu + n)\right]}{e^{\delta t} f(k) - (\mu + n)\,k - u}.$$

In order to obtain a more convenient form of the equations,

we denote $x(t) = k(t)e^{-\delta t}$, $y(t) = c(t)e^{-\delta t}$. We have

$$\frac{\partial f_0}{\partial u}(t, k(t), u(t)) = \frac{e^{-\alpha t}}{c(t)}$$

and the Euler–Lagrange equation can be written as

$$\frac{d}{dt}\frac{e^{-\alpha t}}{c(t)} + \frac{e^{-\alpha t}\left[e^{\delta t}f'(k(t)) - (\mu + n)\right]}{c(t)} = 0 \tag{9.19}$$

$$\frac{-\alpha e^{-\alpha t}}{c(t)} - \frac{c'(t)e^{-\alpha t}}{c^2(t)} + \frac{e^{-\alpha t}\left[e^{\delta t}f'(k(t)) - (\mu + n)\right]}{c(t)} = 0,$$

hence

$$-\alpha - \frac{c'(t)}{c(t)} + e^{\delta t}f'(k(t)) - (\mu + n) = 0.$$

We have

$$y'(t) = c'(t)e^{-\delta t} - \delta e^{-\delta t}c(t),$$

$$\frac{y'(t)}{y(t)} = \frac{c'(t)e^{-\delta t}}{c(t)e^{-\delta t}} - \delta = -\delta - (\alpha + n + \mu) + e^{\delta t}f'(k(t)),$$

hence

$$y'(t) = y(t)\left[e^{\delta t}f'\left(x(t)e^{\delta t}\right) - (\alpha + n + \mu + \delta)\right].$$

We also have

$$x'(t) = -\delta x(t) + e^{-\delta t}k'(t) =$$

$$= -\delta x(t) + e^{-\delta t}\left[e^{\delta t}f\left(x(t)e^{\delta t}\right) - (n + \mu)x(t)e^{\delta t} - y(t)e^{\delta t}\right] =$$

$$= -\delta x(t) + f\left(x(t)e^{\delta t}\right) - (n + \mu)x(t) - y(t).$$

In the special case when $\delta = 0$ (the technical progress is absent), we have $x(t) = k(t)$, $y(t) = c(t)$ and we obtain the system

$$\begin{cases} k' = -(n + \mu)k - c + f(k) \\ c' = c\left[f'(k) - (n + \mu + \alpha)\right] \end{cases} . \tag{9.20}$$

If we choose a linear production function: $f(k) = \gamma k$, then the system **(9.20)** becomes:

$$\begin{cases} k' = [\gamma - (n + \mu)]k - c \\ c' = [\gamma - (n + \mu + \alpha)]c, \end{cases} \tag{9.21}$$

which is a <u>*linear*</u> system. We have

$$c(t) = e^{[\gamma - (n+\mu+\alpha)]t} c_0, \tag{9.22}$$

$$k'(t) = [\gamma - (n + \mu)]k - e^{[\gamma - (n+\mu+\alpha)]t} c_0,$$

hence

$$k(t) = e^{[\gamma - (n+\mu)]t} k_0 - c_0 \int_0^t e^{[\gamma-(n+\mu)](t-s)} \cdot e^{[\gamma-(n+\mu+\alpha)]s} ds =$$

$$= e^{[\gamma - (n+\mu)]t} k_0 - c_0\, e^{[\gamma-(n+\mu)]t} \int_0^t e^{-\alpha s} ds =$$

$$= e^{[\gamma - (n+\mu)]t} k_0 + \frac{c_0}{\alpha} e^{[\gamma-(n+\mu)]t} \left(e^{-\alpha t} - 1\right) =$$

$$= e^{[\gamma - (n+\mu)]t} \left(k_0 - \frac{c_0}{\alpha}\right) + \frac{c_0}{\alpha} e^{[\gamma-(n+\mu+\alpha)]t}. \tag{9.23}$$

Using now the final condition $k(T) = k_T$, it follows

$$e^{[\gamma - (n+\mu)]T} k_0 - \frac{c_0}{\alpha} e^{[\gamma-(n+\mu)]T} \left(1 - e^{-\alpha T}\right) = k_T,$$

hence

$$\frac{c_0}{\alpha} = \frac{k_0 - e^{-[\gamma - (n+\mu)]T} k_T}{1 - e^{-\alpha T}}. \tag{9.24}$$

We determined in this way $c_0$, hence $k(t)$ and $c(t)$ are completely determined from **(9.22)** and **(9.23)**.

Let us remark that, in solving the problem, we ignored the constraint $0 \leq c(t) \leq f(k(t))$. From **(9.22)** we see that, if $c_0 > 0$, then $c(t) > 0$ for any $t$. According to **(9.24)**, this condition is satisfied if

$$k_T < k_0\, e^{[\gamma - (n+\mu)]T}. \tag{9.25}$$

As concerns the inequality $c(t) \le f(k(t))$, it becomes

$c(t) \le \gamma\, k(t)$, i.e.

$$\frac{c_0}{\gamma} e^{[\gamma-(n+\mu+\alpha)]t} \le \frac{c_0}{\alpha} e^{[\gamma-(n+\mu+\alpha)]t} + \left(k_0 - \frac{c_0}{\alpha}\right) e^{[\gamma-(n+\mu)]t}.$$

The above condition is automatically satisfied if $\gamma \ge \alpha$ and $k_0 > \frac{c_0}{\alpha}$; but

$$k_0 - \frac{c_0}{\alpha} = \frac{e^{-[\gamma-(n+\mu)]T} k_T - e^{-\alpha T} k_0}{1 - e^{-\alpha T}},$$

hence the condition $k_0 > \frac{c_0}{\alpha}$ is equivalent to

$$k_T > e^{[\gamma-(n+\mu+\alpha)]T} k_0. \tag{9.26}$$

From **(9.25)** and **(9.26)** we deduce that if $k_T$ satisfies the inequalities:

$$e^{[\gamma-(n+\mu+\alpha)]T} \le \frac{k_T}{k_0} \le e^{[\gamma-(n+\mu)]T}$$

then the constraint $0 \le c(t) \le \gamma\, k(t)$ is satisfied for any $t$.

We remark that the situation $k(T) = k_0\, e^{[\gamma-(n+\mu)]T}$ would occur if $c(t)$ vanished for all $t$. Indeed, in this case, the equation **(9.15)** would have the form

$$k'(t) = [\gamma - (n + \mu)]\, k(t). \tag{9.15'}$$

The solution of this linear equation is $k(t) = k_0\, e^{[\gamma-(n+\mu)]t}$, hence, for $t = T$, we would obtain $k(T) = k_0\, e^{[\gamma-(n+\mu)]T}$.

## 9.3 A Discrete – Time Variational Problem

We shall discuss now a discrete analogue of the simplest problem of the Calculus of Variations.

Let $L_t : G \subset \mathrm{R}^n \to \mathrm{R}$, $t = 0,1,\ldots,T-1$, be given $\mathscr{C}^1$ functions. Let $x_0 \in G$, $x_T \in G$ be also given ($G$ is an open connected set).

We state the following problem: find a sequence $x_t \in G$, $t = 1,\ldots,T-1$, minimizing the scalar function

$$\sum_{t=0}^{T-1} L_t(x_t, x_{t+1}),$$

where $x_0$ and $x_T$ are the given ones.

We shall look for necessary optimality conditions.

Let $\widetilde{x}_1,\ldots,\widetilde{x}_{T-1}$ be an optimal sequence; let $\varepsilon$ be with $|\varepsilon|$ small enough and let $v_1,\ldots,v_{T-1}$ be arbitrary vectors in $\mathrm{R}^n$.

Take $x_t = \widetilde{x}_t + \varepsilon v_t$ and define

$$I(\varepsilon) = \sum_{t=0}^{T-1} L_t(\widetilde{x}_t + \varepsilon v_t, \widetilde{x}_{t+1} + \varepsilon v_{t+1}),\ v_0 = 0,\ v_T = 0.$$

Since $G$ is open, it follows that $I(\varepsilon)$ is well defined for $|\varepsilon|$ small enough.

We remark that, for $\varepsilon = 0$, $I$ has a minimal value and $L_t$ being $\mathscr{C}^1$ functions, it follows that $I'(0) = 0$.

By the chain rule, we have:

$$I'(\varepsilon) = \sum_{t=0}^{T-1} \left[ \frac{\partial L_t}{\partial x}(\widetilde{x}_t + \varepsilon v_t, \widetilde{x}_{t+1} + \varepsilon v_{t+1}) v_t + \right.$$
$$\left. + \frac{\partial L_t}{\partial y}(\widetilde{x}_t + \varepsilon v_t, \widetilde{x}_{t+1} + \varepsilon v_{t+1}) v_{t+1} \right]$$

(we denoted by $\frac{\partial L_t}{\partial y}$ the partial derivative with respect to the second argument).

We deduce that, if $\tilde{x}_t$, $t = 0,1,\dots,T-1$, is an optimal sequence, then

$$\sum_{t=0}^{T-1}\left[\frac{\partial L_t}{\partial x}(\tilde{x}_t,\tilde{x}_{t+1})v_t + \frac{\partial L_t}{\partial y}(\tilde{x}_t,\tilde{x}_{t+1})v_{t+1}\right] = 0$$

for an arbitrary sequence $\{v_t\}$ with $v_0 = 0$, $v_T = 0$.

The above condition may be written as

$$\sum_{t=1}^{T-1}\left[\frac{\partial L_t}{\partial x}(\tilde{x}_t,\tilde{x}_{t+1})v_t + \frac{\partial L_{t-1}}{\partial y}(\tilde{x}_{t-1},\tilde{x}_t)v_t\right] = 0.$$

Since the sequence $v_t$, $t = 1,\dots,T-1$ is arbitrary, we obtain finally that the optimal sequence $\tilde{x}_t$, $t = 0,1,\dots,T$, satisfies

$$\frac{\partial L_t}{\partial x}(\tilde{x}_t,\tilde{x}_{t+1}) + \frac{\partial L_{t-1}}{\partial y}(\tilde{x}_{t-1},\tilde{x}_t) = 0, \quad t = 1,\dots,T-1,$$

$\tilde{x}_0 = x_0$, $\tilde{x}_T = x_T$.

We have thus obtained the following result:

**Theorem 9.3** *The optimal sequence* $\tilde{x}_t$, $t = 0,1,\dots,T$, *solves the two-point discrete- time boundary value problem*

$$\frac{\partial L_t}{\partial x}(x_t,x_{t+1}) + \frac{\partial L_{t-1}}{\partial y}(x_{t-1},x_t) = 0$$

*with given* $x_0$ *and* $x_T$.

**Remark** If the above equations can be solved with respect to the argument $x_{t+1}$, then we obtain an explicit recurrence of the form

$$x_{t+1} = F_t(x_{t-1},x_t).$$

## 9.4 An Application

We shall apply the theoretical results of **9.3** to a simple model for optimizing the budget of a household.

We assume that a family possesses an initial capital $k_0$. If this capital consists of a bank deposit, then it grows in geometric progression: $k_{t+1} = (1+i)\,k_t$, where $i$ is the interest rate. We also assume that, at each period $t$, the family has to withdraw a part $c_t$, to cover its consumption expenses. Eventually, the evolution of the capital will be described by

$$k_{t+1} = (1+i)\,k_t - c_t\,. \tag{9.27}$$

We assume that the utility of consumption is measured by the function $\mathscr{U}(c) = \ln c$ and that the family's objective is to maximize the global utility over the period between $t = 0$ and $T$:

$$\max J(c) = \sum_{t=0}^{T-1} \ln c_t\,. \tag{9.28}$$

We also assume that the family intends to have, at time $T$, a final capital $k_T$.

Thus, we have the following problem: find the sequence

$k_1, \ldots, k_{T-1}$ satisfying $k_{t+1} = (1+i)\,k_t - c_t$;

$$\max J(c) = \sum_{t=0}^{T-1} \ln c_t\,, \text{ where } k_0, k_T \text{ are given.}$$

We remark that a problem with the same structure would have been obtained if we considered the discrete analogue of the economic growth model discussed in **9.2** (with logarithmic utility function and linear production function).

The problem described above fits the general scheme discussed in **9.3**, if we write

$$J(c) = \sum_{t=0}^{T-1} L(k_t, k_{t+1}) = -\sum_{t=0}^{T-1} \ln\left[- k_{t+1} + (1+i)k_t\right].$$

According to the general theory, the necessary optimality condition is

$$\frac{\partial L}{\partial x}\left(\tilde{k}_t, \tilde{k}_{t+1}\right) + \frac{\partial L}{\partial y}\left(\tilde{k}_{t-1}, \tilde{k}_t\right) = 0 \text{ for } t = 1, \ldots, T-1,$$

where $\frac{\partial L}{\partial x}$ stands for the partial derivative with respect to the first argument and $\frac{\partial L}{\partial y}$, for the partial derivative with respect to the second argument.

In our case, the above condition becomes:

$$\frac{1+i}{-\tilde{k}_{t+1} + (1+i)\tilde{k}_t} - \frac{1}{(1+i)\tilde{k}_{t-1} - \tilde{k}_t} = 0\,,\ t = 1, \ldots, T-1, \text{ i.e.}$$

$$(1+i)^2\, \tilde{k}_{t-1} - 2(1+i)\,\tilde{k}_t + \tilde{k}_{t+1} = 0\,. \qquad \textbf{(9.29)}$$

The optimal evolution of capital is given by the solution of the above equation, which satisfies the conditions $\tilde{k}_0 = k_0$ and $\tilde{k}_T = k_T$.

Equation **(9.29)** is a linear equation with constant coefficients; its characteristic equation is

$$q^2 - 2q(1+i) + (1+i)^2 = 0,$$

which has the multiple root $q_1 = q_2 = 1 + i$, hence the general solution of **(9.29)** will be of the form $(\alpha t + \beta)(1+i)^t$.

To determine the coefficients $\alpha$ and $\beta$, we use the initial and the final condition. Thus, $\beta = k_0$ and $\alpha = \frac{1}{T}\left((1+i)^{-T} k_T - k_0\right)$.

Thus, the optimal evolution is described by

$$\tilde{k}_t = (1+i)^t \left[\frac{t}{T}\left((1+i)^{-T} k_T - k_0\right) + k_0\right] =$$

$$= (1+i)^t \left[\frac{t}{T}(1+i)^{-T} k_T + k_0\left(1 - \frac{t}{T}\right)\right].$$

It can be seen immediately that $k_t > 0$, for any $t$.

In its turn,

$$\tilde{c}_t = (1+i)\tilde{k}_t - \tilde{k}_{t+1} = \frac{1}{T}(1+i)^t \left[k_0 - (1+i)^{-T} k_T\right].$$

It is reasonable to require that $0 \le \tilde{c}_t \le (1+i)\tilde{k}_t$, whatever $t$. The constraint $\tilde{c}_t \le (1+i)\tilde{k}_t$ is automatically satisfied, since we have shown that $\tilde{k}_{t+1} = (1+i)\tilde{k}_t - \tilde{c}_t > 0$. The constraint $\tilde{c}_t \ge 0$ reduces to the condition $k_T \le (1+i)^T k_0$.

## 9.5 Unrestricted Optimal Control Problem in Discrete Time

A natural generalization of the problem described in **9.3** is the following.

Let $f_t : G_1 \times G_2 \subset \mathrm{R}^n \times \mathrm{R}^m \to \mathrm{R}^n$,

$L_t : G_1 \times G_2 \to \mathrm{R}$, $t = 0, 1, \ldots T-1$, and $g : G_1 \to \mathrm{R}$ be $\mathscr{C}^1$ functions. Consider the dynamics

$$x_{t+1} = f_t(x_t, u_t);\ x_0 \in G_1 \text{ given: } x_0 = \hat{x}, \tag{9.30}$$

and the cost

$$J(u) = g(x_T) + \sum_{t=0}^{T-1} L_t(x_t, u_t). \tag{9.31}$$

We look for a sequence $u_t$, $t = 0, 1, \dots T-1$ such that $J(u)$ is minimum with $x_t$ defined by **(9.30)**.

We shall consider again the necessary conditions for optimality. Let $\tilde{u}_t$, $t = 0, 1, \dots, T-1$, be the optimal control and $\tilde{x}_t$, $t = 1, \dots, T$, be the corresponding optimal solution.

Take $\varepsilon \in \mathrm{R}$ with $|\varepsilon|$ small enough and $v_t$, $t = 0, 1, \dots T-1$, an arbitrary sequence of vectors in $\mathrm{R}^m$.

Consider the control $u_t = \tilde{u}_t + \varepsilon v_t$ and denote by $x_t(\varepsilon)$ the corresponding solution defined by

$$x_{t+1}(\varepsilon) = f_t(x_t(\varepsilon), \tilde{u}_t + \varepsilon v_t), \; x_0(\varepsilon) = \hat{x}. \tag{9.32}$$

Denote $I(\varepsilon) = g(x_T(\varepsilon)) + \sum_{t=0}^{T-1} L_t(x_t(\varepsilon), \tilde{u}_t + \varepsilon v_t)$.

We remark that $I(\varepsilon)$ is well defined for $|\varepsilon|$ small enough and that $\varepsilon = 0$ leads to the minimal value. Since $g$ and $L_t$, $t = 0, 1, \dots T-1$ are $\mathscr{C}^1$, $I(\cdot)$ is differentiable and we deduce that $I'(0) = 0$.

By the chain rule, we have

$$I'(\varepsilon) = \frac{\partial g}{\partial x}(x_T(\varepsilon)) \frac{\partial x_T(\varepsilon)}{\partial \varepsilon} + \sum_{t=0}^{T-1} \left[ \frac{\partial L_t}{\partial x}(x_t(\varepsilon), \tilde{u}_t + \varepsilon v_t) \frac{\partial x_t(\varepsilon)}{\partial \varepsilon} + \right.$$
$$\left. + \frac{\partial L_t}{\partial u}(x_t(\varepsilon), \tilde{u}_t + \varepsilon v_t) v_t \right].$$

Herefrom, we obtain the necessary optimality conditions in the form

$$\frac{\partial g}{\partial x}(\tilde{x}_T)\frac{\partial x_T(0)}{\partial \varepsilon} + \sum_{t=0}^{T-1}\left[\frac{\partial L_t}{\partial x}(\tilde{x}_t,\tilde{u}_t)\frac{\partial x_t(0)}{\partial \varepsilon} + \frac{\partial L_t}{\partial u}(\tilde{x}_t,\tilde{u}_t)v_t\right] \quad \textbf{(9.33)}$$

Let us denote $z_t = \dfrac{\partial x_t(\varepsilon)}{\partial \varepsilon}$.

If we differentiate equation **(9.32)** with respect to $\varepsilon$ at $\varepsilon = 0$, then we deduce

$$z_{t+1} = \frac{\partial f_t}{\partial x}(\tilde{x}_t,\tilde{u}_t)z_t + \frac{\partial f_t}{\partial u}(\tilde{x}_t,\tilde{u}_t)v_t, \quad z_0 = 0,$$

(in the formulae above, we have used the fact that $x_t(0) = \tilde{x}_t$, a consequence of **(9.32)**).

To transform in a convenient way the necessary conditions **(9.33)**, we introduce a sequence of row vectors $p_t$, $t = T-1,\ldots,0$, defined by

$$p_t = p_{t+1}\frac{\partial f_t}{\partial x}(\tilde{x}_t,\tilde{u}_t) + \frac{\partial L_t}{\partial x}(\tilde{x}_t,\tilde{u}_t),$$

$$p_T = \frac{\partial g}{\partial x}(\tilde{x}_T). \quad \textbf{(9.34)}$$

Postmultiplying relation **(9.34)** by $z_t$, we obtain

$$p_t z_t = p_{t+1}\frac{\partial f_t}{\partial x}(\tilde{x}_t,\tilde{u}_t)z_t + \frac{\partial L_t}{\partial x}(\tilde{x}_t,\tilde{u}_t)z_t.$$

On the other hand,

$$\frac{\partial f_t}{\partial x}(\tilde{x}_t,\tilde{u}_t)z_t = z_{t+1} - \frac{\partial f_t}{\partial u}(\tilde{x}_t,\tilde{u}_t)v_t,$$

hence

$$p_t z_t = p_{t+1}z_{t+1} - p_{t+1}\frac{\partial f_t}{\partial u}(\tilde{x}_t,\tilde{u}_t)v_t + \frac{\partial L_t}{\partial x}(\tilde{x}_t,\tilde{u}_t)z_t.$$

It follows that

$$\frac{\partial L_t}{\partial x}(\tilde{x}_t,\tilde{u}_t)z_t = p_t z_t - p_{t+1}z_{t+1} + p_{t+1}\frac{\partial f_t}{\partial u}(\tilde{x}_t,\tilde{u}_t)v_t,$$

hence

$$\sum_{t=0}^{T-1}\frac{\partial L_t}{\partial x}(\tilde{x}_t,\tilde{u}_t)z_t = p_0 z_0 - p_T z_T + \sum_{t=0}^{T-1}p_{t+1}\frac{\partial f_t}{\partial u}(\tilde{x}_t,\tilde{u}_t)v_t.$$

Replacing this sum into **(9.33)** and taking into account the conditions

$$z_0 = 0 \text{ and } p_T = \frac{\partial g}{\partial x}(\tilde{x}_T),$$

we obtain the necessary optimality conditions

$$\sum_{t=0}^{T-1}\left[p_{t+1}\frac{\partial f_t}{\partial u}(\tilde{x}_t,\tilde{u}_t)+\frac{\partial L_t}{\partial u}(\tilde{x}_t,\tilde{u}_t)\right]v_t = 0 .$$

Since the sequence $v_t$ is arbitrary, we deduce the following result:

**Theorem 9.4** *The optimal pair* $(\tilde{x}_t,\tilde{u}_t)$, *together with the auxiliary sequence* $p_t$, *solve the following problem*:

$$x_{t+1} = f_t(x_t,u_t),\ x_0 = \hat{x},$$

$$p_t = p_{t+1}\frac{\partial f_t}{\partial x}(x_t,u_t)+\frac{\partial L_t}{\partial x}(x_t,u_t),\quad p_T = \frac{\partial g}{\partial x}(x_T),$$

$$p_{t+1}\frac{\partial f_t}{\partial u}(x_t,u_t)+\frac{\partial L_t}{\partial u}(x_t,u_t) = 0 .$$

**Remark 1.** If the last equations can be solved with respect to $u_t$, then $(x_t, p_t)$ solve a specific discrete-time two-point boundary value problem.

**Remark 2.** The problem considered in this section can be viewed also as a generalization of the linear-quadratic problem in discrete time, on finite horizon.

## 9.6 An Application

We consider again the problem of optimizing the budget of a household (see **9.4**).

The evolution of the household's capital is the same as in **9.4**:

$$k_{t+1} = (1+i)\,k_t - c_t \text{, with } k_0 \text{ given.} \tag{9.35}$$

This time, we do not impose a final value of capital, but we modify the objective function, by including a term containing that final value.

Thus, we aim to maximize the function

$$J(c) = \rho\, k(T) + (1-\rho)\sum_{t=0}^{T-1} \ln c_t \text{, with } 0 < \rho < 1. \tag{9.36}$$

The model defined by **(9.35)** and **(9.36)** fits the general scheme described in **9.5**, if we take

$$f(k,c) = (1+i)\,k - c\,,$$

$$g(k) = -\,\rho\, k\,,$$

$$L(k,c) = (\rho - 1)\ln c\,.$$

In our problem, $k$ stands for the state variable and $c$ - for the control variable.

The equation defining the auxiliary variable $p_t$ will be:

$$p_t = (1+i)\,p_{t+1} \text{, with final condition } p_T = -\,\rho\,. \tag{9.37}$$

The solution of equation **(9.37)** is $p_t = (1+i)^{-t}\,p_0$.

To determine $p_0$, we use the final condition. We have

$$p_T = -\,\rho = (1+i)^{-T}\,p_0,$$

hence

$$p_0 = -\,\rho\,(1+i)^T \text{, and } p_t = -\,\rho\,(1+i)^{T-t}.$$

The necessary optimality condition will be

$$-\,p_{t+1} + (\rho - 1)\frac{1}{\tilde{c}_t} = 0 \text{, i.e. } \tilde{c}_t = \frac{1-\rho}{\rho}(1+i)^{t-T}. \tag{9.38}$$

Substituting this value into **(9.35)**, we get the evolution for $\tilde{k}_t$:

$$\tilde{k}_{t+1} = (1+i)\,\tilde{k}_t - \frac{1-\rho}{\rho}(1+i)^{t-T},\ \tilde{k}_0 = k_0\,. \tag{9.39}$$

The general solution of the affine equation **(9.39)** is the sum of the general solution $\bar{k}_t = (1+i)^t q$ of the associated linear equation and a particular solution $\hat{k}_t$ of the affine equation.

To find the solution $\hat{k}_t$, we assume it to be of the form

$$\hat{k}_t = (1+i)^t (a\,t + b).$$

Substituting into **(9.39)**, we get

$$\left[a(t+1)+b\right](1+i)^{t+1} = (a\,t+b)(1+i)^{t+1} - \frac{1-\rho}{\rho}(1+i)^{t-T}.$$

Simplifying by $(1+i)^t$, we obtain

$$a = -\frac{1-\rho}{\rho}(1+i)^{-(T+1)},$$

hence

$$\hat{k}_t = \left(b - \frac{1-\rho}{\rho}\frac{t}{(1+i)^{T+1}}\right)(1+i)^t.$$

We deduce that

$$\tilde{k}_t = \bar{k}_t + \hat{k}_t = (1+i)^t\left[m - \frac{1-\rho}{\rho}\frac{t}{(1+i)^{T+1}}\right],$$

where $m = q + b$. We have $\tilde{k}_0 = m = k_0$, hence

$$\tilde{k}_t = (1+i)^t\left[k_0 - \frac{1-\rho}{\rho}\frac{t}{(1+i)^{T+1}}\right].$$

The natural requirement that $\tilde{k}_t$ be positive for all $t$ leads to the condition $k_0 \geq \dfrac{1-\rho}{\rho}\dfrac{T}{(1+i)^{T+1}}$.

If this condition is satisfied, then we also have $\tilde{c}_t \leq (1+i)\tilde{k}_t$ for all $t$, since $(1+i)\tilde{k}_t - \tilde{c}_t = \tilde{k}_{t+1} \geq 0$.

The condition $\tilde{c}_t \geq 0$ is automatically satisfied, as it can be seen from **(9.38)**

From **(9.38)** we can also remark that $\tilde{c}_t$ is an increasing function of $t$.

## 9.7 Optimization with Linear Dynamics and Linear Cost. Continuous Time

We consider the general problem, in which the dynamics is linear, with variable coefficients:

$$x' = A(t)\,x + B(t)\,u\,,\ x(t_0) = x_0\,, \tag{9.40}$$

and the cost is of the form

$$J(u) = \big(q, x(T)\big) + \int_{t_0}^{T} \big[\big(c(t), x(t)\big) + \big(d(t), u(t)\big)\big]\,dt\,. \tag{9.41}$$

In order that the problem be well formulated, we introduce <u>*constraints*</u> upon the admissible controls; we consider admissible only the controls whose values belong to a given set $U$ and we admit that the control functions $u(\cdot)$ be only *piece-wise continuous*, with possible *jumps*.

We look for those admissible controls which minimize the cost.

Begin by transforming the expression of the cost, by means of an auxiliary function.

Consider the function $p$ defined for $t < T$ and the Cauchy problem

$$p' = -\,A^*(t)\,p + c(t),\ p(T) = -q\,. \tag{9.42}$$

From the above formula, we have

$$c(t) = p'(t) + A^*(t)\,p(t)\,.$$

Substituting this expression into **(9.41)**, we have:

$$J(u) = -\,\big(p(T), x(T)\big) + \int_{t_0}^{T} \Big[\Big(p'(t) + A^*(t)\,p(t), x(t)\Big) +$$

$$+ (d(t), u(t))\big] dt = - (p(T), x(T)) +$$

$$+ \int_{t_0}^{T} \big[(p'(t), x(t)) + (p(t), A(t)\, x(t)) + (d(t), u(t))\big] dt =$$

$$= - (p(T), x(T)) + \int_{t_0}^{T} \big[(p'(t), x(t)) + (p(t), x'(t) - B(t)\, u(t)) +$$

$$+ (d(t), u(t))\big] dt = - (p(T), x(T)) +$$

$$+ \int_{t_0}^{T} \left[\frac{d}{dt}(p(t), x(t)) + \big(d(t) - B^*(t)\, p(t), u(t)\big)\right] dt =$$

$$= - \big(p(t_0), x_0\big) + \int_{t_0}^{T} \big(d(t) - B^*(t)\, p(t), u(t)\big)\, dt\,.$$

In this way, the problem reduced to that of finding a piece-wise continuous function $u$, such that $u(t) \in U$ for $t \in [t_0, T]$ and $\int_{t_0}^{T} \big(d(t) - B^*(t)\, p(t), u(t)\big)\, dt$ be minimized.

Obviously, this last condition will be satisfied if at each $t$ for which u is continuous, the scalar product

$$\big(d(t) - B^*(t)\, p(t), u(t)\big)$$

will be minimized.

In this way, we reduced our *dynamic* minimization problem to a *static* minimization problem, depending on the parameter $t$.

To get a better understating of the situation, let us assume that $U$ is characterized by the simple inequalities $\underline{u}_i \le u^i \le \overline{u}_i$, where $u^i$, $i = 1, \ldots, m$ are the components of vector $u$.

We have to minimize, under these constraints, the function

$$\sum_{i=1}^{m} \left[ d^i(t) - \sum_{j=1}^{n} b^i_j(t)\, p^j(t) \right] u^i(t);$$

it can be seen that we get the minimum value by choosing

$$u^i(t) = \underline{u}^i \text{ if } d^i(t) - \sum_{j=1}^{n} b^i_j(t)\, p^j(t) > 0\,,$$

$$u^i(t) = \overline{u}^i \text{ if } d^i(t) - \sum_{j=1}^{n} b^i_j(t)\, p^j(t) < 0\,.$$

The above situation makes clear the reason for admitting jumps for the function $u$; indeed, the solution described above has jumps at the points $t$ such that

$$d^i(t) - \sum_{j=1}^{n} b^i_j(t)\, p^j(t) = 0\,.$$

In general, if the set $U$ is a convex polyhedron (described by linear inequalities and equalities), then the problem of determining the optimal value for the control $u(t)$ reduces, for each fixed $t$, to a *linear programming problem* (minimization of a linear function with linear constraints). Since the *data* of the linear programming problem depend on a parameter $t$, we have a *parametric linear programming problem*.

When $U$ is a convex polyhedron, it is known that, for each $t$, the minimum value is obtained if $u(t)$ is a vertex of the polyhedron.

When we have a *scalar* control and $U$ is an interval, then the function $u$ minimizing the cost takes values only at the boundary of the interval.

Richard Bellman and his co-workers called this type of control "*bang-bang*"; this term was extended to all situations in which the control is piece-wise constant, taking values in the vertices of the polyhedron $U$.

To end this section, let us remark that, in the above discussion, we have shown directly a procedure for determining an optimal control.

We could have shown, by using a special class of variations, that any optimal control must be a solution of the parametric programming problem described above. We do not insist here on this aspect, since we shall be able to deduce the conclusion from a general theory of necessary optimality conditions, in Chapter 10.

## 9.8 Some Microeconomic Models

### 9.8.1 Optimization of the Maintenance of an Equipment

We denote by $V(t)$ the value of the equipment at time $t$ and we admit that the evolution of this value is described by the law

$$V'(t) = -\, a(t)\, V(t) + b(t)\, m(t),$$

where $a(t)$ represents the coefficient of wear and tear, and $b(t)\, m(t)$ is the effect of maintenance activities ($m(t)$ being the expenses for maintenance). To a value $V(t)$ of the equipment, we associate an income $R(t) = r\, V(t)$.

The profit over the time interval $[\, t_0, T\,]$ is given by

$$B = -\, V_0 + V(T)\, e^{-\delta(T-t_0)} + \int_{t_0}^{T} [r\, V(t) - m(t)]\, e^{-\delta t}\, dt .$$

Here $V_0$ represents the initial value, $V(T)\, e^{-\delta(T-t_0)}$ is the discounted value of the equipment at time T and the integral represents the "discounted sum" of net income (the difference between gross income $R(t)$ and the expenses for maintenance $m(t)$).

The problem consists in choosing a maintenance policy which ensures the maximization of profit.

Since in **9.7** we discussed a minimization problem, we shall define

$$J(u) = -\, x(T)\, e^{-\delta(T-t_0)} + \int_{t_0}^{T} [u(t) - r\, x(t)]\, e^{-\delta t}\, dt\,,$$

$$x' = -\, a(t)\, x + b(t)\, u\,;\ x(t_0) = V_0\,;\ 0 \le u \le M\,.$$

The problem fits the general scheme. We have

$$q = -\, e^{-\delta(T-t_0)},\ d(t) = e^{-\delta t},\ c(t) = -r\, e^{-\delta t},$$

$$p'(t) = a(t)\, p(t) - r\, e^{-\delta t},\ p(T) = e^{-\delta(T-t_0)}.$$

The optimal control is obtained by minimizing

$\left[e^{-\delta t} - b(t)\, p(t)\right] u$ on the set $0 \le u \le M$.

Thus, we obtain

$$\tilde{u}(t) = 0 \text{ if } b(t)\, p(t) < e^{-\delta t} \text{ and } \tilde{u}(t) = M \text{ if } b(t) p(t) > e^{-\delta t}.$$

The optimal behaviour is of "bang-bang" type. While $b(t)\, p(t) < e^{-\delta t}$, nothing is spent for maintenance; when $b(t) p(t) > e^{-\delta t}$, then maximum effort is made for maintenance.

### 9.8.2 A Financial Policy Model at Enterprise Level

We denote by $K(t)$ the capital and by $N(t)$ the debt at time t. The evolution speed of liquidities is given by $r\, K(t) - \delta\, N(t)$, the first term representing the income associated with the capital $K(t)$ and the second–the payments.

Since the capital increment is distributed for investment and dividends, we have

$$r\, K(t) - \delta\, N(t) + N'(t) = I(t) + D(t)\,.$$

Considering also the usual equation for the evolution of capital:

$$K' = -\,a\,K(t) + I(t),$$

we obtain the following system of differential equations:

$$K' = -\,a\,K + I;$$

$$N' = -\,r\,K + \delta\,N + I + D.$$

The state of the system is described by the level of capital and the level of debt. The control (decision) variables are the flow of investments and the flow of dividends; these flows are restricted by the constraints

$$0 \le I \le I_{\max}, \quad 0 \le D \le D_{\max}.$$

Our objective-function will represent a "compromise" between income and debt:

$$J(I,D) = \rho\left[\alpha\,N(T) - \beta\,K(T)\right] +$$

$$+ \int_0^T \left[\alpha\,N(t) - \beta\,K(t) - \gamma\,D(t)\right] e^{\varepsilon t} dt.$$

We intend to *minimize* this function. This problem fits the general scheme described in **9.7**, if we take

$$A = \begin{pmatrix} -a & 0 \\ -r & \delta \end{pmatrix}, \quad B = \begin{pmatrix} 1 & 0 \\ 1 & 1 \end{pmatrix};$$

$$q = \begin{pmatrix} -\rho\beta \\ \rho\alpha \end{pmatrix}, \quad c(t) = \begin{pmatrix} -\beta\,e^{\varepsilon t} \\ \alpha\,e^{\varepsilon t} \end{pmatrix}, \quad d(t) = \begin{pmatrix} 0 \\ -\gamma\,e^{\varepsilon t} \end{pmatrix}.$$

The system for $p$ is:

$$\begin{pmatrix} p_1'(t) \\ p_2'(t) \end{pmatrix} = \begin{pmatrix} a & r \\ 0 & -\delta \end{pmatrix} \begin{pmatrix} p_1(t) \\ p_2(t) \end{pmatrix} + \begin{pmatrix} -\beta\,e^{\varepsilon t} \\ \alpha\,e^{\varepsilon t} \end{pmatrix}, \quad \begin{pmatrix} p_1(T) \\ p_2(T) \end{pmatrix} = \begin{pmatrix} \rho\beta \\ -\rho\alpha \end{pmatrix}.$$

We have

$$B^* p(t) = \begin{pmatrix} 1 & 1 \\ 0 & 1 \end{pmatrix} \begin{pmatrix} p_1(t) \\ p_2(t) \end{pmatrix} = \begin{pmatrix} p_1(t) + p_2(t) \\ p_2(t) \end{pmatrix},$$

$$d(t) - B^* p(t) = \begin{pmatrix} -p_1(t) - p_2(t) \\ -\gamma e^{\varepsilon t} - p_2(t) \end{pmatrix}.$$

The parametric linear programming problem is

$$\min\left[\left(-p_1(t) - p_2(t)\right) u_1(t) - \left(\gamma e^{\varepsilon t} + p_2(t)\right) u_2(t)\right],$$

$$0 \le u_1(t) \le I_{\max}, \quad 0 \le u_2(t) \le D_{\max};$$

or

$$\max\left[\left(p_1(t) + p_2(t)\right) u_1(t) + \left(\gamma e^{\varepsilon t} + p_2(t)\right) u_2(t)\right],$$

with the same constraints.

Since $u_1$ and $u_2$ are varying independently, we have:

$$\tilde{u}_1(t) = 0 \text{ if } p_1(t) + p_2(t) < 0,$$

$$\tilde{u}_1(t) = I_{\max} \text{ if } p_1(t) + p_2(t) > 0,$$

$$\tilde{u}_2(t) = 0 \text{ if } \gamma e^{\varepsilon t} + p_2(t) < 0,$$

$$\tilde{u}_2(t) = D_{\max} \text{ if } \gamma e^{\varepsilon t} + p_2(t) > 0.$$

The problem can be solved explicitly. We have

$$p_2'(t) = -\delta p_2(t) + \alpha e^{\varepsilon t}, \quad p_2(T) = -\rho\alpha,$$

hence

$$p_2(t) = -\rho\alpha e^{-\delta(t-T)} + \int_T^t e^{-\delta(t-\tau)} \alpha e^{\varepsilon\tau} d\tau =$$

$$= -\rho\alpha e^{\delta(T-t)} + \alpha e^{-\delta t} \frac{1}{\delta+\varepsilon}\left[e^{(\delta+\varepsilon)t} - e^{(\delta+\varepsilon)T}\right];$$

it follows that we can describe explicitly the optimal choice for $u_2$, i.e. for $D$. Further,

$$p_1'(t) = a\, p_1(t) + r\, p_2(t) - \beta e^{\varepsilon t}, \quad p_1(T) = \rho\beta,$$

and, having the explicit expression for $p_2$, we can also determine $p_1$; finally, we also obtain an explicit form for the optimal investment policy.

## 9.9 Optimization with Linear Dynamics and Linear Cost. Discrete Time

We consider now the discrete analogue of the problem discussed in **9.7**.

The dynamics is described by

$$x_{t+1} = A_t\, x_t + B_t\, u_t \tag{9.43}$$

and the cost is

$$J(u) = (q, x_T) + \sum_{t=t_0}^{T-1} \left[ (c_t, x_t) + (d_t, u_t) \right]. \tag{9.44}$$

As in the continuous-time case, we also consider the constraint $u_t \in U$.

As in the continuous-time case, we shall transform the expression of the cost, using the system

$$p_t = A_t^*\, p_{t+1} + c_t, \quad p_T = q\,.$$

The sequence $p_t$ can be determined by backward recurrence, for $t < T$. Using this sequence, we have

$$c_t = p_t - A_t^*\, p_{t+1}$$

and, substituting into **(9.44)**, we obtain

$$J(u) = (p_T, x_T) + \sum_{t=t_0}^{T-1} \left[ \left( p_t - A^* p_{t+1}, x_t \right) + (d_t, u_t) \right] =$$

$$= (p_T, x_T) + \sum_{t=t_0}^{T-1} \left[ (p_t, x_t) - (p_{t+1}, A_t x_t) + (d_t, u_t) \right] =$$

$$= (p_T, x_T) + \sum_{t=t_0}^{T-1} \left[ (p_t, x_t) - (p_{t+1}, x_{t+1} - B_t u_t) + (d_t, u_t) \right] =$$

$$= (p_T, x_T) + \sum_{t=t_0}^{T-1} \left[ (p_t, x_t) - (p_{t+1}, x_{t+1}) \right] +$$

$$+ \sum_{t=t_0}^{T-1} \left( d_t + B_t^* p_{t+1}, u_t \right) = \left( p_{t_0}, x_{t_0} \right) + \sum_{t=t_0}^{T-1} \left( d_t + B_t^* p_{t+1}, u_t \right).$$

This formula shows that the minimum value of the cost is obtained if, for each $t$, $u_t$ minimizes the scalar product

$$\left( d_t + B_t^* p_{t+1}, u \right) \text{ for } u \in U.$$

As in **9.7**, if $U$ is a convex polyhedron, then we have to solve, at each step, a *linear programming problem*.

However, we remark that the discrete-time problem may be considered, from the beginning, as a large size linear programming problem, with variables $x_{t_0}, \dots x_T$, $u_{t_0}, \dots u_{T-1}$; the dynamic equations appear as *linear equality constraints*.

Our reasoning took into account the *dynamic* structure of the problem, reducing the large size linear programming problem to a sequence of problems of smaller size.

In the special case when the control constraints are of the form $\underline{u}^i \leq u^i \leq \overline{u}^i$, the i-th component of the optimal control is

$$\tilde{u}_t^i = \underline{u}^i \text{ if } \left( d_t + B_t^* p_{t+1} \right)^i > 0$$

and

$$\tilde{u}_t^i = \overline{u}^i \text{ if } \left( d_t + B_t^* p_{t+1} \right)^i < 0.$$

## 9.10 Applications

**9.10.1** We consider the discrete-time variant of the model for the optimization of the maintenance of an equipment (see **9.8.1**).

The evolution of the value $V_t$ of the equipment is described by

$$V_{t+1} - V_t = - a_t V_t + b_t m_t ;$$

the profit is given by

$$B = - V_0 + V_T s^{T-t_0} + \sum_{t=t_0}^{T-1} (r V_t - m_t) s^t , \text{ where } 0 < s < 1.$$

In order to maximize the profit, we minimize the function

$$J(u) = - s^{T-t_0} x_T + \sum_{t=t_0}^{T-1} (u_t - r x_t) s^t ,$$

with

$$x_{t+1} = (1 - a_t) x_t + b_t u_t ; \; 0 \le u_t \le M .$$

We have

$$q = - s^{T-t_0} , \; c_t = - r s^t , \quad d_t = s^t ,$$

$$p_t = (1 - a_t) p_{t+1} - r s^t , \quad p_T = - s^{T-t_0} .$$

We can obtain simple explicit formulae, if $a_t$ is constant.

**9.10.2** The discrete-time variant of the model of financial policy of an enterprise (see **9.8.2**) is the following:

$$K_{t+1} - K_t = - a K_t + I_t$$

$$N_{t+1} - N_t = - r K_t + \delta N_t + I_t + D_t ,$$

$$0 \le I_t \le I^{\max} , \quad 0 \le D_t \le D^{\max} ,$$

$$\min_{I,D}\left[\rho(\delta N_T - \beta K_T) + \sum_{t=0}^{T-1}(\delta N_t - \beta K_t - \gamma D_t)s^t\right].$$

We have

$$A_t = \begin{pmatrix} 1-a & 0 \\ -r & 1+\delta \end{pmatrix},\ B_t = \begin{pmatrix} 1 & 0 \\ 1 & 1 \end{pmatrix},$$

$$q = \begin{pmatrix} -\rho\beta \\ \rho\alpha \end{pmatrix},\ c_t = \begin{pmatrix} -\beta s^t \\ \alpha s^t \end{pmatrix},\ d_t = \begin{pmatrix} 0 \\ -\gamma s^t \end{pmatrix}.$$

The system for the vector $p_t$ is:

$$p_t = \begin{pmatrix} 1-a & -r \\ 0 & 1+\delta \end{pmatrix} p_{t+1} + \begin{pmatrix} -\beta \\ \alpha \end{pmatrix} s^t,\ p_T = \rho\begin{pmatrix} -\beta \\ \alpha \end{pmatrix}.$$

As in the continuous-time case, the vector $p_t$ can be determined explicitly. We have

$$p_t^2 = (1+\delta)\,p_{t+1}^2 + \alpha s^t,\ p_T^2 = \alpha\rho. \tag{9.45}$$

To find a particular solution, we assume $\hat{p}_t^2 = h s^t$; we have $h s^t = (1+\delta) h s^{t+1} + \alpha s^t$, wherefrom

$$h = \frac{\alpha}{1 - s(1+\delta)}$$

and the particular solution is

$$\hat{p}_t^2 = \frac{\alpha s^t}{1 - s(1+\delta)}.$$

The general solution of the associated linear equation is

$$\tilde{p}_t^2 = v(1+\delta)^{T-t},$$

hence the general solution of equation **(9.45)** will be of the form

$$p_t^2 = \frac{\alpha s^t}{1 - s(1+\delta)} + v(1+\delta)^{T-t}.$$

Taking into account the condition $p_T^2 = \alpha\rho$, we deduce

$$\rho\alpha = v + \frac{\alpha s^T}{1 - s(1+\delta)},$$

wherefrom

$$v = \alpha\left[\rho - \frac{s^T}{1 - s(1+\delta)}\right]$$

and

$$p_t^2 = \frac{\alpha\, s^t}{1 - s(1+\delta)} + \alpha\left[\rho - \frac{s^T}{1 - s(1+\delta)}\right](1+\delta)^{T-t}.$$

Having this explicit formula for $p_t^2$, we can find the explicit expression for $p_t^1$ and, then, the explicit structure of the optimal controls $\tilde{u}_t^1$, $\tilde{u}_t^2$.

## 9.11 Notes and References

The mathematical topics considered in this chapter are rather simple and well known; it is however difficult to indicate specific references for the simple situations described.

What concerns the economic models, they are also spread in different places. The models discussed in **9.8** and **9.10** are inspired from [10].

For economic growth optimization models, a classical reference may be considered Cass [6].

CHAPTER 10

# General Problem of Optimal Control

## 10.1 Problem Statement. General Theorems.

### 10.1.1 The Optimal Control Problem

The optimization problems studied in the preceeding chapters can be joined in a general framework. We shall describe this framework.

The problem is defined by the open, connected sets $I \subset R$, $G_1 \subset R^n$, $G_2 \subset R^m$, $B \subset R^s$, $U \subset G_2$ and by the functions

$f: I \times G_1 \times G_2 \times B \to R^n$,

$L_\gamma: I \times G_1 \times G_2 \times B \to R$, $\gamma = 0, 1, \dots, l$,

$g_\gamma: B \to R$, $\gamma = 0, 1, \dots, l$,

$\alpha: B \to G_1$, $\omega: B \to G_1$, $t_0: B \to I$, $T: B \to I$.

A <u>control</u> is a pair $(b, u)$, where $b \in B$ and

$u: [t_0(b), T(b)] \to U$ is a piece-wise continuous function.

We associate to any control the dynamics of state variables, described by $x' = f(t, x, u(t), b)$ and by the initial conditions $x(t_0(b)) = \alpha(b)$.

We also associate to the control $(b, u)$ the numbers

$$I_\gamma(b,u) = g_\gamma(b) + \int_{t_0(b)}^{T(b)} L_\gamma(t, x(t), u(t), b)\, dt,$$

where $x(t)$ is the solution of the above initial-value problem.

A control $(b, u)$ is <u>*admissible*</u> if the solution of the corresponding Cauchy problem is defined on the whole interval $[t_0(b), T(b)]$, satisfies the final condition $x(T(b)) = \omega(b)$ and the constraints

$$I_\gamma(b,u) \le 0 \text{ for } \gamma = 1, \dots, l',$$

$$I_\gamma(b,u) = 0 \text{ for } \gamma = l' + 1, \dots, l.$$

A control $(\tilde{b}, \tilde{u})$ is <u>*optimal*</u> if it is admissible and if

$$I_0(\tilde{b}, \tilde{u}) \le I_0(b, u)$$

for any admissible control $(b, u)$.

To get a better understanding of these concepts, it is useful to consider some special cases.

**a)** The functions $t_0$, $T$, $\alpha$, are constant, $l = 0$, $B = G_1$, $\omega$ is the identity mappine $\omega(b) = b$, $f$ and $L_0$ do not depend on $b$.

In this case, the problem is independent of $b$; a control is a piece-wise continuous function $u: [t_0, T] \to U$.

To a given control $u$, we associate the evolution given by

$$x' = f(t, x, u(t)),\ x(t_0) = \alpha$$

and the cost

$$I_0(u) = g_0(x(T)) + \int_{t_0}^{T} L_0(t, x(t), u(t))\, dt .$$

The control is admissible if the corresponding solution is defined on the whole interval $[t_0, T]$.

We remark that the problem described above is the continuous-time analogue of the problem discussed in **9.5**.

We also remark that the optimization problem with linear dynamics and quadratic cost fits the above scheme, with

$$f(t,x,u) = A\,x + B\,u,\ g_0(x) = (G\,x, x),$$

$$L_0(t,x,u) = (Q\,x, x) + \mu(u,u),$$

$$G_1 = \mathrm{R}^n,\ G_2 = \mathrm{R}^m,\ U = \mathrm{R}^m .$$

Another special case corresponds to a linear cost.

The problems with $L_0 \equiv 0$ are called *problems with terminal cost*, while those with $g_0 = 0$ are called *Lagrange type problems*.

**b)** We admit again that $t_0$, $T$, $\alpha$ are constant, $l = 0$, and $f$, $L_0$ are independent of $b$; we assume $\omega$ constant and $g_0 = 0$.

In this case, a control is a piece-wise continuous function $u: [t_0, T] \to U$ and it is admissible if the solution of the Cauchy problem $x' = f(t,x,u(t))$, $x(t_0) = \alpha$ is defined on $[t_0, T]$ and satisfies $x(T) = \omega$.

The control is optimal if it minimizes the cost

$$\int_{t_0}^{T} L_0(t, x(t), u(t))\, dt$$

within the class of admissible controls.

The simplest problem of Calculus of Variations fits this case, with $f(t,x,u) = u$.

**c)** We assume again that $t_0$, $\alpha$, $\omega$ are constant, $l = 0$, $f$ and $L_0$ are independent of $b$. We also assume that $T$ is the identity mapping $T(b) = b$. This means that the final time is now free and the solution has to reach, under the action of the control, the target $\omega$. If $g_0 = 0$, $L_0 = 1$, then an optimal solution is a solution that reaches the final point $\omega$ in the shortest time.

### 10.1.2 Necessary optimality conditions

For the general problem formulated above, we shall obtain *necessary optimality conditions*, characterizations of the optimal solution which, in some cases, can lead to its identification.

We shall now state the two fundamental results, one concerning the case of fixed horizon and the other concerning the case of free initial and final time.

**Theorem 10.1** *Let $t_0$, $T$ be constant, $(\tilde{b}, \tilde{u})$ – an optimal control and $\tilde{x}$ – the corresponding solution of the problem:*

$$x' = f(t, x, \tilde{u}), \; x(t_0) = \alpha(\tilde{b}).$$

*Then there exist the numbers $\lambda_0, \lambda_1, \ldots, \lambda_{l+n}$ (Lagrange multipliers), at least one of them non-zero, such that:*

**1)** $\lambda_\gamma \geq 0$ *for* $\gamma = 0, 1, \ldots, l$ *and moreover,* $\lambda_\gamma I_\gamma(\tilde{b}, \tilde{u}) = 0$ *for* $1 \leq \gamma \leq l$.

**2)** *Let $p$ be the solution of the system*

$$p' = -A^*(t)\, p + \sum_{\gamma=0}^{l} \lambda_\gamma c_\gamma(t), \; p_j(T) = -\lambda_{l+j},$$

*where*

$$A(t) = \frac{\partial f}{\partial x}\left(t, \tilde{x}(t), \tilde{u}(t), \tilde{b}\right),$$

$$c_\gamma(t) = \frac{\partial L_\gamma}{\partial x}\left(t, \tilde{x}(t), \tilde{u}(t), \tilde{b}\right) \text{ (written as column-vector).}$$

*We denote*

$$H(t, x, u, b, p) = \big(p, f(t, x, u, b)\big) - \sum_{\gamma=0}^{l} \lambda_\gamma L_\gamma(t, x, u, b).$$

*Then, at any point t where* $\tilde{u}$ *is continuous, we have*

$$H\left(t, \tilde{x}(t), \tilde{u}(t), \tilde{b}, p(t)\right) \geq H\left(t, \tilde{x}(t), u, \tilde{b}, p(t)\right) \text{ for any } u \in U.$$

This property of the optimal control is known as "*Pontryagin's Maximum Principle*".

**3)** *The following relations hold:*

$$\sum_{\gamma=0}^{l} \lambda_\gamma \frac{\partial g_\gamma}{\partial b^\sigma}\left(\tilde{b}\right) + \left(p(T), \frac{\partial \omega}{\partial b^\sigma}\left(\tilde{b}\right)\right) - \left(p(t_0), \frac{\partial \alpha}{\partial b^\sigma}\left(\tilde{b}\right)\right) -$$

$$- \int_{t_0}^{T} \frac{\partial H}{\partial b^\sigma}\left(t, \tilde{x}(t), \tilde{u}(t), \tilde{b}, p(t)\right) dt = 0, \quad \sigma = 1, 2, \ldots, s.$$

These relations are called "*transversality conditions*".

It is instructive to see what gives this theorem in some special cases.

**a)** We assume $\alpha$ constant, $l = 0$, $B = G_1$, $\omega$–the identity mapping ($\omega(b) = b$), $f$ independent of $b$, $L_0 \equiv 0$, $g(b) = (c, b)$.

Condition **1)** implies $\lambda_0 \geq 0$.

The equation for $p$ is $p' = -A^*(t)p$, with final conditions $p_j(T) = -\lambda_j$. The function $H$ becomes

$$H(t, x, u, p) = (p, f(t, x, u)),$$

while the transversality conditions become

$$\lambda_0 \frac{\partial g_0}{\partial b^\sigma}\left(\tilde{b}\right) + p_\sigma(T) = 0\,, \text{ hence } \lambda_0\, c_\sigma - \lambda_\sigma = 0\,.$$

It follows $\lambda_0 > 0$, since otherwise we would have all $\lambda_\sigma = 0$. We may choose $\lambda_0 = 1$ and, then, we have $\lambda_j = -\, c_j$.

If the dynamics is linear: $f(t,x,u) = A(t)\,x + B(t)\,u$, then $p$ can be determined independently of the optimal control and the condition given by the Maximum Principle becomes:

$$H\big(t, \tilde{x}(t), \tilde{u}(t), p(t)\big) = \big(p(t), A(t)\,\tilde{x}(t) + B(t)\,\tilde{u}(t)\big) \geq$$

$$\geq H\big(t, \tilde{x}(t), u, p(t)\big) = \big(p(t), A(t)\,\tilde{x}(t) + B(t)\,u\big)$$

for any $u \in U$. This condition reduces to

$$\big(B^*(t)\,p(t), \tilde{u}(t)\big) \geq \big(B^*(t)\,p(t), u\big)$$

for any $u \in U$ and any continuity point $t$ for $\tilde{u}$. In this way, the optimal control is characterized, at any continuity point, as a solution of a static optimization problem.

**b)** In the simplest problem of the Calculus of Variations, we have $t_0$, $T$ constant, $g_0 = 0$, $l = 0$, $L_0$ independent of $b$; the functions $\alpha$ and $\omega$ are constant, and $f(t,x,u) = u$.

The equation for $p$ becomes $p' = \lambda_0 \dfrac{\partial L_0}{\partial x}\big(t, \tilde{x}(t), \tilde{u}(t)\big)$.

The Maximum Principle gives

$$\big(p(t), \tilde{u}(t)\big) - \lambda_0\, L_0\big(t, \tilde{x}(t), \tilde{u}(t)\big) \geq \big(p(t), u\big) - \lambda_0\, L_0\big(t, \tilde{x}(t), u\big)$$

for any $u$.

Since we have no constraint on $u$, we may apply the Fermat theorem, which leads to

$$p(t) - \lambda_0 \frac{\partial L_0}{\partial u}\big(t, \tilde{x}(t), \tilde{u}(t)\big) = 0\,.$$

In this case, the transversality conditions provide no information.

From the equation satisfied by $p$, it follows

$$p(t) = p(t_0) + \lambda_0 \int_{t_0}^{t} \frac{\partial L_0}{\partial x}(s, \tilde{x}(s), \tilde{u}(s))\, ds = \lambda_0 \frac{\partial L_0}{\partial u}(t, \tilde{x}(t), \tilde{u}(t)).$$

If we had $\lambda_0 = 0$, then it would follow that $p(t) \equiv 0$, hence $p(T) = 0$, therefore all multipliers would vanish.

It follows that $\lambda_0 > 0$ and we deduce

$$\frac{\partial L_0}{\partial u}(t, \tilde{x}(t), \tilde{u}(t)) - \int_{t_0}^{t} \frac{\partial L_0}{\partial x}(s, \tilde{x}(s), \tilde{u}(s))\, ds = const.$$

By differentiating with respect to $t$ the above formula, we rediscover the Euler–Lagrange equations.

**c)** In the linear-quadratic optimization problem, the system satisfied by $p$ becomes

$$p'(t) = -\, A^*(t)\, p(t) + 2\, \lambda_0\, Q(t)\, \tilde{x}(t);$$

the transversality conditions give: $2\, \lambda_0\, G\, \tilde{x}(T) + p(T) = 0$, hence

$$p(T) = -\, 2\, \lambda_0\, G\, \tilde{x}(T).$$

Again, it can be seen that $\lambda_0 > 0$ and, for $\lambda_0 = \frac{1}{2}$, we rediscover the system we had in **7.1**.

The Maximum Principle gives

$$(p(t), A(t)\, \tilde{x}(t) + B(t)\, \tilde{u}(t)) - \frac{1}{2}\left[(Q(t)\, \tilde{x}(t), \tilde{x}(t)) + \mu\, (\tilde{u}(t), \tilde{u}(t))\right] \geq$$

$$\geq (p(t), A(t)\, \tilde{x}(t) + B(t)\, u) - \frac{1}{2}\left[(Q(t)\, \tilde{x}(t), \tilde{x}(t)) + \mu\, (u, u)\right],$$

what reduces to

$$\left(B^*(t)\, p(t), \tilde{u}(t)\right) - \frac{\mu}{2}(\tilde{u}(t), \tilde{u}(t)) \geq \left(B^*(t)\, p(t), u\right) - \frac{\mu}{2}(u, u)$$

for any $u$. Since we have no constraints on $u$, the above relation implies:

$$B^*(t)\,p(t) - \mu\,\widetilde{u}(t) = 0\,, \quad \text{or} \quad \widetilde{u}(t) = \frac{1}{\mu}\,B^*(t)\,p(t)\,,$$

the same result that we obtained directly in **7.1**.

Let us make the important remark that, in the general problem considered, we admitted final condition, as well as inequality and equality constraints of the form

$$I_\gamma \le 0\,,\ \gamma = 1,\dots,l'\,,\ I_\gamma = 0\,,\ \gamma = l'+1,\dots,l\,,$$

but we did not admit *state* constraints.

However, such constraints occur frequently in economic problems; they are neglected at a first stage and, after the solution is obtained, we have to check whether it is valid from the viewpoint of economic interpretations.

We shall now state the general result in the case when $t_0$ and $T$ are no longer assumed to be constant.

**Theorem 10.2** *Let* $\left(\widetilde{b},\widetilde{u}\right)$ *be the optimal control and* $\widetilde{x}$ *– the corresponding optimal solution. Then there exist the numbers* $\lambda_0 \ge 0, \lambda_1,\dots,\lambda_{l+n+1}$, *at least one of them being non-zero, such that*

**1)** $\lambda_\gamma \ge 0$ *for* $1 \le \gamma \le l$, $\lambda_\gamma\, I_\gamma\left(\widetilde{u},\widetilde{b}\right) = 0\,,\ 1 \le \gamma \le l$;

**2)** *Let p be the solution of the problem*

$$p' = -\,A^*(t)\,p + \sum_{\gamma=0}^{l} \lambda_\gamma\, c_\gamma(t)\,,\quad p_j\left(T\left(\widetilde{b}\right)\right) = -\,\lambda_{l+j}\,,$$

*where A and* $c_\gamma$ *are the same as in* Theorem 10.1.

*Let H be defined as in* Theorem 10.1. *Then*

$$H\left(t,\widetilde{x}(t),\widetilde{u}(t),\widetilde{b},p(t)\right) \ge H\left(t,\widetilde{x}(t),u,\widetilde{b},p(t)\right)$$

*for any* $u \in U$ *and any continuity point t* of $\widetilde{u}$.

*Moreover,*

$$H\left(t,\tilde{x}(t),\tilde{u}(t),\tilde{b},p(t)\right)=\lambda_{n+l+1}-$$

$$-\int_t^{T(\tilde{b})}\frac{\partial H}{\partial t}\left(\tau,\tilde{x}(\tau),\tilde{u}(\tau),\tilde{b},p(\tau)\right)d\tau .$$

**3)** *The following relations hold:*

$$\sum_{\gamma=0}^{l}\lambda_\gamma\frac{\partial g_\gamma}{\partial b^\sigma}\left(\tilde{b}\right)+\left(p\left(T\left(\tilde{b}\right)\right),\frac{\partial\omega}{\partial b^\sigma}\left(\tilde{b}\right)\right)-\left(p\left(t_0\left(\tilde{b}\right)\right),\frac{\partial\alpha}{\partial b^\sigma}\left(\tilde{b}\right)\right)-$$

$$-H\left(T\left(\tilde{b}\right),\tilde{x}\left(T\left(\tilde{b}\right)\right),\tilde{u}\left(T\left(\tilde{b}\right)\right),\tilde{b},p\left(T\left(\tilde{b}\right)\right)\right)\frac{\partial T}{\partial b^\sigma}\left(\tilde{b}\right)+$$

$$+H\left(t_0\left(\tilde{b}\right),\tilde{x}\left(t_0\left(\tilde{b}\right)\right),\tilde{u}\left(t_0\left(\tilde{b}\right)\right),\tilde{b},p\left(t_0\left(\tilde{b}\right)\right)\right)\frac{\partial t_0}{\partial b^\sigma}\left(\tilde{b}\right)-$$

$$-\int_{t_0(\tilde{b})}^{T(\tilde{b})}\frac{\partial H}{\partial b^\sigma}\left(t,\tilde{x}(t),\tilde{u}(t),\tilde{b},p(t)\right)dt=0 \quad \textit{for } \sigma=1,2,\dots,s .$$

Let us see how does this theorem work in some special cases.

**a)** We assume that $l=0$, $f$ and $L_0$ do not depend on $t$ and on $b$, $t_0$ and $\alpha$ are constant and $T(b)=b$.

The problem defining the function $p$ becomes:

$$p'=-A^*(t)\,p+\lambda_0\,c_0(t),\quad p_j\left(\tilde{T}\right)=-\lambda_j$$

and the Maximum Principle gives

$$\left(p(t),f(\tilde{x}(t),\tilde{u}(t))\right)-\lambda_0\,L_0(\tilde{x}(t),\tilde{u}(t))\geq$$

$$\geq\left(p(t),f(\tilde{x}(t),u)\right)-\lambda_0\,L_0(\tilde{x}(t),u)$$

for any $u\in U$ and any continuity point $t$ of $\tilde{u}$; moreover,

$$\left(p(t),f(\tilde{x}(t),\tilde{u}(t))\right)-\lambda_0\,L_0(\tilde{x}(t),\tilde{u}(t))=\lambda_{n+1}\,.$$

The transversality conditions become
$$\lambda_0 \frac{\partial g_0}{\partial b^\sigma}\left(\tilde{b}\right) = 0, \quad H\left(\tilde{x}\left(\tilde{T}\right), \tilde{u}\left(\tilde{T}\right)\right) = 0.$$

Since we have, for any $t$, $H(\tilde{x}(t), \tilde{u}(t)) = \lambda_{n+1}$, we deduce that $\lambda_{n+1} = 0$, hence $H(\tilde{x}(t), \tilde{u}(t)) = 0$.

**b)** In the standard <u>*minimum time problem*</u> we have, moreover, $g_0 = 0$, $L_0 \equiv 1$. In this case, the conditions become:

$$p' \neq -A^*(t)\,p, \quad (p(t), f(\tilde{x}(t), \tilde{u}(t))) \geq (p(t), f(\tilde{x}(t), u))$$

for any $u \in U$ and any continuity point $t$ of $\tilde{u}$;

moreover, $(p(t), f(\tilde{x}(t), \tilde{u}(t)))$ is a positive constant and $p$ does not vanish identically.

To end this section, let us remark that, for the sake of simplicity, we did not specify the smoothness assumptions concerning the functions $f$, $L_\gamma$, $g_\gamma$, $\alpha$, $\omega$, $T$, $b$; from the statements, it can be seen that $f$ and $L_\gamma$ are assumed to be $\mathscr{C}^1$ with respect to the variables on the second, third and fourth positions and continuous ( or piece-wise continuous) with respect to the variable on the first position; the other functions are assumed to be $\mathscr{C}^1$. All these assumptions will be pointed out in the course of the proofs.

## 10.2 Optimum Capital Accumulation under the Minimum Time Objective

We shall apply the results described in **10.1** to a two-sector economic growth model. The first sector uses capital and labour and produces capital goods; the second sector produces consumer goods and uses capital and labour. We use the following notations:

$K_1$ – capital goods used in the first sector;

$L_1$ – labour employed in the first sector;

$K_2$ – capital used in the second sector;

$L_2$ – labour employed in the second sector.

The production functions are

$$Q_i = F_i(K_i, L_i), \ (i = 1, 2) \tag{10.1}$$

We assume that $F_i$ are homogenous of first degree and concave, hence

$$F_i(K_i, L_i) = L_i f_i\left(\frac{K_i}{L_i}\right), \ f_i' > 0, \ f_i'' < 0, \ (i = 1, 2).$$

We assume that capital goods are not "shiftable" after they have been placed in either sector. Capital–labour substitution prevails within each sector, but only labour can move freely from one sector to the other.

We assume that total labour $L$ grows at a constant rate $n$:

$$L = e^{nt} \tag{10.2}$$

and that capital goods depreciate at a constant rate $\mu$.

Since $K_1' + \mu K_1$ is total gross investment in the capital goods sector and $K_2' + \mu K_2$ is total gross investment in the consumer goods sector, we have

$$F_1 = (K_1' + \mu K_1) + (K_2' + \mu K_2). \tag{10.3}$$

We define

$$\frac{K_1' + \mu K_1}{F_1(K_1, L_1)} = w, \quad \frac{L_1}{L} = v, \quad k_1 = \frac{K_1}{L}, \quad k_2 = \frac{K_2}{L}. \tag{10.4}$$

We have

$$k_1' = \frac{K_1'}{K_1'} k_1 - \frac{L'}{L} k_1, \quad k_2' = \frac{K_2'}{K_2} k_2 - \frac{L'}{L} k_2 .$$

Taking into account **(10.2)**, **(10.3)**, **(10.4)**, we obtain the following evolutions for $k_1$ and $k_2$:

$$\begin{cases} k_1' = w\, v\, f_1\left(\dfrac{k_1}{v}\right) - (\mu + n)\, k_1 \\ k_2' = (1-w)\, v\, f_1\left(\dfrac{k_1}{v}\right) - (\mu + n)\, k_2 \end{cases} . \qquad \textbf{(10.5)}$$

We remark that, by definition, $0 \le w \le 1$ and $0 \le v \le 1$.

We shall use the following convention: if $v = 0$, then we set $F_1(k_1, v) = 0$ and $w = 0$. We also exclude the case $k_1(0) = 0$.

Our objective will be to minimize the time to economic maturity, i.e. the time $T$ needed to reach a given level $\left(k_1^T, k_2^T\right)$.

**Remark** For economic reasons, the levels $k_1^T, k_2^T$ are chosen so that to satisfy the conditions

$$k_1^T < \frac{f_1\left(k_1^T\right)}{n + \mu}, \quad k_2^T < \frac{f_1\left(k_1^T\right)}{n + \mu}.$$

We shall not insist here on the meaning of these conditions.

The model described above can be viewed as a minimum time problem; it fits the scheme described in **10.1** for this problem. We have

$$x = \begin{pmatrix} k_1 \\ k_2 \end{pmatrix}, \quad u = \begin{pmatrix} w \\ v \end{pmatrix},$$

$$f(x,u) = \begin{pmatrix} w\, v\, f_1\left(\dfrac{k_1}{v}\right) - (\mu + n)\, k_1 \\ (1-w)\, v\, f_1\left(\dfrac{k_1}{v}\right) - (\mu + n)\, k_2 \end{pmatrix}.$$

The set $U$ is the product $[0,1]\times[0,1]\subset \mathrm{R}^2$,

$$A(t)=\frac{\partial f}{\partial x}(\tilde{x}(t),\tilde{u}(t))=\begin{pmatrix}\tilde{w}(t)\,f_1'\left(\frac{\tilde{k}_1}{\tilde{v}}\right)-(\mu+n) & 0\\ (1-\tilde{w}(t))\,f_1'\left(\frac{\tilde{k}_1}{\tilde{v}}\right) & -(\mu+n)\end{pmatrix};$$

$$A^*(t)=\begin{pmatrix}\tilde{w}(t)\,f_1'\left(\frac{\tilde{k}_1}{\tilde{v}}\right)-(\mu+n) & (1-\tilde{w}(t))\,f_1'\left(\frac{\tilde{k}_1}{\tilde{v}}\right)\\ 0 & -(\mu+n)\end{pmatrix}.$$

The system that gives $p(t)$ is

$p'=-A^*(t)p,\quad p_j\left(\tilde{T}\right)=-\lambda_j$, i.e.

$$p_1'(t)=-\left[\tilde{w}\,f_1'\left(\frac{\tilde{k}_1}{\tilde{v}}\right)-(\mu+n)\right]p_1(t)-(1-\tilde{w})\,f_1'\left(\frac{\tilde{k}_1}{\tilde{v}}\right)p_2$$

$$p_2'(t)=(\mu+n)\,p_2(t)$$

$$p_1\left(\tilde{T}\right)=-\lambda_1,\quad p_2\left(\tilde{T}\right)=-\lambda_2.$$

The Maximum Principle gives

$\left(p(t),f(\tilde{x}(t),\tilde{u}(t))\right)\geq\left(p(t),f(\tilde{x}(t),u)\right)$, i.e.

$$p_1(t)\left[\tilde{w}(t)\,\tilde{v}(t)\,f_1\left(\frac{\tilde{k}_1(t)}{\tilde{v}(t)}\right)-(\mu+n)\,\tilde{k}_1(t)\right]+$$

$$+\,p_2(t)\left[(1-\tilde{w}(t))\,\tilde{v}(t)\,f_1\left(\frac{\tilde{k}_1(t)}{\tilde{v}(t)}\right)-(\mu+n)\,\tilde{k}_2(t)\right]\geq$$

$$\geq p_1(t)\left[w\,v\,f_1\left(\frac{\tilde{k}_1(t)}{v}\right)-(\mu+n)\,\tilde{k}_1(t)\right]+$$

$$+\,p_2(t)\left[(1-w)\,v\,f_1\left(\frac{\tilde{k}_1(t)}{v}\right)-(\mu+n)\,\tilde{k}_2(t)\right]$$

or

$$p_1(t)\left[\tilde{w}(t)\,\tilde{v}(t)\,f_1\left(\frac{\tilde{k}_1(t)}{\tilde{v}(t)}\right) - w\,v\,f_1\left(\frac{\tilde{k}_1(t)}{v}\right)\right] +$$

$$+\ p_2(t)\left[(1-\tilde{w}(t))\,\tilde{v}(t)\,f_1\left(\frac{\tilde{k}_1(t)}{\tilde{v}(t)}\right) - (1-w)\,v\,f_1\left(\frac{\tilde{k}_1(t)}{v}\right)\right] \geq 0$$

for any $0 \leq w \leq 1$, $0 \leq v \leq 1$ and any $t$ which is a continuity point for $\tilde{w}(t)$ and $\tilde{v}(t)$.

Moreover, $\left(p(t), f(\tilde{x}(t),\tilde{u}(t))\right)$ is a positive constant and $p(t)$ does not vanish identically.

Using the above results, some qualitative properties of the optimal strategy can be derived. We shall state, without proofs, some of these properties. The proofs can be found in [17 ].

**Theorem 10.3** *Let* $(\tilde{w}(t),\tilde{v}(t))$ *be the optimal strategy. Then*

**a)** $\tilde{w}(t) = 0$ *or* $\tilde{w}(t) = 1$

**b)** *If* $p_1(t) \neq 0$ *and* $p_2(t) \neq 0$, *then* $v(t) = 0$ *or* $v(t) = 1$.

*If* $p_1(t) = 0$ *or* $p_2(t) = 0$, *then* $v(t) = 0$ *or* $v(t) = 1$ *or* $v(t)$ *is undetermined.*

**Theorem 10.4** *Let* $(\tilde{w}(t),\tilde{v}(t))$ *be the optimal strategy. Then*

**1)** *There are only two possibilities for the* $\tilde{w}(t)$ *strategy:*

**a)** *There is no switch in the* $\tilde{w}(t)$ *strategy.*

**b)** *There is only one switch in the* $\tilde{w}(t)$ *strategy and it must have the form*

$$\tilde{w}(t) = \begin{cases} 1, & 0 \leq t \leq t^* \\ 0, & t^* \leq t \leq \tilde{T} \end{cases},$$

*where* $\tilde{T}$ *is the optimum.*

2) *In addition to the results of* Theorem 10.3, *we can say that if for some interval* $[t_0, t_1]$ *we have* $\tilde{w}(t) = 0$, $\tilde{v}(t) = 0$ *with* $p_1(t) = p_2(t)$, *then there can be a switch in the* $\tilde{v}(t)$ *strategy at* $t_1$ *from* $\tilde{v}(t) = 0$ *into* $\tilde{v}(t) = 1$.

A detailed analysis of this model can be found in the paper of M. Kurz [ 17].

## 10.5 Reduction of Problems with Free Initial and Final Time to Problems on Fixed Horizon

We associate to the given problem with free initial and final time the problem on fixed horizon described by

$$y'(s) = f(\xi(s), y(s), v(s), b)\,\eta(s), \quad \xi'(s) = \eta(s),$$

$$y(0) = \alpha(b), \ \xi(0) = t_0(b), \ y(1) = \omega(b), \ \xi(1) = T(b);$$

the controls are <u>*triples*</u> $(b, v, \eta)$ with $\eta > 0$, $v(t) \in U$, $b \in B$; the state variables are $(y, \xi)$. The functions $L_\gamma$ are replaced by

$$\tilde{L}_\gamma(\xi, y, v, b, \eta) = L_\gamma(\xi, y, v, b)\,\eta.$$

Let $(\tilde{b}, \tilde{u})$ be an optimal control for the given problem and $\tilde{x}$ - the corresponding optimal evolution.

We define the control $(\tilde{b}, \tilde{v}, \tilde{\eta})$ by

$$\tilde{\eta}(s) = T(\tilde{b}) - t_0(\tilde{b}), \quad \tilde{v}(s) = \tilde{u}\left[s\,T(\tilde{b}) + (1 - s)\,t_0(\tilde{b})\right].$$

We shall prove that this is an optimal control for the associated problem. In this way, applying Theorem 10.1 to the associated problem, we shall deduce Theorem 10.2.

The evolution $\left(\tilde{y}, \tilde{\xi}\right)$ associated to the control $\left(\tilde{b}, \tilde{v}, \tilde{\eta}\right)$ is given by the relations

$$\tilde{y}(s) = \tilde{x}\left[s\,T\left(\tilde{b}\right) + (1 - s)\,t_0\left(\tilde{b}\right)\right],$$

$$\tilde{\xi}(s) = s\,T\left(\tilde{b}\right) + (1 - s)\,t_0\left(\tilde{b}\right);$$

indeed, we have

$$\tilde{y}(0) = \tilde{x}\left(t_0\left(\tilde{b}\right)\right) = \alpha\left(\tilde{b}\right), \quad \tilde{y}(1) = \tilde{x}\left(T\left(\tilde{b}\right)\right) = \omega\left(\tilde{b}\right),$$

$$\tilde{\xi}(0) = t_0\left(\tilde{b}\right), \; \tilde{\xi}(1) = T\left(\tilde{b}\right), \quad \tilde{\xi}'(s) = T\left(\tilde{b}\right) - t_0\left(\tilde{b}\right) = \tilde{\eta}(s),$$

$$\tilde{y}'(s) = \tilde{x}'\left[s\,T\left(\tilde{b}\right) + (1 - s)\,t_0\left(\tilde{b}\right)\right]\left[T\left(\tilde{b}\right) - t_0\left(\tilde{b}\right)\right] =$$

$$= f\left(s\,T\left(\tilde{b}\right) + (1 - s)\,t_0\left(\tilde{b}\right), \tilde{x}\left[s\,T\left(\tilde{b}\right) + (1 - s)\,t_0\left(\tilde{b}\right)\right],\right.$$

$$\left.\tilde{u}\left[s\,T\left(\tilde{b}\right) + (1 - s)\,t_0\left(\tilde{b}\right)\right], \tilde{b}\right)\left[T\left(\tilde{b}\right) - t_0\left(\tilde{b}\right)\right] =$$

$$= f\left(\tilde{\xi}(s), \tilde{y}(s), \tilde{v}(s), \tilde{b}\right)\tilde{\eta}(s).$$

Further, we have

$$\tilde{I}_\gamma\left(\tilde{b}, \tilde{v}, \tilde{\eta}\right) = g_\gamma\left(\tilde{b}\right) + \int_0^1 \tilde{L}_\gamma\left(\tilde{\xi}(s), \tilde{y}(s), \tilde{v}(s), \tilde{b}, \tilde{\eta}(s)\right) ds -$$

$$= g_\gamma\left(\tilde{b}\right) + \int_0^1 L_\gamma\left(\tilde{\xi}(s), \tilde{x}\left(\tilde{\xi}(s)\right), \tilde{u}\left(\tilde{\xi}(s)\right), \tilde{b}\right)\tilde{\xi}'(s)\,ds =$$

$$= g_\gamma\left(\tilde{b}\right) + \int_{t_0\left(\tilde{b}\right)}^{T\left(\tilde{b}\right)} L_\gamma\left(t, \tilde{x}(t), \tilde{u}(t), \tilde{b}\right) dt.$$

From the above formulae it follows that $\left(\tilde{b}, \tilde{v}, \tilde{\eta}\right)$ is an admissible control, since $\left(\tilde{b}, \tilde{u}\right)$ is admissible. We shall prove that $\left(\tilde{b}, \tilde{v}, \tilde{\eta}\right)$ is even optimal.

Let $(b, v, \eta)$ be an arbitrary admissible control; $\eta(s) > 0$ and $\xi'(s) = \eta(s)$ imply that $\xi$ is strictly increasing and, since it is also continuous, it follows that it is *invertible*; let

$\zeta:[t_0(b), T(b)] \to [0, 1]$ be the inverse.

We have $\xi[\zeta(t)] = t$, $\zeta[\xi(s)] = s$ for $t \in [t_0(b), T(b)]$, $s \in [0, 1]$.

We define $u(t) = v[\zeta(t)]$, $x(t) = y[\zeta(t)]$.

We show that $(b, u)$ is an admissible control for the problem we started with and $x$ is the corresponding evolution.

We remark that $x$ is defined on $[t_0(b), T(b)]$,

$$x(t_0(b)) = y[\zeta(t_0(b))] = y(0) = \alpha(b),$$

$$x(T(b)) = y[\zeta(T(b))] = y(1) = \omega(b),$$

$$x'(t) = y'(\zeta(t))\,\zeta'(t) = f(\xi(\zeta(t)), y(\zeta(t)), v(\zeta(t)), b)\,\eta(\zeta(t))\,\zeta'(t) =$$

$$= f(t, x(t), u(t), b),$$

since $\zeta'(t) \cdot \xi'(\zeta(t)) = 1$ and $\xi'(\zeta(t)) = \eta(\zeta(t))$. Further,

$$I_\gamma(b, u) = g_\gamma(b) + \int_{t_0(b)}^{T(b)} L_\gamma(t, x(t), u(t), b)\, dt =$$

$$= g_\gamma(b) + \int_0^1 L_\gamma(\xi(s), x(\xi(s)), u(\xi(s)), b)\, \xi'(s)\, ds =$$

$$= g_\gamma(b) + \int_0^1 L_\gamma(\xi(s), y(s), v(s), b)\, \eta(s)\, ds = \tilde{I}_\gamma(b, v, \eta).$$

From the fact that $(b, v, \eta)$ is admissible, it follows now that $(b, u)$ is admissible.

Since $(\tilde{b}, \tilde{u})$ is optimal for the given problem, we have $I_0(\tilde{b}, \tilde{u}) \le I_0(b, u)$; the above equalities show that

$\tilde{I}_0(\tilde{b}, \tilde{v}, \tilde{\eta}) \le \tilde{I}_0(b, v, \eta)$, hence $(\tilde{b}, \tilde{v}, \tilde{\eta})$ is optimal.

We are now in position to deduce Theorem 10.2 from Theorem 10.1.

Since $\left(\tilde{b},\tilde{v},\tilde{\eta}\right)$ is optimal for the associated problem with fixed horizon, from Theorem 10.1 it follows that there exist multipliers $\lambda_0, \lambda_1, \ldots, \lambda_{l+n+1}$, at least one of which is non-zero, such that

**1)** $\lambda_\gamma \geq 0$ for $\gamma = 0,\ldots,l'$, $\lambda_\gamma \, \tilde{I}_\gamma\left(\tilde{b},\tilde{v},\tilde{\eta}\right) = 0$ for $1 \leq \gamma \leq l'$.

**2)** We define

$$\tilde{A}(s) = \begin{pmatrix} \frac{\partial f}{\partial x}\left(\tilde{\xi}(s), \tilde{y}(s), \tilde{v}(s), \tilde{b}\right)\tilde{\eta} & \frac{\partial f}{\partial t}\left(\tilde{\xi}(s), \tilde{y}(s), \tilde{v}(s), \tilde{b}\right)\tilde{\eta} \\ 0 & 0 \end{pmatrix}$$

$$\tilde{c}_\gamma(s) = \begin{pmatrix} \frac{\partial L_\gamma}{\partial x}\left(\tilde{\xi}(s), \tilde{y}(s), \tilde{v}(s), \tilde{b}\right)\tilde{\eta} \\ \frac{\partial L_\gamma}{\partial t}\left(\tilde{\xi}(s), \tilde{y}(s), \tilde{v}(s), \tilde{b}\right)\tilde{\eta} \end{pmatrix}, \quad \gamma = 0, 1, \ldots, l.$$

We define $\tilde{p}$ as a solution of the problem

$$\tilde{p}'(s) = -\tilde{A}^*(s)\,\tilde{p}(s) + \sum_{\gamma=0}^{l} \lambda_\gamma\, \tilde{c}_\gamma(s), \quad \tilde{p}_j(1) = -\lambda_{l+j}.$$

We define

$$\tilde{H}(\xi, y, v, \eta, \tilde{p}, b) = \left(\tilde{p}, \tilde{f}(\xi, y, v, \eta, b)\right) - \sum_{\gamma=0}^{l} \lambda_\gamma\, L_\gamma(\xi, y, v, b)\,\eta,$$

where

$$\tilde{f}(\xi, y, v, \eta, b) = \begin{pmatrix} f(\xi, y, v, b)\,\eta \\ \eta \end{pmatrix}.$$

The following inequality holds:

$$\tilde{H}\left(\tilde{\xi}(s), \tilde{y}(s), \tilde{v}(s), \tilde{\eta}, \tilde{b}, \tilde{p}(s)\right) \geq \tilde{H}\left(\tilde{\xi}(s), \tilde{y}(s), v, \eta, \tilde{b}, \tilde{p}(s)\right)$$

for any $v \in U$, $\eta > 0$.

3) $$\sum_{\gamma=0}^{l} \lambda_\gamma \frac{\partial g_\gamma}{\partial b^\sigma}(\widetilde{b}) + \left(\widetilde{p}(1), \frac{\partial \widetilde{\omega}}{\partial b^\sigma}(\widetilde{b})\right) - \left(\widetilde{p}(0), \frac{\partial \widetilde{\alpha}}{\partial b^\sigma}(\widetilde{b})\right) -$$

$$- \int_0^1 \frac{\partial \widetilde{H}}{\partial b^\sigma}\left(\widetilde{\xi}(\tau), \widetilde{y}(\tau), \widetilde{v}(\tau), \widetilde{\eta}(\tau), \widetilde{p}(\tau), \widetilde{b}\right) d\tau = 0, \ \sigma = 1, \ldots, s,$$

where we denoted $\widetilde{\omega} = \begin{pmatrix} \omega \\ T \end{pmatrix}$, $\widetilde{\alpha} = \begin{pmatrix} \alpha \\ t_0 \end{pmatrix}$.

We remark that an additional multiplier appeared, because the dimension of the state space describing the evolution increased.

To make the above conditions more explicit, we write $\widetilde{p} = \begin{pmatrix} \hat{p} \\ \rho \end{pmatrix}$ and we deduce

$$\hat{p}'(s) = -\left[\frac{\partial f}{\partial x}\left(\widetilde{\xi}(s), \widetilde{y}(s), \widetilde{v}(s), \widetilde{b}\right) \widetilde{\eta}\right]^* \hat{p}(s) +$$

$$+ \sum_{\gamma=0}^{l} \lambda_\gamma \frac{\partial L_\gamma}{\partial x}\left(\widetilde{\xi}(s), \widetilde{y}(s), \widetilde{v}(s), \widetilde{b}\right) \widetilde{\eta},$$

$$\rho'(s) = -\left(\hat{p}(s), \frac{\partial f}{\partial t}\left(\widetilde{\xi}(s), \widetilde{y}(s), \widetilde{v}(s), \widetilde{b}\right) \widetilde{\eta}\right) +$$

$$+ \sum_{\gamma=0}^{l} \lambda_\gamma \frac{\partial L_\gamma}{\partial t}\left(\widetilde{\xi}(s), \widetilde{y}(s), \widetilde{v}(s), \widetilde{b}\right) \widetilde{\eta},$$

$$\hat{p}_j(1) = -\lambda_{l+j}, \ j = 1, \ldots, n, \quad \rho(1) = -\lambda_{l+n+1}.$$

The maximum condition becomes:

$$\left(\hat{p}(s), f\left(\widetilde{\xi}(s), \widetilde{y}(s), \widetilde{v}(s), \widetilde{b}\right) \widetilde{\eta}\right) + \rho(s)\, \widetilde{\eta} -$$

$$- \sum_{\gamma=0}^{l} \lambda_\gamma L_\gamma\left(\widetilde{\xi}(s), \widetilde{y}(s), \widetilde{v}(s), \widetilde{b}\right) \widetilde{\eta} \geq$$

$$\geq \left(\hat{p}(s), f\left(\widetilde{\xi}(s), \widetilde{y}(s), v, \widetilde{b}\right) \eta\right) + \rho(s)\, \eta - \sum_{\gamma=0}^{l} \lambda_\gamma L_\gamma\left(\widetilde{\xi}(s), \widetilde{y}(s), v, \widetilde{b}\right) \eta$$

for any $v \in U$, $\eta > 0$.

Taking, in the above inequality, $v = \tilde{v}(s)$, we deduce

$$\left[\left(\hat{p}(s), f\left(\tilde{\xi}(s), \tilde{y}(s), \tilde{v}(s), \tilde{b}\right)\right) + \rho(s) - \right.$$

$$\left. - \sum_{\gamma=0}^{l} \lambda_\gamma L_\gamma\left(\tilde{\xi}(s), \tilde{y}(s), \tilde{v}(s), \tilde{b}\right)\right](\tilde{\eta} - \eta) \geq 0, \text{ for any } \eta > 0.$$

This implies that

$$\left(\hat{p}(s), f\left(\tilde{\xi}(s), \tilde{y}(s), \tilde{v}(s), \tilde{b}\right)\right) + \rho(s) -$$

$$- \sum_{\gamma=0}^{l} \lambda_\gamma L_\gamma\left(\tilde{\xi}(s), \tilde{y}(s), \tilde{v}(s), \tilde{b}\right) \equiv 0$$

Taking, in the maximum condition, $\eta = \tilde{\eta}$, we obtain

$$\left(\hat{p}(s), f\left(\tilde{\xi}(s), \tilde{y}(s), \tilde{v}(s), \tilde{b}\right)\right) - \sum_{\gamma=0}^{l} \lambda_\gamma L_\gamma\left(\tilde{\xi}(s), \tilde{y}(s), \tilde{v}(s), \tilde{b}\right) \geq$$

$$\geq \left(\hat{p}(s), f\left(\tilde{\xi}(s), \tilde{y}(s), v, \tilde{b}\right)\right) - \sum_{\gamma=0}^{l} \lambda_\gamma L_\gamma\left(\tilde{\xi}(s), \tilde{y}(s), v, \tilde{b}\right)$$

for any $v \in U$.

We write these relations at the point $\tilde{\xi}^{-1}(t)$; we have

$$\tilde{y}\left(\tilde{\xi}^{-1}(t)\right) = \tilde{x}(t), \quad \tilde{v}\left(\tilde{\xi}^{-1}(t)\right) = \tilde{u}(t);$$

we denote $p(t) = \hat{p}\left(\tilde{\xi}^{-1}(t)\right)$, $\beta(t) = \rho\left(\tilde{\xi}^{-1}(t)\right)$. We obtain

$$\left(p(t), f\left(t, \tilde{x}(t), \tilde{u}(t), \tilde{b}\right)\right) + \beta(t) - \sum_{\gamma=0}^{l} \lambda_\gamma L_\gamma\left(t, \tilde{x}(t), \tilde{u}(t), \tilde{b}\right) \equiv 0$$

$$\left(p(t), f\left(t, \tilde{x}(t), \tilde{u}(t), \tilde{b}\right)\right) - \sum_{\gamma=0}^{l} \lambda_\gamma L_\gamma\left(t, \tilde{x}(t), \tilde{u}(t), \tilde{b}\right) \geq$$

$$\geq \left(p(t), f\left(t, \tilde{x}(t), v, \tilde{b}\right)\right) - \sum_{\gamma=0}^{l} \lambda_\gamma L_\gamma\left(t, \tilde{x}(t), v, \tilde{b}\right)$$

for any $v \in U$; the relations are valid at any $t$ which is a continuity point for $\tilde{u}$.

Further, we have

$$p'(t) = \hat{p}'\left(\tilde{\xi}^{-1}(t)\right)\frac{d}{dt}\tilde{\xi}^{-1}(t) =$$

$$= \hat{p}'\left(\tilde{\xi}^{-1}(t)\right)\frac{1}{\tilde{\xi}'\left(\tilde{\xi}^{-1}(t)\right)} = \hat{p}'\left(\tilde{\xi}^{-1}(t)\right)\frac{1}{\tilde{\eta}}$$

and $\beta'(t) = \rho'\left(\tilde{\xi}^{-1}(t)\right)\frac{1}{\tilde{\eta}}$. We deduce

$$p'(t) = -\left[\frac{\partial f}{\partial x}\left(t, \tilde{x}(t), \tilde{u}(t), \tilde{b}\right)\right]^{*} p(t) + \sum_{\gamma=0}^{l} \lambda_\gamma \frac{\partial L_\gamma}{\partial x}\left(t, \tilde{x}(t), \tilde{u}(t), \tilde{b}\right),$$

$$\beta'(t) = -\left(\frac{\partial f}{\partial t}\left(t, \tilde{x}(t), \tilde{u}(t), \tilde{b}\right), p(t)\right) + \sum_{\gamma=0}^{l} \lambda_\gamma \frac{\partial L_\gamma}{\partial t}\left(t, \tilde{x}(t), \tilde{u}(t), \tilde{b}\right),$$

$$p_j\left(T\left(\tilde{b}\right)\right) = \hat{p}_j(1) = -\lambda_{l+j}, \quad \beta\left(T\left(\tilde{b}\right)\right) = \rho(1) = -\lambda_{l+n+1}.$$

It follows that

$$\beta(t) = -\lambda_{l+n+1} + \int_t^{T(\tilde{b})} \frac{\partial H}{\partial t}\left(\tau, \tilde{x}(\tau), \tilde{u}(\tau), \tilde{b}, p(\tau)\right) d\tau.$$

Thus, the maximum conditions become

$$H\left(t, \tilde{x}(t), \tilde{u}(t), \tilde{b}, p(t)\right) + \beta(t) \equiv 0,$$

$$H\left(t, \tilde{x}(t), \tilde{u}(t), \tilde{b}, p(t)\right) \geq H\left(t, \tilde{x}(t), u, \tilde{b}, p(t)\right)$$

for any $u \in U$ and any continuity point $t$ of $\tilde{u}$.

The transversality conditions become

$$\sum_{\gamma=0}^{l} \lambda_\gamma \frac{\partial g_\gamma}{\partial b^\sigma}\left(\tilde{b}\right) + \left(p\left(T\left(\tilde{b}\right)\right), \frac{\partial \omega}{\partial b^\sigma}\left(\tilde{b}\right)\right) + \beta\left(T\left(\tilde{b}\right)\right)\frac{\partial T}{\partial b^\sigma}\left(\tilde{b}\right) -$$

$$-\left(p\left(t_0\left(\tilde{b}\right)\right), \frac{\partial \alpha}{\partial b^\sigma}\left(\tilde{b}\right)\right) - \beta\left(t_0\left(\tilde{b}\right)\right)\frac{\partial t_0}{\partial b^\sigma}\left(\tilde{b}\right) -$$

$$-\int_0^1 \frac{\partial \widetilde{H}}{\partial b^\sigma}\left(\widetilde{\xi}(\tau),\widetilde{y}(\tau),\widetilde{v}(\tau),\widetilde{\eta},\widetilde{b},\widetilde{p}(\tau)\right)d\tau = 0 .$$

But $\widetilde{H}\left(\xi,\widetilde{y},\widetilde{v},\widetilde{\eta},\widetilde{b},\widetilde{p}\right) = H\left(\xi,\widetilde{y},\widetilde{v},\widetilde{b},\hat{p}\right)\widetilde{\eta} + \rho\,\widetilde{\eta}$, hence

$$\frac{\partial \widetilde{H}}{\partial b^\sigma}\left(\widetilde{\xi},\widetilde{y},\widetilde{v},\widetilde{\eta},\widetilde{b},\widetilde{p}\right) = \frac{\partial H}{\partial b^\sigma}\left(\widetilde{\xi},\widetilde{y},\widetilde{v},\widetilde{b},\hat{p}\right)\widetilde{\eta}.$$

Performing the change of variables $\tau = \widetilde{\xi}^{-1}(t)$, we obtain

$$\int_0^1 \frac{\partial \widetilde{H}}{\partial b^\sigma}\left(\widetilde{\xi}(\tau),\widetilde{y}(\tau),\widetilde{v}(\tau),\widetilde{\eta},\widetilde{b},\widetilde{p}(\tau)\right)d\tau =$$

$$= \int_{t_0(\widetilde{b})}^{T(\widetilde{b})} \frac{\partial H}{\partial b^\sigma}\left(t,\widetilde{x}(t),\widetilde{u}(t),\widetilde{b},p(t)\right)dt .$$

We also have

$$\beta\left(T\left(\widetilde{b}\right)\right) = -H\left(T\left(\widetilde{b}\right),\widetilde{x}\left(T\left(\widetilde{b}\right)\right),\widetilde{u}\left(T\left(\widetilde{b}\right)\right),\widetilde{b},p\left(T\left(\widetilde{b}\right)\right)\right),$$

$$\beta\left(t_0\left(\widetilde{b}\right)\right) = -H\left(t_0\left(\widetilde{b}\right),\widetilde{x}\left(t_0\left(\widetilde{b}\right)\right),\widetilde{u}\left(t_0\left(\widetilde{b}\right)\right),\widetilde{b},p\left(t_0\left(\widetilde{b}\right)\right)\right).$$

Taking into account the above considerations, we obtain the relations in the statement of Theorem 10.2.

## 10.4 An Abstract Multiplier Rule

We begin with a definition.

**Definition 10.5** *Let us consider* $J_\rho : \mathscr{C} \to \mathrm{R}$, $\rho = 0, 1, \dots, m$, $\mathscr{C}$ *– an unspecified set and* $\widetilde{w} \in \mathscr{C}$. *A set* $K$ *of vectors from* $\mathrm{R}^{m+1}$ *is called* <u>*derivative set*</u> *for* $\{J_\rho\}$ *at* $\widetilde{w}$ *if, for any collection of vectors* $k_1, k_2, \dots, k_N$ *from* $K$ *there exists a function* $w : [0,\delta]^N \to \mathscr{C}$, *such that* $w(0) = \widetilde{w}$ *and*

$$J_\rho(w(\varepsilon)) - J_\rho(\widetilde{w}) = \sum_{j=1}^{N} \varepsilon_j k_j^\rho + o(\varepsilon) \text{ for } \varepsilon \in [0,\delta]^N .$$

**Theorem 10.6** *(Hestenes' multiplier rule)*

*Let $K$ be a derivative set for $\{J_\rho\}$ at $\tilde{w} \in \mathscr{C}$. If $\tilde{w}$ is a minimum for the function $J_0$ in the class of elements $w \in \mathscr{C}$ satisfying $J_\rho(w) \leq 0$, $\rho = 1, 2, \ldots, l'$, $J_\rho(w) = 0$, $\rho = l' + 1, \ldots, m$, then there exist multipliers $\lambda_0 \geq 0$, $\lambda_1, \ldots, \lambda_m$, at least one of them being non-zero, such that, for any $k \in K$, $\sum_{\rho=0}^{m} \lambda_\rho k^\rho \geq 0$.*

*Moreover, $\lambda_\rho \geq 0$ for $1 \leq \rho \leq l'$ and $\lambda_\rho J_\rho(\tilde{w}) = 0$.*

To prove this theorem, we need several auxiliary results.

**Lemma 10.7** *Let $K$ be a derivative set, $K^*$ – the convex cone generated by $K$ (the set of vectors of the form $\sum_{i=1}^{N} \alpha_i k_i$, $\alpha_i \geq 0$, $k_i \in K$). Then $K^*$ is also a derivative set.*

***Proof*** Let $k_1^*, k_2^*, \ldots, k_r^*$ be a finite set of vectors from $K^*$, hence $k_j^* = \sum_{i=1}^{N} \alpha_{ij} k_i$, $\alpha_{ij} \geq 0$, $k_i \in K$; since $K$ is a derivative set, we can associate to the vectors $k_i$ a function $w : [0, \delta]^N \to \mathscr{C}$, such that $w(0) = \tilde{w}$, $J_\rho(w(\varepsilon)) - J_\rho(\tilde{w}) = \sum_{j=1}^{N} \varepsilon_j k_j^\rho + o(\varepsilon)$.

We consider the function $\varepsilon : [0, \delta']^r \to [0, \delta]^N$, defined by $\varepsilon_i(\eta_1, \ldots, \eta_r) = \sum_{j=1}^{r} \alpha_{ij} \eta_j$, and the function $w^* : [0, \delta']^r \to \mathscr{C}$, defined by $w^*(\eta) = w(\varepsilon(\eta))$.

We deduce that $w^*(0) = w(\varepsilon(0)) = w(0) = \tilde{w}$ and

$$J_\rho\big(w^*(\eta)\big) - J_\rho(\tilde{w}) = \sum_{j=1}^{N} \varepsilon_j(\eta)\, k_j^\rho + o(\varepsilon(\eta)) = \sum_{l=1}^{r} \big(k_l^*\big)^\rho\, \eta_l + o(\eta),$$

hence $K^*$ is a derivative set.

Taking into account this lemma, we may prove the abstract multiplier rule by assuming, from the beginning, that the derivative set $K$ is a convex cone.

We shall denote by $K^-$ a cone whose non-zero vectors are such that $k^0 < 0$, $k^\gamma < 0$ for $1 \le \gamma \le l'$ if $J_\gamma(\tilde{w}) = 0$, $k^\gamma = 0$ for $l' < \gamma \le m$.

Further, we denote by $K^+$ the set of vectors of the form $k - k^-$, where $k \in K$, $k^- \in K^-$. The set $K^+$ is a convex cone; indeed, if $k^+ \in K^+$, then we can write

$$k^+ = k - k^-, \quad \alpha k^+ = \alpha k - \alpha k^-$$

and

$$\alpha k \in K,\ \alpha k^- \in K^- \text{ for } \alpha > 0,\ \alpha k^+ = 0 \text{ for } \alpha = 0.$$

Further,

$$k_1^+ + k_2^+ = k_1 + k_2 - \big(k_1^- + k_2^-\big),$$

hence

$$k_1^+ + k_2^+ \in K^+ \text{ if } k_1^+ \in K^+,\ k_2^+ \in K^+.$$

**Lemma 10.8** *If there exists $k \in \mathrm{R}^{m+1}$ which does not belong to $K^+$, then there exist multipliers with the properties stated in* Theorem 10.6; *among $\lambda_1, \dots, \lambda_m$, at least one is non-zero,*

$$\lambda_\rho \ge 0 \text{ for } 0 \le \rho \le l',\ \lambda_\rho J_\rho(\tilde{w}) = 0 \text{ for } 1 \le \rho \le l',$$

$$\sum_{\rho=0}^{m} \lambda_\rho k^\rho \ge 0 \text{ for any vector } k \in K.$$

***Proof*** Since $K^+$ is convex, its interior coincides with the interior of its closure; since $K^+$ does not coincide with the whole space, its closure does not coincide with the whole space. Let $\bar{k}$ be the vector with the first co-ordinate equal to $-1$ and the other co-ordinates equal to zero. If this vector does not belong to the closure of $K^+$, then we choose $k_0 = \bar{k}$. If $\bar{k}$ belongs to the closure of $K^+$, then we choose as $k_0$ an arbitrary vector of unit length, which is not in the closure of $K^+$; such a vectoı exists, since this closure is a cone which does not coincide with the whole space.

Further, we choose $k_1$ as the vector from the closure of $K^+$ such that $|k_1 - k_0| =$ *the distance from* $k_0$ *to the closure of* $K^+$; such a vector exists and it is unique for any closed convex set. For $k$ in the closure of $K^+$, we have $k_1 + tk$ located in the closure of $K^+$, for all $t \geq 0$ (the property of a convex cone).

We consider the function $g(t,k) = \frac{1}{2}|k_1 + tk - k_0|^2$. It is defined for $t \geq 0$ and has a minimum at $t = 0$.

For $k = k_1$, the point $k_1 + tk = (1+t)k_1$ belongs to the closure of $K^+$ for any $t > -1$, hence $t = 0$ is an interior point and, therefore, $\frac{\partial g}{\partial t}(0, k_1) = 0$.

On the other hand, for any $k$ we have

$$\frac{\partial g}{\partial t}(0,k) = (k_1 - k_0, k) \geq 0;$$

indeed,

$$g(t,k) \geq g(0,k) \text{ for any } t \geq 0,$$

$$\frac{g(t,k) - g(0,k)}{t} \geq 0 \text{ for any } t \geq 0$$

and hence, taking the limit, we have $\dfrac{\partial g}{\partial t}(0,k) \geq 0$.

We define $\lambda_\rho = k_1^\rho - k_2^\rho$ and we show that these numbers have the properties required. The condition $(k_1 - k_0, k) \geq 0$ becomes $\sum \lambda_\rho k^\rho \geq 0$ and it is satisfied for any k in the closure of $K^+$; since $K$ is contained in the closure of $K^+$, we deduce that

$$\sum \lambda_\rho k^\rho \geq 0 \text{ for any } k \in K.$$

The vector $\bar{k}_\gamma$, defined by $\bar{k}_\gamma^\gamma = 1$, $\bar{k}_\gamma^\rho = 0$ for $\rho \neq \gamma$, $\gamma = 0, 1, \ldots, l'$, belongs to the closure of $K^+$, since $-\bar{k}_\gamma$ belongs to the closure of $K^-$ and $0$ belongs to $K$; it follows that

$$\sum \lambda_\rho \bar{k}_\gamma^\rho \geq 0,$$

hence $\lambda_\gamma \geq 0$ for $\gamma = 0, 1, \ldots, l'$.

If $J_\rho(\tilde{w}) < 0$ for an index $\rho \in \{1, \ldots, l'\}$, then the corresponding co-ordinate can be chosen in an arbitrary way and $-\bar{k}_\rho$ belongs to the closure of $K^+$; we deduce that $-\lambda_\rho \geq 0$ and, since $\lambda_\rho \geq 0$, we have $\lambda_\rho = 0$.

The lemma is thus proved.

**Remark** When $\bar{k} = -\bar{k}_0$ belongs to the closure of $K^+$, then it follows $-\lambda_0 \geq 0$, hence $\lambda_0 = 0$. If $\bar{k}$ does not belong to the closure of $K^+$, then we take $k_0 = \bar{k}$ and we have

$$0 < |k_1 - k_0|^2 = 2(k_1 - k_0, k_1) + |k_0|^2 - |k_1|^2 = 1 - |k_1|^2,$$

hence $|k_1| < 1$, therefore $|k_1|^0 < 1$ and $\lambda_0 = k_1^0 - k_0^0 = k_1^0 + 1 > 0$.

We deduce that, in order that $\lambda_0 > 0$, it is necessary and sufficient that the vector having the first co-ordinate $-1$ and the other co-ordinates $0$ does not belong to the closure of $K^+$.

**Lemma 10.9** *If* $K^{+} = \mathrm{R}^{m+1}$, *then there exist* $N = m - l' + 1$ *vectors* $k_1, \dots, k_N$ *in* $K$ *such that* $\bar{k} = k_1 + \dots + k_N \in K^{-}$ *and* $\operatorname{rank}\left(k_j^{\gamma}\right) = N - 1 = m - l'$, *for* $\gamma = l' + 1, \dots, m$, $j = 1, \dots, N$.

***Proof*** We choose the vectors $k_1, \dots, k_{N-1}$ in $K^{+}$ such that $\det\left(k_j^{\gamma}\right) \neq 0$; we choose $k_N^{\gamma}$ such that $k_1^{\gamma} + \dots + k_{N-1}^{\gamma} + k_N^{\gamma} = 0$ for $\gamma = l' + 1, \dots, m$ and $k_N^{\gamma}$ arbitrary for $0 \leq \gamma \leq l'$.

Such $k_N^{\gamma}$ will belong to $K^{+}$, since $K^{+}$ is the whole space.

From $k_j \in K^{+}$ it follows $k_j = k_j' - k_j^{-}$, $k_j' \in K$, $k_j^{-} \in K^{-}$; since, according to the definition of $K^{-}$, we have $\left(k_j^{-}\right)^{\gamma} = 0$ for $\gamma = l' + 1, \dots, m$, we deduce that the vectors $k_j'$ have the same components as $k_j$, for $l' < \gamma < m$, hence they have the same properties as $k_j$. For this reason, we may assume, from the beginning, that $k_j \in K$.

We show that there exists $k' \neq 0$ belonging to $K^{-} \cap K$.

We consider $k \neq 0$, $k \in K^{-}$; since $K^{+} = \mathrm{R}^{m+1}$, it follows $k \in K^{+}$, hence $k = k' - k''$, $k' \in K$, $k'' \in K^{-}$, hence $k' = k + k''$ and $k' \in K^{-}$, hence $k' \in K^{-} \cap K$. Assuming $k' = 0$, we would have $k = -k''$, $k \in K^{-}$, $k'' \in K^{-}$, hence $k \in K^{-}$ and $-k \in K^{-}$; this is not possible, since $k \neq 0$.

Thus, we have shown that there exist

$$k_1, \dots, k_N \in K \text{ with } \det\left(k_j^{\gamma}\right) \neq 0,$$

$$\gamma = l' + 1, \dots, m, \quad j = 1, \dots, N - 1, \quad N = m - l' + 1,$$

$$k_1^{\gamma} + \dots + k_N^{\gamma} = 0 \text{ for } \gamma = l' + 1, \dots, m$$

and there exists $k' \in K^{-} \cap K$, $k' \neq 0$.

For $\varepsilon > 0$ small enough, we have

$k^* = k' + \varepsilon\,(k_1 + \ldots + k_N) \in K^- \cap K$.

On the one hand, $k_j \in K$ for $j = 1,\ldots,N$ and $k' \in K$ imply that $k^* \in K$; on the other hand, the last components of $k^*$ are equal to zero (as those of $k'$ and of $k_1 + \ldots + k_N$), while the components of $k^*$ that correspond to strictly negative components of $k'$ are still negative, if $\varepsilon$ is small enough, hence $k^* \in K^-$.

We define $k_j^* = \dfrac{1}{N} k' + \varepsilon\, k_j$; since $k' \in K$, $k_j \in K$, we have $k_j^* \in K$ and

$$\sum_{j=1}^{N} k_j^* = k' + \varepsilon \sum_{j=1}^{N} k_j = k^* \in K^- .$$

Since $k' \in K^-$, it follows that its last components are equal to zero and hence $\left(k_j^*\right)^\gamma = \varepsilon\, k_j^\gamma$, $j = 1,\ldots,N$, $\gamma = l' + 1,\ldots,m$, therefore $\operatorname{rank}\left(\left(k_j^*\right)^\gamma\right) = \operatorname{rank}\left(k_j^\gamma\right) = N - 1$.

We deduce that the vectors $k_j^*$ have the properties in the statement.

**Lemma 10.10** **(An implicit functions theorem)**

*Let $G_\alpha$, $\alpha = 1,\ldots,m$ be continuous functions, defined for $|y| \le r$, $0 \le t \le \varepsilon$. We assume that*

$$G_\alpha(y,0) = \sum_{i=1}^{n} a_{\alpha i}\, y^i, \ \alpha = 1,\ldots,m \ \textit{and that} \ \operatorname{rank}\left(a_{\alpha i}\right) = m .$$

*We define* $N(t) = \max\limits_{|y| \le r} \left| G(y,t) - G(y,0) \right|$.

*Then there exist a constant $M$ and a function* $y\colon [\,0,\delta\,] \to R^n$

*such that*

$$y(0) = 0, \ |y(t)| \le MN(t) \le r,$$

$$G_\alpha(y(t),t) = 0, \ \lim_{t\to 0_+} y(t) = y(0) = 0.$$

***Proof*** Let $a_{\beta j}$, $\beta = m+1,\dots,n$, $j = 1,\dots,n$ be chosen such that the matrix $A = (a_{\beta j})$ be non-singular, $\beta = 1,\dots,n$, $j = 1,\dots,n$ (the system of linearly independent vectors has been completed to form a basis). We define

$$G_\beta(y,t) = \sum_{j=1}^{n} a_{\beta j}\, y^j, \ \beta = m+1,\dots,n.$$

The system of functions $G_1,\dots,G_n$ obtained in this way satisfies the conditions stated, with the same function $N(t)$. We choose $M$ such that $|A^{-1}y| \le M|y|$.

The function $F$ defined by $F(y,t) = y - A^{-1}G(y,t)$ satisfies

$$|F(y,t)| \le MN(t) \text{ for } |y| \le r,$$

since

$$F(y,t) = A^{-1}(A\,y - G(y,t)) = A^{-1}[G(y,0) - G(y,t)].$$

We choose $\delta$ such that $MN(t) \le r$ for $0 \le t \le \delta$; this is possible, since $\lim_{t\to 0_+} N(t) = 0$. It follows that, for $t \in [0,\delta]$, the mapping $F$ maps the ball with radius $MN(t)$ in itself; on the basis of Brouwer's fixed point theorem, there exists $y(t)$ such that $F(y(t),t) = y(t)$ and $|y(t)| \le MN(t)$.

Taking into account the definition of the function $F$, it follows that $G(y(t),t) = 0$ and the lemma is proved.

***Proof*** **of Theorem 10.6**

According to the Lemma 10.8, it suffices to prove that $K^+ \neq \mathrm{R}^{m+1}$.

If we had $K^+ = \mathrm{R}^{m+1}$ then, according to Lemma 10.9, we could choose vectors $k_1, \dots, k_N$ with the properties stated in that lemma. From the definition of the derivative set it follows that there exists a function

$$w : [0, \delta]^N \to C \quad \text{with } w(0) = \tilde{w},$$

$$J_\rho(w(\varepsilon)) - J_\rho(\tilde{w}) = \sum_{j=1}^{N} \varepsilon_j k_j^\rho + o(\varepsilon).$$

We choose $\bar{k}^j = 1$ for $j = 1, \dots, N$; $\sum_{j=1}^{N} k_j^\rho \bar{k}^j = \bar{k}^\rho$.

According to Lemma 10.9, $\bar{k} \in K^-$.

For $\delta'$ small enough, $\varepsilon_j = t\bar{k}^j + t y^j$, $t \in [0, \delta']$, $|y| \leq 1$ satisfy $0 \leq \varepsilon_j \leq \delta$. We define

$$F_\rho(y, t) = \frac{1}{t}\left[J_\rho(w(\varepsilon)) - J_\rho(\tilde{w})\right] \text{ for } 0 < t \leq \delta,$$

$$F_\rho(y, 0) = \bar{k}^\rho + \sum_{j=1}^{N} k_j^\rho y^j.$$

We have

$$F_\rho(y, t) = \frac{1}{t}\left[J_\rho(w(\varepsilon)) - J_\rho(\tilde{w})\right] = \frac{1}{t}\left[\sum_{j=1}^{N} \varepsilon_j k_j^\rho + o(\varepsilon)\right] =$$

$$= \frac{1}{t}\sum_{j=1}^{N} t\left(\bar{k}^j + y^j\right) k_j^\rho + \frac{o(t)}{t} = \sum_{j=1}^{N} \left(\bar{k}^j + y^j\right) k_j^\rho + \frac{o(t)}{t} =$$

$$= \bar{k}^\rho + \sum_{j=1}^{N} k_j^\rho y^j + \frac{o(t)}{t} = F_\rho(y, 0) + \frac{o(t)}{t},$$

hence $F_\rho(y,t) - F_\rho(y,0) = O(t)$, uniformly with respect to $y$ in $|y| \le 1$.

Since $\bar{k}^\rho = 0$ for $\rho = l'+1,\dots,m$ $\left(\bar{k} \in K^-\right)$ and

$\operatorname{rank}\left(k_j^\rho\right) = N-1$ for $\rho = l'+1,\dots,m$ (from Lemma 10.9),

the functions $F_{l'+1},\dots,F_m$ satisfy the assumptions of Lemma 10.10.

According to this lemma, there exists $y$ such that

$$\lim_{t\to 0_+} y(t) = 0\,, \quad F_\rho(y(t),t) = 0\,, \quad \rho = l'+1,\dots,m\,.$$

We define

$$\varepsilon(t) = t\left[\bar{k} + y(t)\right], \quad v(t) = w(\varepsilon(t)),$$

$$g_\rho(t) = J_\rho(v(t)) - J_\rho(\tilde{w})\,, \quad \rho = 0, 1,\dots,m\,.$$

We have

$$\lim_{t\to 0_+} \frac{g_\rho(t)}{t} = \lim_{t\to 0_+} F_\rho(y(t),t) = \lim_{t\to 0_+}\left[F_\rho(y(t),0) + O(t)\right] =$$

$$= \lim_{t\to 0_+}\left[\bar{k}^\rho + \sum_{j=1}^{N} k_j^\rho\, y^j(t)\right] = \bar{k}^\rho\,.$$

If $1 \le \gamma \le l'$ and $J_\gamma(\tilde{w}) = 0$ then $\bar{k}^\gamma < 0$ (from the definition of $K^-$), hence $\lim_{t\to 0_+} \frac{g_\gamma(t)}{t} < 0$; therefore, for $t$ small enough, $g_\gamma(t) < 0$.

It follows that $J_\gamma(v(t)) < 0$ for $\gamma$ satisfying $J_\gamma(\tilde{w}) = 0$. If $J_\gamma(\tilde{w}) < 0$, then $\lim_{t\to 0_+} g_\gamma(t) = 0$, hence $\lim_{t\to 0_+} J_\gamma(v(t)) = J_\gamma(\tilde{w}) < 0$, hence, for $t$ small enough, we have again $J_\gamma(v(t)) < 0$.

Therefore, for $t$ small enough, we have $J_\gamma(v(t)) < 0$, whatever $1 \le \gamma \le l'$. On the other hand,

$$J_\rho(v(t)) - J_\rho(\widetilde{w}) = J_\rho(v(t)) \text{ for } \rho = l' + 1, \ldots, m$$

and, from $F_\rho(y(t),t) \equiv 0$ for $\rho = l' + 1, \ldots, m$, it follows that

$$J_\rho(v(t)) = 0, \ \rho = l' + 1, \ldots, m.$$

At the same time, since $\bar{k}^0 < 0$ (according to the definition of $K^-$) and since $\lim\limits_{t \to 0_+} \frac{g_0(t)}{t} = \bar{k}^0 < 0$, it follows that $g_0(t) < 0$ for $t$ small enough, hence $J_0(v(t)) < J_0(\widetilde{w})$.

Thus, we find out that, for $t > 0$ small enough, $v(t)$ satisfies the constraints and gives a smaller value of $J_0$ than the one given by $\widetilde{w}$, what contradicts the optimality of $\widetilde{w}$.

In this way, Theorem 10.6 is proved.

This multiplier rule represents the basic instrument for obtaining the necessary optimality conditions in the general optimization problem we deal with.

However, its applicability is much broader.

As an illustration, we shall show how it works when applied to a standard linear programming problem.

In the linear programming problems, we have

$$J_\rho = \sum_{j=1}^{n} a_{\rho j} u_j - c_\rho, \ u_j \geq 0.$$

Consider $\widetilde{u}_j \geq 0$. The set $K$ of vectors having co-ordinates of the form

$$k^\rho = a_{\rho j}\left(u - \widetilde{u}_j\right)', \ u \geq 0, \ j = 1, \ldots, n$$

is a derivative set for the system of functions $J_\rho$ at $\widetilde{u}$.

Consider $k_1, \ldots, k_N \in K$, $k_\alpha^\rho = a_{\rho j_\alpha}\left(u_\alpha - \tilde{u}_{j_\alpha}\right)$. We define

$u_j(\varepsilon) = \tilde{u}_j$ for $j \neq j_\alpha$,

$$u_j(\varepsilon) = \sum_{j_\beta = j_\alpha}\left(u_\beta - \tilde{u}_{j_\beta}\right)\varepsilon_\beta + \tilde{u}_{j_\alpha} \text{ if } j = j_\alpha .$$

For $\varepsilon_\beta$ small enough, we have $u_j(\varepsilon) \geq 0$. We have

$$J_\rho(u(\varepsilon)) - J_\rho(\tilde{u}) = \sum_{j=1}^{n} a_{\rho j}\left(u_j(\varepsilon) - \tilde{u}_j\right) =$$

$$= \sum_\alpha a_{\rho j_\alpha}\left(u_{j_\alpha}(\varepsilon) - \tilde{u}_{j_\alpha}\right) = \sum_\alpha a_{\rho j_\alpha} \sum_{j_\beta = j_\alpha}\left(u_\beta - \tilde{u}_{j_\beta}\right)\varepsilon_\beta =$$

$$= a_{\rho j_1}\left[\left(u_1 - \tilde{u}_{j_1}\right)\varepsilon_1 + \ldots + \left(u_{k_1} - \tilde{u}_{j_1}\right)\varepsilon_{k_1}\right] + \ldots +$$

$$+ a_{\rho j_N}\left[\left(u_{k_1 + \ldots + k_{s-1} + 1} - \tilde{u}_{j_N}\right)\varepsilon_{k_1 + \ldots + k_{s-1} + 1} + \ldots + \left(u_N - \tilde{u}_N\right)\varepsilon_N\right] =$$

$$= a_{\rho j_1}\left(u_1 - \tilde{u}_{j_1}\right)\varepsilon_1 + \ldots + a_{\rho j_{k_1}}\left(u_{k_1} - \tilde{u}_{j_{k_1}}\right)\varepsilon_{k_1} + \ldots +$$

$$+ a_{\rho j_{k_1 + \ldots + k_{s-1} + 1}}\left(u_{k_1 + \ldots + k_{s-1} + 1} - \tilde{u}_{k_1 + \ldots + k_{s-1} + 1}\right)\varepsilon_{k_1 + \ldots + k_{s-1} + 1} + \ldots +$$

$$+ a_{j_N}\left(u_N - \tilde{u}_N\right)\varepsilon_N = \sum_\alpha a_{\rho j_\alpha}\left(u_\alpha - \tilde{u}_\alpha\right)\varepsilon_\alpha = \sum_\alpha k_\alpha^\rho \varepsilon_\alpha ,$$

what shows that $K$ is indeed a derivative set.

From the general theorem it follows that there exist $\lambda_\gamma$, at least one being non-zero, with

$$\lambda_\gamma \geq 0, \quad 0 \leq \gamma \leq l', \quad \lambda_\gamma J_\gamma = 0, \quad 1 \leq \gamma \leq l',$$

such that

$$\sum_\rho \lambda_\rho k^\rho \geq 0 \text{ for any } k \in K .$$

Thus, we have

$$\sum_\rho \lambda_\rho a_{\rho j}\left(u - \tilde{u}_j\right) \geq 0 \text{ for any } j \text{ and any } u \geq 0.$$

For the indices $j$ satisfying $\widetilde{u}_j = 0$ it follows $\sum_\rho \lambda_\rho a_{\rho j} \geq 0$;

for the indices $j$ such that $\widetilde{u}_j > 0$ it follows $\sum_\rho \lambda_\rho a_{\rho j} = 0$.

In conclusion, if $\{\widetilde{u}_j\}_1^n$ is a minimum for $\sum_{j=1}^n a_{0j} u_j - c_0$ in the class of programs satisfying

$$\sum_{j=1}^n a_{\rho j} u_j \leq c_\rho, \; 1 \leq \rho \leq l',$$

$$\sum_{j=1}^n a_{\rho j} u_j = c_\rho, \; l' < \rho \leq m,$$

then there exist multipliers $\lambda_\rho$ such that

$$\widetilde{u}_j = 0 \text{ if } \sum_\rho \lambda_\rho a_{\rho j} \geq 0$$

and

$$\widetilde{u}_j > 0 \text{ if } \sum_\rho \lambda_\rho a_{\rho j} = 0.$$

This is a known result in the theory of linear programming.

## 10.5 Proof of Theorem 10.1

We begin by performing some transformations.

Let $(\widetilde{b}, \widetilde{u})$ be an optimal control, $\widetilde{x}$ – the corresponding optimal evolution,

$$A(t) = \frac{\partial f}{\partial x}\left(t, \widetilde{x}(t), \widetilde{u}(t), \widetilde{b}\right), \quad c_\gamma(t) = \frac{\partial L_\gamma}{\partial x}\left(t, \widetilde{x}(t), \widetilde{u}(t), \widetilde{b}\right).$$

We define the functions $q_\gamma$ as solutions of the system

$$q_\gamma' = -A^*(t)\, q_\gamma - c_\gamma(t), \quad q_\gamma(T) = 0, \; \gamma = 0, 1, \ldots, l.$$

The system is linear, hence the functions $q_\gamma$ are defined on the whole interval $[t_0, T]$.

We define the functions $\tilde{q}_i$ as solutions of the system

$$\tilde{q}_i' = -A^*(t)\tilde{q}_i, \quad \tilde{q}_i^j(T) = \delta_i^j, \quad \delta_i^j = 0 \text{ if } i \neq j, \quad \delta_i^i = 1.$$

We also define

$$F_\gamma(t,x,u,b) = L_\gamma(t,x,u,b) - \left(c_\gamma(t), x\right) +$$

$$+ \left(q_\gamma(t), f(t,x,u,b) - A(t)x\right),$$

$$G_\gamma(b) = g_\gamma(b) + \left(q_\gamma(t_0), \alpha(b)\right), \quad \gamma = 0, 1, \dots, l,$$

$$F_{l+i}(t,x,u,b) = \left(\tilde{q}_i(t), f(t,x,u,b) - A(t)x\right),$$

$$G_{l+i}(b) = -\omega^i(b) + \left(\tilde{q}_i(t_0), \alpha(b)\right), \quad i = 1, \dots, n,$$

$$J_\rho(b,u) = G_\rho(b) - \int_{t_0}^{T} F_\rho(t, x(t), u(t), b)\, dt, \quad \rho = 0, 1, \dots, l+n.$$

We shall show that $(\tilde{b}, \tilde{u})$ is an optimal control for the problem

$$x' = f(t,x,u,b), \quad x(t_0) = \alpha(b), \quad J_\gamma(b,u) \le 0 \text{ for } \gamma = 1, \dots, l',$$

$$J_\gamma(b,u) = 0 \text{ for } \gamma = l'+1, \dots, l+n, \quad \min J_0.$$

Let $(b,u)$ be an admissible control, $x$ – the corresponding evolution. We have

$$\frac{d}{dt}\left(q_\gamma(t), x(t)\right) = \left(q_\gamma'(t), x(t)\right) + \left(q_\gamma(t), x'(t)\right) =$$

$$= \left(-A^*(t)q_\gamma(t) - c_\gamma(t), x(t)\right) + \left(q_\gamma(t), f(t, x(t), u(t), b)\right) =$$

$$= -\left(c_\gamma(t), x(t)\right) + \left(q_\gamma(t), f(t, x(t), u(t), b) - A(t)x(t)\right) =$$

$$= F_\gamma(t, x(t), u(t), b) - L_\gamma(t, x(t), u(t), b), \quad \gamma = 0, 1, \dots, l.$$

It follows that

$$J_\gamma(b,u) = g_\gamma(b) + \big(q_\gamma(t_0), \alpha(b)\big) +$$

$$+ \int_{t_0}^{T} L_\gamma(t, x(t), u(t), b)\, dt + \int_{t_0}^{T} \frac{d}{dt}\big(q_\gamma(t), x(t)\big)\, dt =$$

$$= I_\gamma(b,u) + \big(q_\gamma(t_0), \alpha(b)\big) + \big(q_\gamma(T), x(T)\big) - \big(q_\gamma(t_0), x(t_0)\big) =$$

$$= I_\gamma(b,u).$$

Further,

$$J_{l+i}(b,u) = -\,\omega^i(b) + \big(\tilde{q}_i(t_0), \alpha(b)\big) +$$

$$+ \int_{t_0}^{T} \big(\tilde{q}_i(t), f(t, x(t), u(t), b) - A(t)\, x(t)\big)\, dt =$$

$$= -\,\omega^i(b) + \big(\tilde{q}_i(t_0), \alpha(b)\big) + \int_{t_0}^{T} \big[\big(\tilde{q}_i(t), x'(t)\big) + \tilde{q}_i'(t)\, x(t)\big]\, dt =$$

$$= -\,\omega^i(b) + \big(\tilde{q}_i(t_0), x(t_0)\big) + \big(\tilde{q}_i(T), x(T)\big) - \big(\tilde{q}_i(t_0), x(t_0)\big) =$$

$$= -\,\omega^i(b) + x^i(T).$$

From these relations we deduce that if $(b,u)$ is admissible for the given problem, then it is also admissible for the modified problem and conversely. Herefrom, it can be seen that if $\big(\tilde{b}, \tilde{u}\big)$ is optimal for the given problem, then it is also optimal for the modified one.

For the modified problem, we shall show that the set $K = K_1 \cup K_0$ is a derivative set at $\big(\tilde{b}, \tilde{u}\big)$, where $K_1$ is the set of vectors of the form

$$k^\rho = F_\rho\big(t, \tilde{x}(t), u, \tilde{b}\big) - F_\rho\big(t, \tilde{x}(t), \tilde{u}(t), \tilde{b}\big), \quad \rho = 0, 1, \dots, l+n,$$

$u \in U$, $t \in [t_0, T]$, such that $\tilde{u}$ is continuous at $t$

and $K_0$ is the set of vectors of the form

$$k^{\rho} = \sum_{\sigma} h^{\sigma} \left[ \frac{\partial G_{\rho}}{\partial b^{\sigma}} \left(\widetilde{b}\right) + \int_{t_0}^{T} \frac{\partial F_{\rho}}{\partial b^{\sigma}} \left(t, \widetilde{x}(t), \widetilde{u}(t), \widetilde{b}\right) dt \right]$$

where $h^{\sigma}$ are arbitrary real numbers.

If we shall apply the general multiplier rule for this derivative set, then we shall obtain directly the statement of Theorem 10.1.

Indeed, we have

a) $\displaystyle\sum_{\rho} \lambda_{\rho} \left[ F_{\rho}\left(t, \widetilde{x}(t), u, \widetilde{b}\right) - F_{\rho}\left(t, \widetilde{x}(t), \widetilde{u}(t), \widetilde{b}\right) \right] \geq 0$

for any $u \in U$ and any $t \in [t_0, T]$ where $\widetilde{u}$ is continuous;

b) $\displaystyle\sum_{\rho} \lambda_{\rho} \sum_{\sigma} h^{\sigma} \left[ \frac{\partial G_{\rho}}{\partial b^{\sigma}} \left(\widetilde{b}\right) + \int_{t_0}^{T} \frac{\partial F_{\rho}}{\partial b^{\sigma}} \left(t, \widetilde{x}(t), \widetilde{u}(t), \widetilde{b}\right) dt \right] \geq 0$

for any real $h^{\sigma}$, what implies

$$\sum_{\rho} \lambda^{\rho} \left[ \frac{\partial G_{\rho}}{\partial b^{\sigma}} \left(\widetilde{b}\right) + \int_{t_0}^{T} \frac{\partial F_{\rho}}{\partial b^{\sigma}} \left(t, \widetilde{x}(t), \widetilde{u}(t), \widetilde{b}\right) dt \right] = 0 \text{ for all } \sigma.$$

We define

$$p(t) = -\sum_{i=1}^{n} \lambda_{l+i}\, \widetilde{q}_i(t) - \sum_{\gamma=0}^{l} \lambda_{\gamma}\, q_{\gamma}(t)$$

and we find out that $p$ satisfies the system in the statement of Theorem 10.1, with final conditions $p_j(T) = -\lambda_{l+j}$. We have

$$\left(p(t), f\left(t, \widetilde{x}(t), u, \widetilde{b}\right)\right) - \sum_{\gamma=0}^{l} \lambda_{\gamma}\, L_{\gamma}\left(t, \widetilde{x}(t), \widetilde{u}(t), \widetilde{b}\right) =$$

$$= -\sum_{\gamma=0}^{l} \lambda_{\gamma}\, F_{\gamma}\left(t, \widetilde{x}(t), u, \widetilde{b}\right) - \sum_{\gamma=0}^{l} \lambda_{\gamma} \left(c_{\gamma}(t), \widetilde{x}(t)\right) -$$

$$- \sum_{\gamma=0}^{l} \lambda_{\gamma} \left(q_{\gamma}(t), A(t)\, \widetilde{x}(t)\right) - \sum_{i=1}^{n} \lambda_{l+i}\, F_{l+i} \left(t, \widetilde{x}(t), u, \widetilde{b}\right) -$$

$$-\sum_{i=1}^{n} \lambda_{l+i} \left(\tilde{q}_i(t), A(t)\, \tilde{x}(t)\right) =$$

$$= -\sum_{\rho=0}^{l+n} F_\rho\left(t, \tilde{x}(t), u, \tilde{b}\right) - \left(p(t), A(t)\, \tilde{x}(t)\right) - \sum_{\gamma=0}^{l} \lambda_\gamma \left(c_\gamma(t), \tilde{x}(t)\right).$$

Property **a)** leads to the inequality

$$H\left(t, \tilde{x}(t), \tilde{u}(t), \tilde{b}, p(t)\right) - H\left(t, \tilde{x}(t), u, \tilde{b}, p(t)\right) \geq 0$$

from the statement of Theorem 10.1.

In a similar way, property **b)** leads to the last conditions in the statement of Theorem 10.1.

In this way, the proof of Theorem 10.1 reduces to the proof of the fact that the set $K$ is a derivative set for the modified problem.

Let $k_0, k_1, \ldots, k_N$ be vectors from $K$, $K = K_1 \cup K_0$. To fix the ideas and for the sake of simplicity, we shall assume that $k_0 \in K_0$, $k_1, \ldots, k_N \in K_1$. Therefore, for $j = 1, \ldots, N$, we have

$$k_j^\rho = -F_j\left(t_j, \tilde{x}\left(t_j\right), \tilde{u}\left(t_j\right), \tilde{b}\right) + F_\rho\left(t_j, \tilde{x}\left(t_j\right), u_j, \tilde{b}\right),\ t_j \in \left[t_0, T\right],$$

$u_j \in U$, $\tilde{u}$ continuous at $t_j$.

We can assume that $t_1 \leq t_2 \leq \ldots \leq t_N$.

We define $T_1 = t_1$, $T_j = t_j + \varepsilon_1 + \ldots + \varepsilon_{j-1}$.

We denote by $M(\varepsilon)$ the set of points $t \in \left[t_0, T\right]$ that do not belong to any interval of the form $\left[T_j, T_j + \varepsilon_j\right)$; we remark that

$$T_{j+1} = t_{j+1} + \varepsilon_1 + \ldots + \varepsilon_j \geq t_j + \varepsilon_1 + \ldots + \varepsilon_j = T_j + \varepsilon_j,$$

hence the intervals considered are disjoint.

We define

$u(t,\varepsilon) = u_j$ for $t \in \left[T_j, T_j + \varepsilon_j\right)$,

$u(t,\varepsilon) = \widetilde{u}(t)$ for $t \in M(\varepsilon)$.

The function $u(t,\varepsilon)$ defined in this way on $\left[t_0, T\right]$ is piece-wise continuous and $u(t,\varepsilon) \in U$.

Let $h^\sigma$ be the numbers that appear in the definition of $k_0$

$$\left( k_0^\rho = \sum_\sigma h^\sigma \left[ \frac{\partial G_\rho}{\partial b^\sigma}\left(\widetilde{b}\right) + \int_{t_0}^{T} \frac{\partial F_\rho}{\partial b^\sigma}\left(t, \widetilde{x}(t), \widetilde{u}(t), \widetilde{b}\right) dt \right] \right).$$

We define $b^\sigma(\varepsilon) = \widetilde{b}^\sigma + \varepsilon_0 h^\sigma$ and we consider the evolution

$x' = f\left(t, x, u(t,\varepsilon), b(\varepsilon)\right)$, with $x\left(t_0\right) = \alpha\left(b(\varepsilon)\right)$.

Since, for $\varepsilon = 0$, the system admits the solution $\widetilde{x}$ defined on the whole interval $\left[t_0, T\right]$, from the *general theorem concerning (global) dependence of solutions on initial conditions and on parameters*, it follows that, for $\varepsilon_0, \varepsilon_1, \ldots, \varepsilon_N$ small enough, the solution, denoted $x(\cdot,\varepsilon)$, will be defined on $\left[t_0, T\right]$ and, moreover, $\lim\limits_{\varepsilon \to 0} x(t,\varepsilon) = \widetilde{x}(t)$.

We define

$$f_\rho(\varepsilon) = \int_{t_0}^{T} \left[F_\rho\left(t, x(t,\varepsilon), u(t,\varepsilon), b(\varepsilon)\right) - F_\rho\left(t, \widetilde{x}(t), \widetilde{u}(t), \widetilde{b}\right)\right] dt +$$
$$+ G_\rho\left(b(\varepsilon)\right) - G_\rho\left(\widetilde{b}\right).$$

We shall prove that the mapping $\varepsilon \mapsto f_\rho(\varepsilon)$ admits partial derivatives at $\varepsilon = 0$ and that $\dfrac{\partial f_\rho}{\partial \varepsilon_j}(0) = k_j^\rho$; then

$$f_\rho(\varepsilon) - f_\rho(0) = \sum_j k_j^\rho \varepsilon_j + o(\varepsilon) \text{ for } \varepsilon_j > 0 \text{ small enough.}$$

Herefrom it will follow that $K$ is a derivative set.

Since $G_\rho$ are $\mathscr{C}^1$ and $b(\varepsilon)$ is linear with respect to $\varepsilon$, it suffices to analyze the term

$$\int_{t_0}^{T}\left[F_\rho\left(t, x(t,\varepsilon), u_j, b(\varepsilon)\right) - F_\rho\left(t, \tilde{x}(t), \tilde{u}(t), \tilde{b}\right)\right]dt =$$

$$= \sum_j \int_{T_j}^{T_j+\varepsilon_j}\left[F_\rho\left(t, x(t,\varepsilon), u_j, b(\varepsilon)\right) - F_\rho\left(t, \tilde{x}(t), \tilde{u}(t), \tilde{b}\right)\right]dt +$$

$$+ \int_{M(\varepsilon)}\left[F_\rho(t, x(t,\varepsilon), \tilde{u}(t), b(\varepsilon)) - F_\rho\left(t, \tilde{x}(t), \tilde{u}(t), \tilde{b}\right)\right]dt.$$

We show first that

$x(t,\varepsilon) = \tilde{x}(t) + O(\varepsilon)$ for $t \in \left[T_j + \varepsilon_j, T_{j+1}\right]$,

$t \in \left[t_0, t_1\right)$, $t \in \left[T_N + \varepsilon_N, T\right]$,

$\tilde{x}(t,\varepsilon) = \tilde{x}\left(t_j\right) + O(\varepsilon)$ for $t \in \left[T_j, T_j + \varepsilon_j\right)$.

On the interval $\left[t_0, t_1\right]$, the functions $x(\cdot,\varepsilon)$ and $\tilde{x}$ satisfy

$$x' = f(t, x, \tilde{u}(t), b(\varepsilon)),\ x(t_0) = \alpha(b(\varepsilon))$$

and

$$x' = f\left(t, x, \tilde{u}(t), \tilde{b}\right),\ x(t_0) = \alpha\left(\tilde{b}\right) \text{ respectively.}$$

The conclusion

$$x(t,\varepsilon) = \tilde{x}(t) + O(\varepsilon) \text{ for } t \in \left[t_0, t_1\right]$$

follows directly from the *theorem on Lipschitz dependence on initial conditions and on parameters*; in fact, under our assumptions, the function $\varepsilon \mapsto x(t,\varepsilon)$ is even differentiable on $\left[t_0, t_1\right]$.

On the interval $\left[t_1, t_1 + \varepsilon_1\right]$, we have

$$x(t,\varepsilon) = x(t_1,\varepsilon) + \int_{t_1}^{t} f(s, x(s,\varepsilon), u_1, b(\varepsilon))\, ds$$

and, since

$$x(t_1,\varepsilon) = \tilde{x}(t_1) + O(\varepsilon), \quad \int_{t_1}^{t} f\big(s, x(s,\varepsilon), u_1, b(\varepsilon)\big)\, ds = O(\varepsilon),$$

we deduce $x(t,\varepsilon) = \tilde{x}(t_1) + O(\varepsilon)$.

We consider now the interval $\left[t_1 + \varepsilon_1, t_2 + \varepsilon_1\right] = \left[T_1 + \varepsilon_1, T_2\right]$.

On this interval, $x(\cdot,\varepsilon)$ is the solution of the problem

$$x' = f\big(t, x, \tilde{u}(t), \tilde{b}\big), \quad x\big(t_1 + \varepsilon_1\big) = \tilde{x}\big(t_1 + \varepsilon_1\big).$$

We deduce again, on the basis of the theorem on Lipschitz dependence on initial conditions and on parameters, that

$$x(t,\varepsilon) = \tilde{x}(t) + O(\varepsilon).$$

Further, for $t \in \left[T_2, T_2 + \varepsilon_2\right] = \left[t_2 + \varepsilon_1, t_2 + \varepsilon_1 + \varepsilon_2\right]$ we have

$$x(t,\varepsilon) = x(t_2 + \varepsilon_1, \varepsilon) + \int_{T_2}^{t} f\big(s, x(s,\varepsilon), u_2, b(\varepsilon)\big)\, ds =$$

$$= x\big(t_2 + \varepsilon_1, \varepsilon\big) + O(\varepsilon) = \tilde{x}\big(t_2 + \varepsilon_1, \varepsilon\big) + O(\varepsilon) = \tilde{x}\big(t_2\big) + O(\varepsilon).$$

In the interval: $\left[T_2 + \varepsilon_2, T_3\right] = \left[t_2 + \varepsilon_1 + \varepsilon_2, t_3 + \varepsilon_1 + \varepsilon_2\right]$,

$x(\cdot,\varepsilon)$ is the solution of the problem

$$x' = f\big(t, x, \tilde{u}(t), b(\varepsilon)\big), \quad x\big(t_2 + \varepsilon_1 + \varepsilon_2\big) = \tilde{x}\big(t_2\big)$$

and $\tilde{x}$ is the solution of the problem

$$x' = f\big(t, x, \tilde{u}(t), \tilde{b}\big), \quad x\big(t_2 + \varepsilon_1 + \varepsilon_2\big) = \tilde{x}\big(t_2 + \varepsilon_1 + \varepsilon_2\big);$$

again, we deduce

$$x(t,\varepsilon) = \tilde{x}(t) + O(\varepsilon).$$

The reasoning continues in the same way. We have

$$\int_{T_j}^{T_j+\varepsilon_j} \left[F_\rho\left(t, x(t,\varepsilon), u_j, b(\varepsilon)\right) - F_\rho\left(t, \tilde{x}(t), \tilde{u}(t), \tilde{b}\right)\right] dt =$$

$$= \varepsilon_j \left[F_\rho\left(t_j, \tilde{x}\left(t_j\right), u_j, \tilde{b}\right) - F_\rho\left(t_j, \tilde{x}\left(t_j\right), \tilde{u}\left(t_j\right), \tilde{b}\right)\right] +$$

$$+ \int_{T_j}^{T_j+\varepsilon_j} \left[F_\rho\left(t, x(t,\varepsilon), u_j, b(\varepsilon)\right) - F_\rho\left(t_j, \tilde{x}\left(t_j\right), u_j, \tilde{b}\right)\right] dt -$$

$$- \int_{T_j}^{T_j+\varepsilon_j} \left[F_\rho\left(t, \tilde{x}(t), \tilde{u}(t), \tilde{b}\right) - F_\rho\left(t_j, \tilde{x}\left(t_j\right), \tilde{u}\left(t_j\right), \tilde{b}\right)\right] dt .$$

But

$$b(\varepsilon) - \tilde{b} = O(\varepsilon), \quad x(t,\varepsilon) - \tilde{x}\left(t_j\right) = O(\varepsilon),$$

$$0 < t - t_j < \varepsilon_1 + \varepsilon_2 + \ldots + \varepsilon_j ,$$

hence $t - t_j = O(\varepsilon)$; it follows

$$\left| F_\rho\left(t, x(t,\varepsilon), u_j, b(\varepsilon)\right) - F_\rho\left(t_j, \tilde{x}\left(t_j\right), u_j, \tilde{b}\right) \right| \le \gamma_1(\varepsilon),$$

with $\lim\limits_{\varepsilon \to 0} \gamma_1(\varepsilon) = 0$ and therefore

$$\left| \int_{T_j}^{T_j+\varepsilon_j} \left[F_\rho\left(t, x(t,\varepsilon), u_j, b(\varepsilon)\right) - F_\rho\left(t_j, \tilde{x}\left(t_j\right), u_j, \tilde{b}\right)\right] dt \right| \le$$

$$\le \varepsilon_j \, \gamma_1(\varepsilon) = o(\varepsilon) .$$

We also have

$$\left| \tilde{x}(t) - \tilde{x}\left(t_j\right) \right| \le L \left| t - t_j \right|, \quad \tilde{u}(t) - \tilde{u}\left(t_j\right) \to 0 \text{ for } t \to t_j ,$$

since $t_j$ is a continuity point for $\tilde{u}$.

We deduce that

$$\left| F_\rho\left(t, \tilde{x}(t), \tilde{u}(t), \tilde{b}\right) - F_\rho\left(t_j, \tilde{x}\left(t_j\right), \tilde{u}\left(t_j\right), \tilde{b}\right) \right| \le \gamma_2(\varepsilon),$$

with $\lim\limits_{\varepsilon \to 0} \gamma_2(\varepsilon) = 0$.

Finally, we obtain

$$\int_{T_j}^{T_j+\varepsilon_j} \left[F_\rho\left(t, x(t,\varepsilon), u_j, b(\varepsilon)\right) - F_\rho\left(t, \tilde{x}(t), \tilde{u}(t), \tilde{b}\right)\right] dt =$$

$$= \varepsilon_j \left[F_\rho\left(t_j, \tilde{x}\left(t_j\right), u_j, \tilde{b}\right) - F_\rho\left(t_j, \tilde{x}\left(t_j\right), \tilde{u}\left(t_j\right), \tilde{b}\right)\right] + \varepsilon_j \gamma_3(\varepsilon),$$

$$\lim_{\varepsilon\to 0} \gamma_3(\varepsilon) = 0.$$

Further, $M(\varepsilon)$ is a union of intervals of the form

$$\left[T_j + \varepsilon_j, T_{j+1}\right) = \left[t_j + \varepsilon_1 + \ldots + \varepsilon_j, t_{j+1} + \varepsilon_1 + \ldots + \varepsilon_j\right),$$

hence $\int_{M(\varepsilon)}$ is a sum of integrals of the form

$$\int_{t_j+\varepsilon_1+\ldots+\varepsilon_j}^{t_{j+1}+\varepsilon_1+\ldots+\varepsilon_j} \left[F_\rho(t, x(t,\varepsilon), \tilde{u}(t), b(\varepsilon)) - F_\rho\left(t, \tilde{x}(t), \tilde{u}(t), \tilde{b}\right)\right] dt.$$

For the interval $\left[t_j + \varepsilon_1 + \ldots + \varepsilon_j, t_{j+1} + \varepsilon_1 + \ldots + \varepsilon_j\right)$, we can write

$$F_\rho(t, x(t,\varepsilon), \tilde{u}(t), b(\varepsilon)) - F_\rho\left(t, \tilde{x}(t), \tilde{u}(t), \tilde{b}\right) =$$

$$= \frac{\partial F_\rho}{\partial x}\left(t, \tilde{x}(t), \tilde{u}(t), \tilde{b}\right)\left[x(t,\varepsilon) - \tilde{x}(t)\right] +$$

$$+ \frac{\partial F_\rho}{\partial b}\left(t, \tilde{x}(t), \tilde{u}(t), \tilde{b}\right)\left[b(\varepsilon) - \tilde{b}\right] +$$

$$+ \gamma_4\left(x(t,\varepsilon), b(\varepsilon)\right)\left(\left|x(t,\varepsilon) - \tilde{x}(t)\right| + \left|b(\varepsilon) - \tilde{b}\right|\right).$$

But, by the definition of $F_\rho$ and of $b(\varepsilon)$,

$$\frac{\partial F_\rho}{\partial x}\left(t, \tilde{x}(t), \tilde{u}(t), \tilde{b}\right) \equiv 0 \text{ and } b(\varepsilon) - \tilde{b} = \varepsilon_0 h,$$

therefore

$$F_\rho(t, x(t,\varepsilon), \tilde{u}(t), b(\varepsilon)) - F_\rho\left(t, \tilde{x}(t), \tilde{u}(t), \tilde{b}\right) =$$

$$= \frac{\partial F_\rho}{\partial b}\left(t, \tilde{x}(t), \tilde{u}(t), \tilde{b}\right)\varepsilon_0 h +$$

$$+ \gamma_4\left(x(t,\varepsilon), b(\varepsilon)\right)\left(\left|x(t,\varepsilon) - \tilde{x}(t)\right| + \left|b(\varepsilon) - \tilde{b}\right|\right) =$$

$$= \varepsilon_0 \frac{\partial F_\rho}{\partial b}\left(t, \tilde{x}(t), \tilde{u}(t), \tilde{b}\right) h + \gamma_5(\varepsilon) O(\varepsilon).$$

Herefrom it follows

$$\int_{t_j+\varepsilon_1+\ldots+\varepsilon_j}^{t_{j+1}+\varepsilon_1+\ldots+\varepsilon_j}\left[F_\rho(t,x(t,\varepsilon),\tilde{u}(t),b(\varepsilon)) - F_\rho\left(t,\tilde{x}(t),\tilde{u}(t),\tilde{b}\right)\right]dt =$$

$$= \varepsilon_0\int_{t_j+\varepsilon_1+\ldots+\varepsilon_j}^{t_{j+1}+\varepsilon_1+\ldots+\varepsilon_j}\frac{\partial F_\rho}{\partial b}\left(t,\tilde{x}(t),\tilde{u}(t),\tilde{b}\right)h\,dt + O(\varepsilon)\,\gamma_6(\varepsilon) =$$

$$= \varepsilon_0\int_{t_j}^{t_{j+1}}\frac{\partial F_\rho}{\partial b}\left(t,\tilde{x}(t),\tilde{u}(t),\tilde{b}\right)h\,dt +$$

$$+ \varepsilon_0\int_{t_j+\varepsilon_1+\ldots+\varepsilon_j}^{t_j}\frac{\partial F_\rho}{\partial b}\left(t,\tilde{x}(t),\tilde{u}(t),\tilde{b}\right)h\,dt +$$

$$+ \varepsilon_0\int_{t_{j+1}}^{t_{j+1}+\varepsilon_1+\ldots+\varepsilon_j}\frac{\partial F_\rho}{\partial b}\left(t,\tilde{x}(t),\tilde{u}(t),\tilde{b}\right)h\,dt + O(\varepsilon)\,\gamma_6(\varepsilon) =$$

$$= \varepsilon_0\int_{t_j}^{t_{j+1}}\frac{\partial F_\rho}{\partial b}\left(t,\tilde{x}(t),\tilde{u}(t),\tilde{b}\right)h\,dt + O(\varepsilon)\,\gamma_7(\varepsilon).$$

Finally, it follows

$$\int_{t_0}^{T}\left[F_\rho(t,x(t,\varepsilon),u(t,\varepsilon),b(\varepsilon)) - F_\rho\left(t,\tilde{x}(t),\tilde{u}(t),\tilde{b}\right)\right]dt =$$

$$= \sum_j \varepsilon_j\left[F_\rho\left(t_j,\tilde{x}(t_j),u_j,\tilde{b}\right) - F_\rho\left(t_j,\tilde{x}(t_j),\tilde{u}(t_j),\tilde{b}\right)\right] +$$

$$+ \varepsilon_0\int_{t_0}^{T}\frac{\partial F_\rho}{\partial b}\left(t,\tilde{x}(t),\tilde{u}(t),\tilde{b}\right)h\,dt + o(\varepsilon),$$

hence

$$f_\rho(\varepsilon) = \varepsilon_0\left[\frac{\partial G_\rho}{\partial b}\left(\tilde{b}\right) + \int_{t_0}^{T}\frac{\partial F_\rho}{\partial b}\left(t,\tilde{x}(t),\tilde{u}(t),\tilde{b}\right)dt\right]h +$$

$$+ \sum_j \varepsilon_j\left[F_\rho\left(t_j,\tilde{x}(t_j),u_j,\tilde{b}\right) - F_\rho\left(t_j,\tilde{x}(t_j),\tilde{u}(t_j),\tilde{b}\right)\right] + o(\varepsilon) =$$

$$= \varepsilon_0\,k_0^\rho + \sum_{j=1}^{N}\varepsilon_j\,k_j^\rho + o(\varepsilon),$$

what ends the proof.

## 10.6 Notes and References

There are several ways to state and prove the necessary optimality conditions in optimal control, starting with the one proposed by Pontryagin and his colleagues [24].

We have used here the version proposed by M. R. Hestenes [11].

The application to economic growth was given by M. Kurz in [17].

# References

1. Allen, R.G.D. (1960) *Mathematical economics.* Macmillan, London.

2. Altăr, M., Arsene, C., Stroe, R. (1975) *Problems concerning identification of macroeconomic production functions for the Romanian economy* (Romanian). Working Paper, Academy of Economic Studies, Bucharest.

3. Brand, L. (1966) *Differential and difference equations.* John Wiley & Sons, New York–London–Sidney.

4. Bródy, A. (1981) *An economy controlled by stocks and profits,* in *"Non-price control"* (edited by J. Kornai and B. Martos). Akadémiai Kiadó, Budapest, pp. 149 – 162.

5. Calvo, G.A., Vegh. C.A. (1990) *Money supply and interest rate policy in a new-Keynesian framework* (Unpublished Manuscript). International Monetary Fund, IMF Working Paper, Dec, 1990.

6. Cass, D. (1965) *Optimum growth in an aggregative model of capital accumulation.* Rev. of Ec. Studies, July 1965.

7. Halanay, A (1966) *Differential equations.* Academic Press, New York.

8. Halanay, A., Judita Samuel (1993) *A monetary policy model.* Economic Computation and Economic Cybernetics Studies and Research, 1 – 4, pp. 5 – 19.

9. Hall, G., Watt, J. M. (editors) (1976) *Modern numerical methods for ordinary differential equations.* Clarendon Press, Oxford.

10. Helmer, J.Y. (1972) *La Commande optimale en economie.* Dunod, Paris.

**11.** Hestenes, M.R. (1965) *On variational theory and optimal control theory*. SIAM J. Control, 3, pp. 23 – 48.

**12.** Hirsch, M.W., Smale, S. (1974) *Differential equations, dynamical systems and linear algebra*. Academic Press, New York and London.

**13.** Ho, Y.C., Bryson, A.E., Baron, M.L. (1965) *Differential games and optimal pursuit–evasion strategies*. IEEE Trans. Automatic – Control, AC–10, pp. 385-389.

**14.** Kalman, R.E., Ho Y.C., Narencha K.S. (1962) *Controllability of linear dynamical systems*. Contrib. Diff. Eqns. 1, pp. 189-213.

**15.** Kalman, R.E. (1960) *Contributions to the theory of optimal control*. Buletin de la Sociedat Mat. Mexicana, Segunda Serie, Vol. 5.1, pp. 102–119.

**16.** Kleinman, D.L. (1968) *On an iterative technique for Riccati equation computations*. IEEE Trans. On Aut. Control, pp. 114– 115.

**17.** Kurz, M. (1965) *Optimal paths of capital accumulation under the minimum time objective*. Econometrica, 1, 1965, pp. 42-66.

**18.** Kwakernaak, H. and Sivan, R. (1972) *Linear optimal control systems*. Wiley Interscience, New York.

**19.** La Salle, J.P. (1981) *The stability and control of discrete processes*. Springer–Verlag, Berlin.

**20.** Lagarias, J.C., Todd, M.J. (editors) (1990) *Mathematical developments arising from linear programming* (Proceedings of a Joint Summer Research Conference held at Bowdoin College, June 25 - July 1, 1988). In *Contemporary Mathematics*, vol 114, American Mathematical Society, Providence, Rhode Island.

**21.** Luenberger, D.G. (1964) *Observing the state of linear systems*. IEEE Trans. On Electr. 8, 1964.

**22.** McNelis, P., Yoshino, N. (1992) *Monetary stabilization with interest rate instruments in Japan: A linear quadratic control analysis.* Bank of Japan, Monetary and Economic Studies, Vol. X, 2, Nov., pp. 79–106.

**23.** Nikaido, H. (1968) *Convex structures and economic theory.* Academic Press, New-York.

**24.** Pontryagin, L.S., Boltyanski, V.G., Gamkrelidze, R.V. and Mishchenko, E.F. (1964) *The mathematical theory of optimal processes.* J. Wiley, New York.

**25.** Popov V.M. (1973) *Hyperstability of control systems.* Springer Verlag.

**26.** Shampine, L.S., Gordon, M.K. (1975) *Computer solution of ordinary differential equations. The initial value problem.* Freeman & Comp., San Francisco.

**27.** Stetter, H.J. (1973) *Analysis of discretization methods for ordinary differential equations.* Springer–Verlag, Berlin–Heidelberg–New-York.

**28.** Varga, A. (1981) *On stabilization algorithms for linear time–invariant systems.* Rev. Roumaine Sci. Tech., Serie Electrotechn. Energ. 26, pp. 115–124.

**29.** Wonham, V.M. (1974) *Linear multivariable control – a geometric approach.* Lecture Notes in Econ. And Math. Systems, 101, Springer–Verlag.

**30.** Wonham, V.M. (1967) *On pole assignment in multi–input controllable linear systems.* IEEE Trans. Aut. Control AC-12, pp. 660–665.

# Index

# MATHEMATICAL MODELLING:
## Theory and Applications

1. M. Křížek and P. Neittaanmäki: *Mathematical and Numerical Modelling in Electrical Engineering*. Theory and Applications. 1996 ISBN 0-7923-4249-6
2. M.A. van Wyk and W.-H. Steeb: *Chaos in Electronics*. 1997 ISBN 0-7923-4576-2
3. A. Halanay and J. Samuel: *Differential Equations, Discrete Systems and Control*. Economic Models. 1997 ISBN 0-7923-4675-0

KLUWER ACADEMIC PUBLISHERS – DORDRECHT / BOSTON / LONDON

MIX
Papier aus verantwortungsvollen Quellen
Paper from responsible sources
FSC® C105338
FSC
www.fsc.org

If you have any concerns about our products,
you can contact us on
**ProductSafety@springernature.com**

In case Publisher is established outside the EU,
the EU authorized representative is:
**Springer Nature Customer Service Center GmbH**
**Europaplatz 3, 69115 Heidelberg, Germany**

Printed by Libri Plureos GmbH
in Hamburg, Germany